Lecture Notes in Geosystems Mathematics and Computing

More information about this series at http://www.springer.com/series/15481

Birkhäuser

Gayaz Khakimzyanov • Denys Dutykh
Zinaida Fedotova • Oleg Gusev

Dispersive Shallow Water Waves

Theory, Modeling, and Numerical Methods

Birkhäuser

Gayaz Khakimzyanov
Institute of Computational Technologies
of the Siberian Branch of the Russian
Academy of Sciences
Novosibirsk, Russia

Denys Dutykh
CNRS - LAMA UMR 5127
Department of Mathematics
University Savoie Mont Blanc
Campus Scientifique
Le Bourget-du-Lac, France

Zinaida Fedotova
Institute of Computational Technologies
of the Siberian Branch of the Russian
Academy of Sciences
Novosibirsk, Russia

Oleg Gusev
Institute of Computational Technologies
of the Siberian Branch of the Russian
Academy of Sciences
Novosibirsk, Russia

Lecture Notes in Geosystems Mathematics and Computing
ISBN 978-3-030-46266-6 ISBN 978-3-030-46267-3 (eBook)
https://doi.org/10.1007/978-3-030-46267-3

Mathematics Subject Classification: 76-XX, 00A71

This book is published under the imprint Birkhäuser, www.birkhauser-science.com by the registered company Springer Nature Switzerland AG
The registered company address is: Gewerbestrasse 11, 6330 Cham, Switzerland

To our teachers: Academicians N.N. Yanenko and Yu.I. Shokin

(GK, ZF, OG)

To Katya, Nicolas and Michel

(DD)

Foreword

The main processes in the ocean (such as circulation, currents, tides, ensembles of eddies) take place on large (planetary) scales and can be described by depth-averaged shallow water-type hydrodynamic equations. The same equations are also quite often used in modelling and mitigation of various natural hazards in seas and rivers such as the storm surges, tsunamis, inundations and dam breaks. These hyperbolic partial differential equations represent a reduced mathematical model as the dependence on the vertical coordinate was ruled out. Thus, unknown functions depend only on two spatial and one temporal coordinates.

The main advantage of this model is that there is a large number of analytical solutions along with rigorous theoretical methods to explain some physical processes happening in the ocean. The first reliable numerical predictions can be already obtained on coarse grids with grid spacing of the order of $\mathcal{O}(10)$ km. However, with the growth of the amount of available DART buoys data along with other types of instrumental measurements and field observation data, their explanation requires more and more accuracy from theoretical models. At some point, predictions based on nonlinear shallow water equations ceased to be satisfactory. Moreover, there is a practical need for simulations with the local grid resolution of up to $\mathcal{O}(10)$ m. Such numerical results can be practically used to elaborate safer evacuation plans for local populations, to design coastal protecting structures and to minimize the overall impact of ocean-related natural hazards onto local economies.

For all these reasons, today we need a new generation of hydrodynamic models, which are able to provide an improved accuracy by better resolving the vertical flow structure. We realized also that a small parameter[1] used in the shallow water wave theory, which is taken to be equal to the ratio of the characteristic water depth to the typical wavelength, is actually not so small in many natural phenomena of practical interest. To give an example, during the catastrophic SUMATRA event of the 26th of December 2004 in the INDIAN ocean, the characteristic wavelength was of the

[1] Small parameters are used in the so-called asymptotic methods to simplify mathematical models.

order of $\mathcal{O}(50)$ km, which provoked a significant accumulation of non-hydrostatic[2] effects on trans-oceanic propagation distances.

These non-hydrostatic effects are caused by the dependence of the wave propagation celerity on its wavelength, and they could be even observed on satellite records during the tsunami boxing day. Moreover, non-hydrostatic pressure effects become almost unavoidable for tsunami-generated waves by landslide and volcanic eruption mechanisms. These waves combine two major difficulties. First of all, they may reach extraordinary heights. For example, in ALASKA in 2015 a wave run-up of 190 m was reported. On the horizontal scale, such waves reach at most only a few kilometres. Thus, the nonlinear shallow water wave theory is not applicable to describe such solutions. That is why various dispersive and non-hydrostatic models have been developed since the second half of the twentieth century.

In this book, these models are referred to as *dispersive shallow water wave hydrodynamics*. The number of various dispersive wave models is quite large as it often happens in higher orders of approximation in perturbation schemes. Many of such models did not possess a solid justification and many desirable properties (such as the energy conservation and GALILEAN invariance). A number of dispersive models have been implemented in numerical codes with the empiric parametrization of important physical processes (i.e. the bottom friction, wave breaking, etc.), which may affect significantly the predictions. The accuracy and stability of employed schemes have not been always studied or clearly specified. It is due to the fact that most of such codes have been developed by practitioners to solve their very specific class of problems. Thus, the transposition of such codes to other types of problems should be done with a particular care.

For all the reasons mentioned hereinabove, the appearance of the present book made me very happy. Here, the authors perform a careful and detailed derivation of all basic nonlinear dispersive equations. Their numerical discretization is studied carefully as well. Moreover, the energy conservation properties of proposed equations discussed in this book may be used to estimate the accuracy of the implemented numerical methods. Moreover, the authors cover both the plane (i.e. CARTESIAN) and the spherical geometries. The latter has become increasingly important because the bathymetry data is available today in geographical (i.e. spherical) coordinates only.

I know personally the authors of this book for many years and I consider them as very strong world-class experts in the theory and numerical methods of geophysical hydrodynamics. The test cases and their results reported in this book should become the standard benchmarks for testing numerical codes, which are being used for the prediction and analysis of ocean-related natural hazards. It is with great pleasure and certitude that I can recommend this book as a golden standard to all the users and developers of new dispersive wave hydrodynamic models and codes. I would like to attract the special attention of those researchers who plan to develop the numerical

[2]As the name 'non-hydrostatic' indicates, these effects are caused by the deviation of the fluid pressure from the hydrostatic distribution across the water column.

codes that will remain in the open-source domain (and, thus, being potentially used by many other researchers and practitioners). This book will be particularly helpful to them. The computational geoscientists may use it as a guideline for the main principles on which modern numerical codes should be based on.

Efim Pelinovsky

Chief Scientist
Institute of Applied Physics
Nizhny Novgorod, Russia

Professor in Applied Mathematics
State Technical University
Nizhny Novgorod, Russia

Adjunct Professor
University of Southern Queensland
Toowoomba, QLD, Australia
May 2019

Preface

The present book is entirely devoted to the development of mathematical models and numerical algorithms to simulate the generation and propagation of long surface water waves, when horizontal components of the velocity vector dominate over the vertical ones. From practical point of view, the main real-world applications of our work include large-scale wave motions in the ocean (such as tsunami waves) and smaller-scale processes in coastal areas. Very often the natural phenomena we consider result from natural hazards such as tsunamigenic earthquakes, landslides, volcano eruptions, etc.

In our work, we opt for the hierarchical approach to the construction of the family of shallow water models. The same approach is also naturally adopted in the development of numerical algorithms specifically designed to solve these equations. The parameter, which allows us to build this hierarchy, is the dominant characteristic scale of the modelled wave propagation process compared to the typical water depth. We pay special attention to weakly and fully nonlinear dispersive wave models. The corresponding hierarchies are derived in globally flat and globally spherical geometries. The proposed models take into account EARTH'S sphericity, rotation, bathymetry and eventual bottom motion. Whenever possible, we provide also the conservative form of the governing equations and we discuss the total mechanical energy balance properties. When the bottom is steady, we recover the exact conservation law of the energy. The GALILEAN invariance of governing equations is discussed as well.

Our numerical approach to the discretization of the governing equations is based on the operator splitting idea, which originated in NOVOSIBIRSK in pioneering works of Academicians N.N. YANENKO and Yu.I. SHOKIN. In other words, in dispersive wave models we separate the nonlinear hyperbolic and elliptic parts. Then, for each operator we apply the most appropriate numerical method: predictor–corrector finite volume scheme for the former and integro-interpolating finite differences for the latter. The adaptivity of the proposed method is achieved, thanks to the moving grid technique. We prove that our scheme is well-balanced[1] even

[1] By well-balancedness, we understand here the exact preservation of the 'lake-at-rest' states.

on moving grids, which is the condition *sine qua non* for the robustness of the numerical model.

Despite all the progress achieved so far, the topic of dispersive wave modelling is far from being finished. Today, the real-time simulation of wave propagation on trans-oceanic scales is impossible without the usage of parallel algorithms. The ability to predict the tsunami wave propagation in real time is crucial for the mitigation of tsunami hazard. Modern tsunami simulation complexes are used to protect people and reduce property losses in the event of a tsunami. On the mathematical side, we find also that fully nonlinear models with enhanced dispersive characteristics have not been sufficiently studied yet by the scientific community. Moreover, the complex geometry of the coastline in certain regions, the interaction of the wave with coral reefs and vegetation and wave propagation (flood) in urban environment still pose challenging modelling problems. We believe that many new phenomena and effects are yet to be discovered in the field of nonlinear water waves.

The present book is based on a series of papers [188–191] that we published previously in *Communications in Computational Physics*. Each paper has been extended and further improved before their inclusion into this book. We take advantage here to acknowledge the *Global Science Press* and the Managing Editor Prof. Tao TANG for organizing a professional peer-review process and giving us the permission to reproduce this work here.

The logical inter-dependence of chapters is schematically presented in the given diagram, where each arrow points to the dependent chapter.

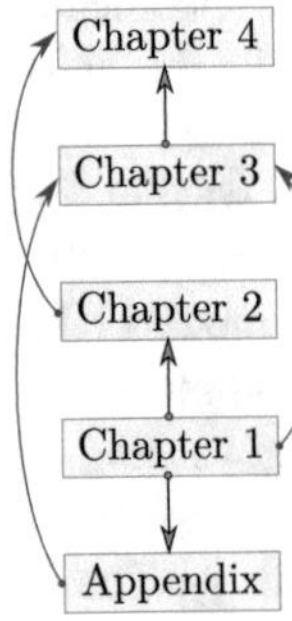

Novosibirsk, Russia
Chambéry, France
Novosibirsk, Russia
Novosibirsk, Russia
May 2020

Gayaz Khakimzyanov
Denys Dutykh
Zinaida Fedotova
Oleg Gusev

Acknowledgements

This work was supported by the CNRS (grant 10-05-91052), the Russian Foundation for Basic Research (grant 12-01-00721a) and by the Russian Science Foundation (grant 14-17-00219). The work of DD has been supported by the French National Research Agency, through the Investments for Future Program (ref. ANR−18−EURE−0016—Solar Academy). DD would like to thank his collaborators, in particular Professors Didier CLAMOND (Université Côte d'Azur, France), Angel DURÁN (Universidad de Valladolid, España) and Dimitrios MITSOTAKIS (Victoria University of Wellington, New Zealand) who kindly shared their precious knowledge with me.

Contents

Acronyms

1D	One-dimensional
2D	Two-dimensional
3D	Three-dimensional
AMR	Adaptive mesh refinement
BBM	Benjamin–Bona–Mahony
BVP	Boundary value problem
CFL	Courant–Friedrichs–Lewy
CPU	Central processing unit
DART	Deep-ocean assessment and reporting of tsunamis
DNS	Direct numerical simulation
FEM	Finite element method
FNWD	Fully nonlinear weakly dispersive
GEBCO	General bathymetric chart of the oceans
GFD	Geophysical fluid dynamics
GN	Green–Naghdi
GPS	Global positioning system
IVP	Initial value problem
KdV	Korteweg–de Vries
MOL	Method of lines
MOST	Method of splitting tsunami
NLS	Nonlinear Schrödinger
NOAA	National oceanic and atmospheric administration
NSWE	Nonlinear shallow water equations
ODE	Ordinary differential equation
PDE	Partial differential equation
RLW	Regularized long wave
SGN	Serre–Green–Naghdi
SOR	Successive over-relaxation
SW	Solitary wave

TVD	Total variation diminishing
USGS	United States Geological Survey
WNBM	Weakly nonlinear base model
WNWD	Weakly nonlinear weakly dispersive

Chapter 1
Model Derivation on a Globally Flat Space

The history of nonlinear dispersive modelling goes back to the end of the nineteenth century [78]. At that time J. Boussinesq [43] proposed (in a footnote on page 360) the celebrated KORTEWEG–DE VRIES (KdV) equation, re-derived later by D. Korteweg and G. de Vries [200]. Of course, J. Boussinesq proposed also the first BOUSSINESQ-type equation [41, 42] as a theoretical explanation of *solitary waves* observed earlier by J. Russell [269]. After this initial active period there was a break in this field until 1950s. The silence was interrupted by the new generation of 'pioneers'—F. Serre [282, 283], C.C. Mei and Le Méhauté [238] and D. Peregrine [264] who derived modern nonlinear dispersive wave models. After this time the modern period started, which can be characterized by the proliferation of journal publications and it is much more difficult to keep track of these records. Subsequent developments can be conventionally divided into two classes:

1. Application and critical analysis of existing models in new (and often more complex) situations;
2. Development of new high-fidelity physical approximate models.

Sometimes both points can be improved in the same publication. We would like to mention that according to our knowledge the first applications of PEREGRINE's model [264] to three-dimensional practical problems were reported in [1, 270].

In parallel, scalar model equations have been developed. They describe the unidirectional wave propagation [117, 257]. For instance, after the above-mentioned KdV equation, its regularized version was proposed first by Peregrine [263], then by Benjamin et al. [29]. Now this equation is referred to as the Regularized Long Wave (RLW) or BENJAMIN–BONA–MAHONY (BBM) equation. In [29] the well-posedness of RLW/BBM equation in the sense of J. HADAMARD was proven as well. Even earlier Whitham [331] proposed a model equation which possesses the dispersion relation of the full EULER equations (it was constructed in an ad-hoc manner to possess this property). It turned out to be an excellent approximation to the EULER equations in certain regimes [246]. Between unidirectional and bi-

G. Khakimzyanov et al., *Dispersive Shallow Water Waves*, Lecture Notes in Geosystems Mathematics and Computing, https://doi.org/10.1007/978-3-030-46267-3_1

directional models there is an intermediate level of scalar equations with second order derivatives in time. Such an intermediate model was proposed, for example, in [195]. Historically, the first BOUSSINESQ-type equation proposed by J. Boussinesq [43] was in this form as well. The main advantage of these models is their simplicity on the one hand, and the ability of providing good quantitative predictions on the other hand.

One possible classification of existing nonlinear dispersive wave models can be made upon the choice of the horizontal velocity variable. Two popular choices were suggested in [264]. Namely, one can use the depth-averaged velocity variable (see e.g. [87, 128, 133, 270, 333, 344]). Usually, such models enjoy nice mathematical properties such as the exact mass conservation equation. The second choice consists in taking the trace of the velocity on a surface defined in the fluid bulk $y = \mathcal{Y}(\mathbf{x}, t)$. Notice that surface $\mathcal{Y}(\mathbf{x}, t)$ may eventually coincide with the free surface [77] or with the bottom $y = -h(\mathbf{x}, t)$ [4, 238]. This technique was used for the derivation of several BOUSSINESQ-type systems with flat bottom, initially in [35] and later in [34, 37] and analysed thoroughly theoretically and numerically in [6–8, 34, 38, 93]. Sometimes the choice of the surface is made in order to obtain a model with improved dispersion characteristics [35, 231, 328]. One of the most popular model of this class is due to O. Nwogu [254] who proposed to use the horizontal velocity defined at $y = \mathcal{Y}(\mathbf{x}) \stackrel{\text{def}}{:=} -\beta h(\mathbf{x}, t)$ with $\beta \approx 0.531$. This result was improved in [293] to $\beta \approx 0.555$ (taking into consideration the shoaling effects as well). However, it was shown later that this theoretical 'improvement' is immaterial when it comes to the description of real sea states [65].

Later, other choices of surface $\mathcal{Y}(\mathbf{x}, t)$ have been proposed. For example, in [182, 226] the surface $\mathcal{Y}(\mathbf{x}, t)$ was chosen to be genuinely unsteady (due to the free surface and/or bottom motion). This choice was motivated by improving also the nonlinear characteristics of the model. Some other attempts can be found in [60, 171, 227, 243, 328]. On the good side of these models we can mention accurate approximation of the dispersion relation up to intermediate water depths and, in some cases, established well-posedness results. On the other side, equations are often cumbersome with unclear mathematical properties (e.g. well-posedness, existence of travelling waves, etc.). Below we shall discuss more closely some of the models of this type.

For another recent complementary review of BOUSSINESQ-type and other nonlinear dispersive models, which discusses also applications and some numerical approaches, we refer to [46] and for a detailed analysis of the theory and asymptotics for the water wave problem we refer to [209].

The main purpose of this book is to propose a uniform derivation procedure to construct long wave approximations in globally flat and globally spherical geometries. This procedure will result in a hierarchical chain of shallow water equations of the first and second approximations having a succession of mathematically and physically substantial properties [188–191, 290].

In the first chapter we attempt to make a literature review on the topic of nonlinear weakly dispersive wave modelling in shallow water environments. This topic is so broad that we apologize in advance if we forgot to mention someone's work. It was not made on purpose. Moreover, we propose a unified modelling framework which encompasses some more or less known models in this field. Namely, we show how several well-known models can be derived from the base model by making judicious choices of dynamic variables and/or their fluxes. We also try to point out some important properties of some model equations that have not attracted so much the attention of the researchers. The second chapter will be devoted to some numerical questions. More precisely, we shall propose an adaptive finite volume discretization of a particular widely used dispersive wave model. The numerical method adaptivity is achieved by moving grid points to the locations where it is needed. The titles of the first two chapters include the wording '*on a globally flat space*'. It means basically that we consider a fluid flow with free surface on a CARTESIAN space, even if some bathymetry variations[1] are allowed, i.e. the bottom is not necessarily flat. The (globally) spherical geometries will be discussed in some detail in Chaps. 3 and 4.

This chapter is organized as follows. In Sect. 1.1 we derive the base model. However, the derivation procedure is quite general and it can be used to derive many other particular models, some of them being well-known and some possibly new. In Sect. 1.2 we propose also a weakly nonlinear version of the base model. Finally, in Sect. 1.4 we outline the main conclusions and perspectives of the present chapter. The present chapter is accompanied also by Appendix A devoted to long wave models based on the potential flow assumption and Appendix B introducing an intermediate class of models between FNWD and WNWD classes.

1.1 Base Model Derivation

First of all we describe the physical problem formulation along with underlying constitutive assumptions. Later on this formulation will be further simplified using the asymptotic (or perturbation) expansions methods [251].

Consider the flow of an ideal incompressible liquid in a physical three-dimensional space. We assume additionally that the fluid is homogeneous (i.e. the density $\rho = \text{const}$) and the gravity acceleration g is constant everywhere.[2] Without any loss of generality from now on we can set $\rho \equiv 1$. For the sake of simplicity, in this chapter we neglect all other forces (such as the CORIOLIS force and friction). Hence, we deal with pure gravity waves.

In order to describe the mathematical model, we introduce a CARTESIAN coordinate system $O\,x_1\,x_2\,y$. The horizontal plane $O\,x_1\,x_2$ coincides with the

[1]The amount of bathymetry variations allowed in our modelling will be discussed in the second chapter of this book.

[2]This assumption is quite realistic since the variation of this parameter around the Earth is less than 1%.

still water level $y = 0$ and the axis $O\,y$ points vertically upwards. By vector $\mathbf{x} = (x_1, x_2)$ we denote the horizontal coordinates. The fluid layer is bounded below by the solid (impenetrable) bottom $y = -h(\mathbf{x}, t)$ and above by the free surface $y = \eta(\mathbf{x}, t)$. The sketch of the fluid domain is schematically shown in Fig. 1.1.

The flow is considered to be completely determined if we find the velocity field $\mathbf{U}(\mathbf{x}, y, t) = \bigl(\mathbf{u}(\mathbf{x}, y, t), v(\mathbf{x}, y, t)\bigr)$ ($\mathbf{u} = (u_1, u_2)$ being the horizontal velocity components) along with the pressure field $p(\mathbf{x}, y, t)$ and the free surface elevation $\eta(\mathbf{x}, t)$, which satisfy the system of EULER equations:

$$\nabla \cdot \mathbf{u} + v_y = 0, \tag{1.1}$$

$$\mathbf{u}_t + (\mathbf{u} \cdot \nabla)\mathbf{u} + v\,\mathbf{u}_y + \nabla p = 0, \tag{1.2}$$

$$v_t + \mathbf{u} \cdot \nabla v + v\,v_y + p_y = -g, \tag{1.3}$$

where $\nabla = (\partial_{x_1}, \partial_{x_2})$ denotes the horizontal gradient operator. The EULER equations are completed with free surface kinematic and dynamic boundary conditions

$$\eta_t + \mathbf{u} \cdot \nabla \eta = v, \quad y = \eta(\mathbf{x}, t), v \tag{1.4}$$

$$p = 0, \quad y = \eta(\mathbf{x}, t). \tag{1.5}$$

Finally, on the bottom we impose the impermeability condition (i.e. the fluid particles cannot penetrate the solid boundary), which states that the normal velocity on the bottom vanishes:

$$h_t + \mathbf{u} \cdot \nabla h + v = 0, \quad y = -h(\mathbf{x}, t). \tag{1.6}$$

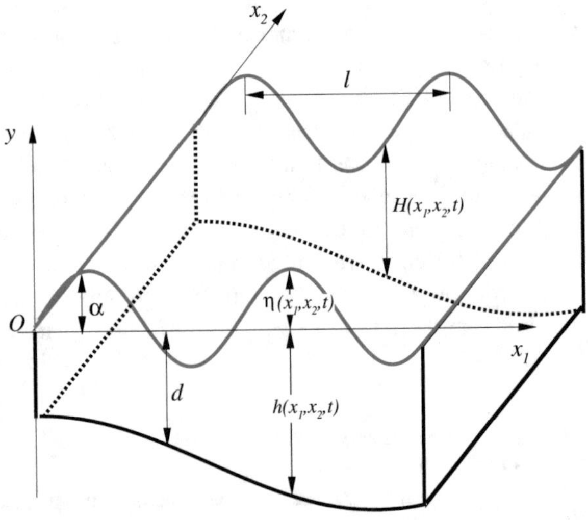

Fig. 1.1 Sketch of the fluid domain

Below we shall discuss also the components of the vorticity vector $\boldsymbol{\omega} = \nabla \times \mathbf{U}$, which are given by

$$\omega_1 = v_{x_2} - u_{2,y},$$
$$\omega_2 = -v_{x_1} + u_{1,y},$$
$$\omega_3 = u_{2,x_1} - u_{1,x_2}.$$

In the derivation of several approximate shallow water models, we shall also employ the integral form of equations (1.1)–(1.3) for arbitrary values of the variable $\zeta \in]-h, \eta[$:

$$\int_{\zeta}^{\eta} (\nabla \cdot \mathbf{u} + v_y)\,\mathrm{d}y = 0, \tag{1.7}$$

$$\int_{\zeta}^{\eta} (\mathbf{u}_t + (\mathbf{u} \cdot \nabla)\mathbf{u} + v\,\mathbf{u}_y + \nabla p)\,\mathrm{d}y = 0, \tag{1.8}$$

$$\int_{\zeta}^{\eta} (v_t + \mathbf{u} \cdot \nabla v + v\,v_y + p_y)\,\mathrm{d}y = -\int_{\zeta}^{\eta} g\,\mathrm{d}y. \tag{1.9}$$

These relations shall be used below for specific or arbitrary values of ζ.

1.1.1 Total Energy Conservation

Let us provide also the total mechanical energy conservation[3] for the full EULER equations (1.1)–(1.3). Each scalar equation in (1.2), (1.3) is multiplied by u_1, u_2 and v correspondingly and the resulting expressions are summed up. In this simple way, one obtains the equation which governs the evolution of the kinetic energy[4] $\mathcal{K} \stackrel{\text{def}}{:=} \frac{1}{2}(|\mathbf{u}|^2 + v^2)$:

$$\mathcal{K}_t + \mathbf{u} \cdot \nabla (\mathcal{K} + p) + v(\mathcal{K} + p)_y + g v = 0. \tag{1.10}$$

It is not difficult to check that the potential energy $\Pi \stackrel{\text{def}}{:=} g\,y$ satisfies the following differential equation:

$$\Pi_t + \mathbf{u} \cdot \nabla \Pi + v\,\Pi_y - g v = 0.$$

[3]A better term would be the 'energy balance' because of the presence of a source term due to gravity.

[4]We remind that for dimensional reasons the constant factor ρ might be included.

Summing up the last equation with (1.10) we obtain the conservation of the total mechanical energy $\mathcal{E} \stackrel{\text{def}}{:=} \mathcal{K} + \Pi$:

$$\mathcal{E}_t + \mathbf{u} \cdot \nabla (\mathcal{E} + p) + v (\mathcal{E} + p)_y = 0 .$$

By using the continuity equation (1.1) multiplied by $\mathcal{E} + p$, we obtain the energy conservation equation in the conservative form:

$$\mathcal{E}_t + \nabla \cdot \big((\mathcal{E} + p)\,\mathbf{u}\big) + \big((\mathcal{E} + p)\,v\big)_y = 0 . \tag{1.11}$$

The last equation, as many other important relations in Fluid Mechanics, admits an elegant integral form. To obtain it, let us integrate Eq. (1.11) over the fluid column and let us use boundary conditions (1.4)–(1.6) which yield:

$$\partial_t \int_{-h}^{\eta} \mathcal{E}\,\mathrm{d}y + \nabla \cdot \int_{-h}^{\eta} (\mathcal{E} + p)\,\mathbf{u}\,\mathrm{d}y + p\,|_{y=-h}\, h_t = 0 . \tag{1.12}$$

The last equation describes the change of the total energy averaged over the water depth. Equation (1.12) becomes a conservation law only in the case of the steady bottom, i.e. $h_t \equiv 0$. The moving bottom effect causes the change of the total energy $\mathcal{E}$ balance. This change is described by Eq. (1.12). Below it will be used to assess the ability of approximate long wave models to describe the total energy evolution during the wave propagation (*cf.* [104]).

1.1.2 Potential Flows

The EULER equations presented above describes the general vortical motion of an ideal incompressible fluid with free surface. If we assume the flow to be irrotational (i.e. $\boldsymbol{\omega} \equiv \mathbf{0}$), then under suitable topological conditions,[5] the velocity potential ϕ exists and the velocity field is given by two following relations:

$$\mathbf{u} = \nabla \phi , \qquad v = \phi_y . \tag{1.13}$$

In this case we obtain the classical potential flow model of the ideal incompressible fluid motion. Its mathematical formulation is detailed below [198, 296]. One has to determine the free surface elevation function η giving the fluid domain shape and the velocity potential ϕ defined inside the fluid domain and which satisfies the LAPLACE equation:

[5]The fluid domain has to be simply connected.

$$\nabla^2 \phi + \phi_{yy} = 0, \qquad \nabla^2 \stackrel{\text{def}}{:=} \partial^2_{x_1 x_1} + \partial^2_{x_2 x_2}. \tag{1.14}$$

The last LAPLACE equation is supplemented by the following boundary conditions:

Kinematic free surface boundary condition:

$$\eta_t + \nabla\phi \cdot \nabla\eta - \phi_y = 0, \qquad y = \eta(\mathbf{x}, t).$$

Dynamic free surface boundary condition:

$$\phi_t + \tfrac{1}{2}|\nabla\phi|^2 + \tfrac{1}{2}\phi_y^2 + g\eta = 0, \qquad y = \eta(\mathbf{x}, t).$$

Bottom impermeability condition:

$$h_t + \nabla\phi \cdot \nabla h + \phi_y = 0, \qquad y = -h(\mathbf{x}, t).$$

Above, we use implicitly the horizontal gradient operator $\nabla \stackrel{\text{def}}{:=} (\partial_{x_1}, \partial_{x_2})$. To obtain a well-posed problem, one must provide also the appropriate initial conditions for the free surface elevation $\eta(\mathbf{x}, 0)$ and for the trace of the potential on the free surface $\phi\big(\mathbf{x}, \eta(\mathbf{x}, 0), 0\big)$.

In this formulation, the fluid pressure p can be easily reconstructed using the so-called CAUCHY–LAGRANGE integral expression:

$$\frac{p}{\rho} = -\Big(\phi_t + \tfrac{1}{2}|\nabla\phi|^2 + \tfrac{1}{2}\phi_y^2 + g y\Big).$$

The fluid velocity field is obtained by simple differentiation of the velocity potential ϕ with respect to the spatial coordinates x_1, x_2, y.

Linearized Equations

During the study of approximate long wave models, the dispersion relation remains the first comparison criterium. Moreover, the properties of numerical schemes can be also discussed in terms of their dispersion relations, see e.g. [214]. That is why we discuss in this section the linear dispersion relation of the classical water wave problem which provides us with the reference solution, which will be approximated below by our modelling and numerical discretizations.

To compute the dispersion relation, we have to linearize the governing equations, presented in the preceding section, to obtain the so-called CAUCHY–POISSON problem [51, 266]. Consider an infinite horizontal strip $\mathbb{R}^2 \times]-d, 0[$ of constant depth $h(\mathbf{x}, t) \equiv d = \text{const} > 0$, which models the linearized (flattened) fluid domain. In this strip we have to solve the LAPLACE equation:

$$\nabla^2 \phi + \phi_{yy} = 0, \qquad (x_1, x_2, y) \in \mathbb{R}^2 \times]-d, 0[. \tag{1.15}$$

The free surface elevation $\eta(\mathbf{x}, t)$ is found from the boundary conditions posed on $y = 0$:

$$\eta_t - \phi_y = 0, \tag{1.16}$$

$$\phi_t + g\,\eta = 0. \tag{1.17}$$

On the solid flat bottom we have the usual impermeability condition:

$$\phi_y = 0, \qquad y = -d. \tag{1.18}$$

It is not difficult to see that Eqs. (1.15)–(1.18) represent the linearized version of the potential flow problem with free surface as described in Sect. 1.1.2.

A particular plane wave solution to the CAUCHY–POISSON problem (1.15)–(1.18) is given by the following formulas:

$$\eta(\mathbf{x}, t) = \alpha_0 \sin(\omega t - \mathbf{k} \cdot \mathbf{x}), \tag{1.19}$$

$$\phi(\mathbf{x}, y, t) = \frac{\alpha_0 g}{\omega \cosh(|\mathbf{k}| d)} \cosh\big(|\mathbf{k}|(y + d)\big) \cos(\omega t - \mathbf{k} \cdot \mathbf{x}), \tag{1.20}$$

where α_0 is the wave amplitude, $\mathbf{k} = (k_1, k_2)$ is the wave vector (i.e. the FOURIER-dual variable to $\mathbf{x}$) and $\omega(\mathbf{k})$ is the wave frequency, which is related to the wave vector $\mathbf{k}$ by the following *dispersion relation*:

$$\omega(\mathbf{k}) = \sqrt{g\,|\mathbf{k}|\,\tanh(|\mathbf{k}|\,d)}. \tag{1.21}$$

The last relation expresses mathematically the necessary condition for the existence of plane wave solutions (1.19), (1.20). It gives us the information regarding the velocity of linear waves, given by the *phase speed*:

$$c_p(\mathbf{k}) \stackrel{\text{def}}{:=} \sqrt{g\,d} \cdot \sqrt{\frac{\tanh(|\mathbf{k}|\,d)}{|\mathbf{k}|\,d}}. \tag{1.22}$$

We remind that the wave length λ and the wave number $|\mathbf{k}|$ are related as $\lambda \equiv \dfrac{2\pi}{|\mathbf{k}|}$. The direct examination of expressions (1.21) and (1.22) suggests the following immediate conclusions:

1. The wave frequency ω and the phase speed c_p do not depend on the direction of vector $\mathbf{k}$, but only on its magnitude $|\mathbf{k}|$ (the isotropy property).
2. From the inequality $\frac{\mathrm{d}c_p}{\mathrm{d}\lambda} > 0$ it follows that longer waves travel faster than the short ones (monotonicity) (see Fig. 1.2).
3. When $\lambda \to 0$ the value of the phase speed c_p vanishes (see also Fig. 1.2).

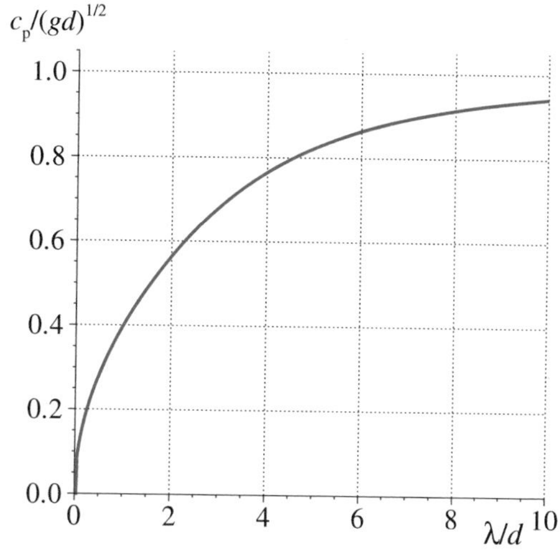

Fig. 1.2 Dependence of the phase velocity on the wavelength in the linearized EULER equations

During the construction of approximate water wave models and their numerical discretizations, the dispersion relation properties of the approximation should be systematically checked against the reference solutions given by Formulas (1.21) and (1.22).

1.1.3 Dimensionless Variables

In order to study the propagation of long gravity waves, we have to scale the governing equations (1.1)–(1.3) along with the boundary conditions (1.4)–(1.6). For this purpose we choose characteristic scales of the flow. Let ℓ, d and α be the typical (wave or basin) length, water depth and wave amplitude correspondingly (they are depicted in Fig. 1.1). Then, dimensionless independent variables can be introduced as follows:

$$x_{1,2}^{*} = \frac{x_{1,2}}{\ell}, \quad y^{*} = \frac{y}{d}, \quad t^{*} = \frac{t}{\ell\,/\sqrt{g\,d}}.$$

The dependent variables are scaled[6] as

[6] We would like to make a comment about the pressure scaling. For dimensional reasons we added in parentheses the fluid density ρ. However, it is not present in governing equations since for an incompressible flow of a homogeneous liquid ρ can be set to the constant 1 without loss of generality.

$$h^* = \frac{h}{d}, \quad \eta^* = \frac{\eta}{\alpha}, \quad p^* = \frac{p}{(\rho)\, g\, d}, \quad \mathbf{u}^* = \frac{\mathbf{u}}{\sqrt{g\, d}}, \quad v^* = \frac{v}{d\, \sqrt{g\, d\, /\, \ell}}.$$

The components of vorticity $\boldsymbol{\omega}$ are scaled as

$$\omega_{1,2}^* = \frac{\omega_{1,2}}{\sqrt{g\, d\, /\, d}}, \qquad \omega_3^* = \frac{\omega_3}{\sqrt{g\, d\, /\, \ell}}.$$

The scaled version of the EULER equations (1.1)–(1.3) read now

$$\nabla \cdot \mathbf{u} + v_y = 0, \tag{1.23}$$

$$\mathbf{u}_t + (\mathbf{u} \cdot \nabla)\, \mathbf{u} + v\, \mathbf{u}_y + \nabla p = 0, \tag{1.24}$$

$$\mu^2 \left(v_t + \mathbf{u} \cdot \nabla v + v\, v_y\right) + p_y = -1, \tag{1.25}$$

where we drop the asterisk symbol $*$ for the sake of notation compactness. Boundary conditions at the free surface similarly become

$$\varepsilon \left(\eta_t + \mathbf{u} \cdot \nabla \eta\right) = v, \quad y = \varepsilon\, \eta\, (\mathbf{x}, t), \tag{1.26}$$

$$p = 0, \quad y = \varepsilon\, \eta\, (\mathbf{x}, t). \tag{1.27}$$

It can be easily checked that the bottom boundary condition (1.6) remains invariant under this scaling. Finally, the scaled components of vorticity $\boldsymbol{\omega}^*$ are

$$\omega_1 = \mu^2\, v_{x_2} - u_{2,\,y},$$

$$\omega_2 = -\mu^2\, v_{x_1} + u_{1,\,y},$$

$$\omega_3 = u_{2,\,x_1} - u_{1,\,x_2}.$$

Above we introduced two important dimensionless parameters:

Nonlinearity $\varepsilon \overset{\text{def}}{:=} \alpha / d$ measures the deviation of waves with respect to the unperturbed water level,

Dispersion $\mu \overset{\text{def}}{:=} d / \ell$ indicates how long the waves are comparing to the mean depth (or equivalently how shallow the water is).

1.1.4 Long Wave Approximation

In shallow water systems the dynamic variables are the total water depth

$$\mathcal{H}\, (\mathbf{x}, t) \overset{\text{def}}{:=} h\, (\mathbf{x}, t) + \varepsilon\, \eta\, (\mathbf{x}, t)$$

and some vector $\bar{\mathbf{u}}(\mathbf{x}, t)$ which is supposed to approximate the horizontal velocity vector of the full model $\mathbf{u}(\mathbf{x}, y, t)$. In many works $\bar{\mathbf{u}}(\mathbf{x}, t)$ is chosen as the trace of the horizontal velocity $\mathbf{u}$ at certain surface $y = \mathcal{Y}_\sigma(\mathbf{x}, t)$ in the fluid bulk [182, 226, 254], i.e.

$$\bar{\mathbf{u}}(\mathbf{x}, t) \stackrel{\text{def}}{:=} \mathbf{u}\bigl(\mathbf{x}, \mathcal{Y}_\sigma(\mathbf{x}, t), t\bigr). \tag{1.28}$$

Another popular choice for the velocity variable consists in taking the depth-averaged velocity [87, 116, 264, 270, 283]:

$$\bar{\mathbf{u}}(\mathbf{x}, t) \stackrel{\text{def}}{:=} \frac{1}{\mathcal{H}(\mathbf{x}, t)} \int_{-h(\mathbf{x}, t)}^{\varepsilon\eta(\mathbf{x}, t)} \mathbf{u}(\mathbf{x}, y, t)\, \mathrm{d}y. \tag{1.29}$$

By applying the mean value theorem [345] to the last integral, we obtain that two approaches are mathematically formally equivalent:

$$\bar{\mathbf{u}}(\mathbf{x}, t) \equiv \mathbf{u}\bigl(\mathbf{x}, \mathcal{Y}_\xi(\mathbf{x}, t), t\bigr).$$

However, this time the surface $y = \mathcal{Y}_\xi(\mathbf{x}, t)$ remains unknown, while above it was explicitly specified. We only know that such surface exists.

Below we shall consider only long wave approximation to the full EULER equations. Namely, we assume that $\bar{\mathbf{u}}(\mathbf{x}, t)$ approximates the true horizontal velocity $\mathbf{u}(\mathbf{x}, y, t)$ to the order $\mathcal{O}(\mu^2)$, i.e.

$$\mathbf{u}(\mathbf{x}, y, t) = \bar{\mathbf{u}}(\mathbf{x}, t) + \mu^2 \tilde{\mathbf{u}}(\mathbf{x}, y, t). \tag{1.30}$$

By integrating the continuity equation (1.23) over the total depth and taking into account boundary conditions (1.6), (1.26) we obtain the mass conservation equation

$$\mathcal{H}_t + \boldsymbol{\nabla}\cdot(\mathcal{H}\bar{\mathbf{u}}) = -\mu^2 \boldsymbol{\nabla}\cdot(\mathcal{H}\mathcal{U}), \tag{1.31}$$

where

$$\mathcal{U}(\mathbf{x}, t) \stackrel{\text{def}}{:=} \frac{1}{\mathcal{H}(\mathbf{x}, t)} \int_{-h(\mathbf{x}, t)}^{\varepsilon\eta(\mathbf{x}, t)} \tilde{\mathbf{u}}(\mathbf{x}, y, t)\, \mathrm{d}y. \tag{1.32}$$

If we choose the variable $\bar{\mathbf{u}}$ to be depth-averaged, then $\mathcal{U}(\mathbf{x}, t) \equiv 0$ and the mass conservation equation (1.31) takes the very familiar form

$$\mathcal{H}_t + \boldsymbol{\nabla}\cdot(\mathcal{H}\bar{\mathbf{u}}) = 0.$$

Integration of equation (1.23) over the vertical coordinate in the limits from $-h(\mathbf{x}, t)$ to y and taking into account the bottom boundary condition (1.6) leads to the following representation for the vertical velocity in the fluid column:

$$v(\mathbf{x},\,y,\,t) \;=\; -\,\mathscr{D}h \;-\; (y\,+\,h)\nabla\cdot\bar{\mathbf{u}} \;+\; \mathcal{O}(\mu^2)\,, \tag{1.33}$$

where for the sake of simplicity we introduced the material (or total, or convective) derivative operator:

$$\mathscr{D}[\,\cdot\,] \stackrel{\text{def}}{:=} [\,\cdot\,]_t \;+\; \bar{\mathbf{u}}\cdot\nabla\,[\,\cdot\,]\,. \tag{1.34}$$

Below the powers of this operator will appear in our computations:

$$\mathscr{D}^k\,[\,\cdot\,] \stackrel{\text{def}}{:=} \underbrace{\mathscr{D}\cdot\mathscr{D}\cdot\ldots\cdot\mathscr{D}}_{k \text{ times}}[\,\cdot\,]\,, \qquad k\,\geqslant\,1\,.$$

We have to express *asymptotically* also the pressure field $p(\mathbf{x},\,y,\,t)$ in terms of the dynamic variables $\bigl(\mathcal{H}(\mathbf{x},\,t),\,\bar{\mathbf{u}}(\mathbf{x},\,t)\bigr)$. Thus, we integrate the vertical momentum equation (1.25) over the vertical coordinate in the limits from y to the free surface:

$$p(\mathbf{x},\,y,\,t) \;=\; \mu^2\int_y^{\varepsilon\eta(\mathbf{x},t)}\Bigl[\mathscr{D}v \;+\; v\,v_y \;+\; \mathcal{O}(\mu^2)\Bigr]\,\mathrm{d}y \;-\; y \;+\; \varepsilon\,\eta(\mathbf{x},\,t)\,. \tag{1.35}$$

The integrand can be expressed in terms of $\mathcal{H}(\mathbf{x},\,t)$ and $\bar{\mathbf{u}}(\mathbf{x},\,t)$ using representation (1.33):

$$\mathscr{D}v \;+\; v\,v_y \;=\; -\,(y\,+\,h)\,\mathscr{R}_1 \;-\; \mathscr{R}_2 \;+\; \mathcal{O}(\mu^2)\,,$$

where we defined

$$\mathscr{R}_1(\mathbf{x},\,t) \stackrel{\text{def}}{:=} \mathscr{D}(\nabla\cdot\bar{\mathbf{u}}) \;-\; (\nabla\cdot\bar{\mathbf{u}})^2\,,$$

$$\mathscr{R}_2(\mathbf{x},\,t) \stackrel{\text{def}}{:=} \mathscr{D}^2\,h\,.$$

Substituting the last result into the integral representation (1.35) and integrating it exactly in y leads to the following expression of the pressure field in the fluid layer:

$$p \;=\; \mathcal{H} \;-\; (y\,+\,h) \;-\; \mu^2\Bigl[\bigl(\mathcal{H} \;-\; (y\,+\,h)\bigr)\,\mathscr{R}_2 \,+ \\ \Bigl(\frac{\mathcal{H}^2}{2} \;-\; \frac{(y\,+\,h)^2}{2}\Bigr)\,\mathscr{R}_1\Bigr] \;+\; \mathcal{O}(\mu^4)\,. \tag{1.36}$$

Notice that this representation does not depend on the expression of the velocity correction $\tilde{\mathbf{u}}(\mathbf{x},\,t)$. If in the last formula we neglect terms of $\mathcal{O}(\mu^4)$ and return to physical variables, we can obtain the pressure reconstruction formula in the fluid bulk:

$$\frac{p}{\rho} = g\Big[\mathcal{H} - (y+h)\Big] - \Big[\big(\mathcal{H} - (y+h)\big)\mathscr{R}_2 + \Big(\frac{\mathcal{H}^2}{2} - \frac{(y+h)^2}{2}\Big)\mathscr{R}_1\Big].$$

We underline the fact that the last formula is accurate to the order $\mathcal{O}(\mu^4)$. This formula will be used in Chap. 2 in order to reconstruct the pressure field under a solitary wave, which undergoes some nonlinear transformations.

In order to obtain an evolution equation for the approximate horizontal velocity $\tilde{\mathbf{u}}(\mathbf{x}, t)$ we integrate over the vertical coordinate equation (1.24):

$$\int_{-h}^{\varepsilon\eta}\big[\mathbf{u}_t + (\mathbf{u}\cdot\nabla)\mathbf{u} + v\,\mathbf{u}_y\big]\,\mathrm{d}y + \nabla\int_{-h}^{\varepsilon\eta} p\,\mathrm{d}y - p|_{y=-h}\cdot\nabla h = 0. \tag{1.37}$$

The pressure variable can be easily eliminated from the last equation using the representation formula (1.36):

$$\nabla\int_{-h}^{\varepsilon\eta} p\,\mathrm{d}y - p|_{y=-h}\cdot\nabla h = \varepsilon\,\mathcal{H}\,\nabla h$$

$$-\,\mu^2\Big[\nabla\big(\tfrac{1}{3}\,\mathcal{H}^3\mathscr{R}_1 + \tfrac{1}{2}\,\mathcal{H}^2\mathscr{R}_2\big) - \mathcal{H}\,\nabla h\,\big(\tfrac{1}{2}\,\mathcal{H}\mathscr{R}_1 + \mathscr{R}_2\big)\Big] + \mathcal{O}(\mu^4).$$

Then, using the representation (1.33) for the vertical velocity v, we can write

$$\int_{-h}^{\varepsilon\eta} v\,\mathbf{u}_y\,\mathrm{d}y = -\mu^2\int_{-h}^{\varepsilon\eta}\big[\mathscr{D}h + (y+h)\nabla\cdot\tilde{\mathbf{u}}\big]\tilde{\mathbf{u}}_y\,\mathrm{d}y + \mathcal{O}(\mu^4)$$

$$= -\,\mu^2(\mathscr{D}h)\cdot\tilde{\mathbf{u}}\Big|_{y=-h}^{y=\varepsilon\eta} - \mu^2\,\nabla\cdot\tilde{\mathbf{u}}\underbrace{\int_{-h}^{\varepsilon\eta}(y+h)\tilde{\mathbf{u}}_y\,\mathrm{d}y}_{(*)} + \mathcal{O}(\mu^4).$$

The integral $(*)$ can be computed using integration by parts

$$\int_{-h}^{\varepsilon\eta}(y+h)\,\tilde{\mathbf{u}}_y\,\mathrm{d}y = \mathcal{H}\cdot\tilde{\mathbf{u}}\Big|^{y=\varepsilon\eta} - \mathcal{H}\mathcal{U}.$$

Combining together these results, we obtain the following asymptotic formula

$$\frac{1}{\mu^2}\int_{-h}^{\varepsilon\eta} v\,\mathbf{u}_y\,\mathrm{d}y = (\mathscr{D}h)\cdot\tilde{\mathbf{u}}\Big|_{y=-h} - \big[\mathscr{D}h + \mathcal{H}\,\nabla\cdot\tilde{\mathbf{u}}\big]\,\tilde{\mathbf{u}}\Big|^{y=\varepsilon\eta}$$

$$+\,\mathcal{H}\mathcal{U}\,\nabla\cdot\tilde{\mathbf{u}} + \mathcal{O}(\mu^2).$$

Finally, we take care of convective terms

$$\int_{-h}^{\varepsilon\eta} \big[\mathbf{u}_t + (\mathbf{u}\cdot\nabla)\,\mathbf{u}\big]\,\mathrm{d}y = \int_{-h}^{\varepsilon\eta} \mathscr{D}\bar{\mathbf{u}}\,\mathrm{d}y + \mu^2 \int_{-h}^{\varepsilon\eta} \mathscr{D}\tilde{\mathbf{u}}\,\mathrm{d}y +$$

$$\mu^2 \int_{-h}^{\varepsilon\eta} (\tilde{\mathbf{u}}\cdot\nabla)\,\bar{\mathbf{u}}\,\mathrm{d}y + \mathcal{O}(\mu^4) = \mathcal{H}\,\mathscr{D}\bar{\mathbf{u}} +$$

$$\mu^2\Big[\mathscr{D}[\mathcal{H}\mathcal{U}] - \mathscr{D}[\varepsilon\,\eta]\cdot\tilde{\mathbf{u}}\big|^{y=\varepsilon\eta} - \mathscr{D}h\cdot\tilde{\mathbf{u}}\big|_{y=-h} + \mathcal{H}\,(\mathcal{U}\cdot\nabla)\,\bar{\mathbf{u}}\Big] + \mathcal{O}(\mu^4).$$

Finally, we obtain

$$\int_{-h}^{\varepsilon\eta} \big[\mathbf{u}_t + (\mathbf{u}\cdot\nabla)\,\mathbf{u} + v\,\mathbf{u}_y\big]\,\mathrm{d}y = \mathcal{H}\,\mathscr{D}\bar{\mathbf{u}} - \mu^2\,\underbrace{\big[\mathscr{D}\mathcal{H} + \mathcal{H}\,\nabla\cdot\bar{\mathbf{u}}\big]}_{(**)}\cdot\,\tilde{\mathbf{u}}\big|^{y=\varepsilon\eta}$$

$$+\,\mu^2\Big[\mathscr{D}[\mathcal{H}\mathcal{U}] + \mathcal{H}\,(\mathcal{U}\cdot\nabla)\,\bar{\mathbf{u}} + \mathcal{H}\mathcal{U}\,\nabla\cdot\bar{\mathbf{u}}\Big].$$

From the mass conservation equation (1.31) we have

$$\mathscr{D}\mathcal{H} + \mathcal{H}\,\nabla\cdot\bar{\mathbf{u}} = -\mu^2\,\nabla\cdot(\mathcal{H}\mathcal{U}) = \mathcal{O}(\mu^2).$$

Thus, the term (**) can be asymptotically neglected. As a result we have

$$\int_{-h}^{\varepsilon\eta} \big[\mathbf{u}_t + (\mathbf{u}\cdot\nabla)\,\mathbf{u} + v\,\mathbf{u}_y\big]\,\mathrm{d}y =$$

$$\mathcal{H}\,\mathscr{D}\bar{\mathbf{u}} + \mu^2\Big[\mathscr{D}[\mathcal{H}\mathcal{U}] + \mathcal{H}\,(\mathcal{U}\cdot\nabla)\,\bar{\mathbf{u}} + \mathcal{H}\mathcal{U}\,\nabla\cdot\bar{\mathbf{u}}\Big] + \mathcal{O}(\mu^4).$$

Substituting all these intermediate results into depth-integrated horizontal momentum equation (1.37), we obtain the required evolution equation for $\bar{\mathbf{u}}$:

$$\bar{\mathbf{u}}_t + (\bar{\mathbf{u}}\cdot\nabla)\,\bar{\mathbf{u}} + \varepsilon\,\nabla\,\eta = \frac{\mu^2}{\mathcal{H}}\Big[\nabla\big(\tfrac{1}{3}\,\mathcal{H}^3\mathscr{R}_1 + \tfrac{1}{2}\,\mathcal{H}^2\mathscr{R}_2\big) - \mathcal{H}\,\nabla\,h\,\big(\tfrac{1}{2}\,\mathcal{H}\mathscr{R}_1 + \mathscr{R}_2\big)\Big]$$

$$-\,\frac{\mu^2}{\mathcal{H}}\Big[\mathscr{D}[\mathcal{H}\mathcal{U}] + \mathcal{H}\,(\mathcal{U}\cdot\nabla)\,\bar{\mathbf{u}} + \mathcal{H}\mathcal{U}\,\nabla\cdot\bar{\mathbf{u}}\Big]. \qquad (1.38)$$

The last equation may look complicated. However, it can be rewritten in a simpler way by pointing out explicitly the non-hydrostatic pressure effects. It turns out that it is advantageous to introduce the depth-integrated (but not depth-averaged) pressure:

$$\mathscr{P}(\mathcal{H},\,\bar{\mathbf{u}}) \stackrel{\text{def}}{:=} \int_{-h}^{\varepsilon\eta} p\,\mathrm{d}y = \frac{\mathcal{H}^2}{2} - \mu^2\Big(\tfrac{1}{3}\,\mathcal{H}^3\mathscr{R}_1 + \tfrac{1}{2}\,\mathcal{H}^2\mathscr{R}_2\Big). \qquad (1.39)$$

We introduce also the pressure trace $\check{p}$ at the bottom:

$$\check{p}(x,\,t) \stackrel{\text{def}}{:=} p|_{y=-h} = \mathcal{H} - \mu^2\Big(\tfrac{1}{2}\,\mathcal{H}^2\mathcal{R}_1 + \mathcal{H}\mathcal{R}_2\Big).$$

Using these new variables equation (1.38) becomes

$$\bar{\mathbf{u}}_t + (\bar{\mathbf{u}}\cdot\nabla)\,\bar{\mathbf{u}} + \frac{\nabla\mathcal{P}}{\mathcal{H}} = \frac{\check{p}\,\nabla h}{\mathcal{H}} - \frac{\mu^2}{\mathcal{H}}\Big[(\mathcal{H}\mathcal{U})_t + (\bar{\mathbf{u}}\cdot\nabla)(\mathcal{H}\mathcal{U}) + \mathcal{H}(\mathcal{U}\cdot\nabla)\,\bar{\mathbf{u}} + \mathcal{H}\mathcal{U}\,\nabla\cdot\bar{\mathbf{u}}\Big].$$

The derived system of equations admits an elegant conservative form:[7]

$$\mathcal{H}_t + \nabla\cdot[\mathcal{H}\mathbf{U}] = 0, \tag{1.40}$$

$$(\mathcal{H}\mathbf{U})_t + \nabla\cdot\Big[\mathcal{H}\bar{\mathbf{u}}\otimes\mathbf{U} + \mathcal{P}(\mathcal{H},\,\bar{\mathbf{u}})\cdot\mathbb{I} + \mu^2\mathcal{H}\mathcal{U}\otimes\bar{\mathbf{u}}\Big] = \check{p}\,\nabla h, \tag{1.41}$$

where we introduced a new velocity variable $\mathbf{U} \stackrel{\text{def}}{:=} \bar{\mathbf{u}} + \mu^2\,\mathcal{U}$ and $\mathbb{I} \in \mathrm{Mat}_{\,2\times 2}(\mathbb{R})$ is the identity matrix. Operator $\otimes$ is the tensorial product, i.e. for two vectors $\mathbf{u} \in \mathbb{R}^m$ and $\mathbf{v} \in \mathbb{R}^n$

$$\mathbf{u}\otimes\mathbf{v} \stackrel{\text{def}}{:=} (u_i\cdot v_j)_{1\leqslant j\leqslant n}^{1\leqslant i\leqslant m} \in \mathrm{Mat}_{\,m\times n}(\mathbb{R}).$$

From now on Eqs. (1.40), (1.41) will be referred to as the *base model* of our study. In order to close the last system of equations (1.40), (1.41), we have to express the variable $\mathcal{U}$ in terms of other dynamic variables $\mathcal{H}(\mathbf{x},\,t)$ and $\bar{\mathbf{u}}(\mathbf{x},\,t)$. Several popular choices will be discussed below. Notice also that *nowhere* in the derivation above the flow irrotationality was assumed.

Remark 1.1 Notice that taking formally the limit $\mu \to 0$ in Eqs. (1.40), (1.41) yields straightforwardly the well-known Nonlinear Shallow Water (NSW or SAINT-VENANT) Equations [84]. Thus, our base model satisfies the BOHR *correspondence principle*.[8] This property is crucial for robust physical wave modelling in coastal environments. Indeed, a wave approaching continental shelf undergoes nonlinear transformations: the water depth is decreasing and the wave amplitude grows, which

[7] This form becomes truly conservative (in the sense of hyperbolic conservation laws) only on the flat bottom, i.e. $h(\mathbf{x},\,t) = h_0 = \text{const} \Rightarrow \nabla h \equiv \mathbf{0}$.

[8] This principle was formulated by Niels Bohr [33]. Loosely speaking, this principle states that Quantum Mechanics reproduces Classical Mechanics in the limit of large quantum numbers. Correspondingly, a nonlinear dispersive model should describe correctly the propagation of non-dispersive waves in the limit when the dispersion vanishes.

often leads to the formation of undular bores. The model has to follow these transformations. Mathematically it means that the model equations should encompass a range of physical regimes varying from fairly shallow water to intermediate depths [143]. There exists an option of coupling different hydrodynamic models as it was done, e.g. in [224]. However, the coupling represents a certain number of difficulties, e.g.

- Boundary conditions at artificial interfaces?
- How to determine automatically the physical regime?
- Dynamic evolution and handling of model applicability areas...

Consequently, in this book we let the physical model to do this work for us.

Energy Conservation

We would like to raise the question of energy conservation in nonlinear dispersive wave models. The full EULER equations naturally have this property. So, it is a priori natural to require that a good approximation to EULER equations conserves the energy as well [133]. An energy conservation equation can be established for the base model (1.40), (1.41) for some choices of the variable $\mathcal{U}\,(\mathcal{H},\ \bar{\mathbf{u}})$. For instance, the classical SGN model discussed in the following section enjoys this property (it corresponds to the choice $\mathcal{U}\ \equiv\ \mathbf{0}$). On moving bottoms this property was discussed in [133]. Here we provide only the final result, i.e. the total energy equation for SGN model on a general moving bottom:[9]

$$(\mathcal{H}\,\mathscr{E})_t\;+\;\boldsymbol{\nabla}\cdot\Bigl[\,\mathcal{H}\,\bar{\mathbf{u}}\,\Bigl(\mathscr{E}\;+\;\frac{\mathscr{P}}{\mathcal{H}}\Bigr)\Bigr]\;=\;-\check{p}\,h_t\,, \tag{1.42}$$

where the total energy $\mathscr{E}$ is defined as

$$\mathscr{E}\;\stackrel{\mathrm{def}}{:=}\;\tfrac{1}{2}\,|\,\bar{\mathbf{u}}\,|^{2}\;+\;\tfrac{1}{6}\,\mathcal{H}^{2}\,(\boldsymbol{\nabla}\cdot\bar{\mathbf{u}})^{2}\;+\;\tfrac{1}{2}\,\mathcal{H}\,(\mathscr{D}h)\,(\boldsymbol{\nabla}\cdot\bar{\mathbf{u}})\;+\;\tfrac{1}{2}\,(\mathscr{D}h)^{2}\;+\;\frac{g}{2}\,(\mathcal{H}\;-\;2\,h)\,.$$

For other choices of the closure $\mathcal{U}\,(\mathcal{H},\ \bar{\mathbf{u}})$ this question of energy conservation has to be studied separately.

Remark 1.2 Recently, Clamond et al. [74] proposed a dispersion-improved SGN-type model which enjoys the energy conservation property. The method employed in that study is the variational approach: the preservation of the variational structure is crucial for the preservation of several invariants.

[9] Of course, this equation becomes a conservation law only when the bottom is static (but not necessarily flat).

Galilean Invariance

The same questions can be raised about the GALILEAN invariance property as well. This property is of fundamental importance for any mathematical model that provides a physically sound description of water waves (stemming from Classical Mechanics and Classical Physics). Some thoughts and tentative corrections can be found in [96, 112]. The base model (1.40), (1.41) is GALILEAN invariant under reasonable assumptions on the closure velocity vector $\mathcal{U}$.

GALILEAN invariance principle states that all mechanical laws are the same in any *inertial* frame of reference [208]. Consequently, the mathematical form of governing equations should be the same as well. It was proposed by Galilei in [138]. Consider the horizontal GALILEAN boost transformation between two inertial frames of reference:

$$\mathbf{x}' = \mathbf{x} + \mathbf{C}\,t\,, \qquad y' = y\,, \qquad t' = t\,, \tag{1.43}$$

where $\mathbf{C}$ is a constant motion speed of the new coordinate system (with primes) relatively to the initial one (without primes). Notice that scalar quantities such as $\mathcal{H}(\mathbf{x},\,t)$ and $h(\mathbf{x},\,t)$ remain invariant since they are defined as distances between two points and distances are preserved by the GALILEAN transformation (1.43). Let us see how the horizontal velocity variable changes under the GALILEAN transformation:

$$\mathbf{u}(\mathbf{x},\,y,\,t) \overset{\text{def}}{:=} \frac{\mathrm{d}\mathbf{x}}{\mathrm{d}t} = \frac{\mathrm{d}\mathbf{x}'}{\mathrm{d}t} - \mathbf{C} \overset{\text{def}}{=:} \mathbf{u}'(\mathbf{x}',\,y',\,t') - \mathbf{C}\,.$$

It is not difficult to understand that the same transformation rule applies to $\bar{\mathbf{u}}(\mathbf{x},\,t)$ regardless if it is defined as a trace or depth-averaged velocity:

$$\bar{\mathbf{u}} = \bar{\mathbf{u}}' - \mathbf{C}\,.$$

Indeed, the last claim is obvious for the case of the trace operator. Let us check it for the depth-averaging operator:

$$\bar{\mathbf{u}}(\mathbf{x},\,t) \overset{\text{def}}{:=} \frac{1}{\mathcal{H}} \int_{-h}^{\varepsilon\eta} \mathbf{u}\,\mathrm{d}y = \frac{1}{\mathcal{H}'} \int_{-h'}^{\varepsilon\eta'} \bigl(\mathbf{u}' - \mathbf{C}\bigr)\,\mathrm{d}y' =$$

$$\frac{1}{\mathcal{H}'} \int_{-h'}^{\varepsilon\eta'} \mathbf{u}'\,\mathrm{d}y' - \mathbf{C} \overset{\text{def}}{=:} \bar{\mathbf{u}}'(\mathbf{x}',\,t') - \mathbf{C}\,.$$

If the velocity $\bar{\mathbf{u}}(\mathbf{x},\,t)$ is defined in a different way, its transformation rule has to be studied separately. From the definition (1.32) it follows that the velocity correction $\mathcal{U}$ should remain invariant under the GALILEAN boost (1.43) (since it is defined as a difference of two velocities):

$$\mathcal{U}' \equiv \mathcal{U}. \tag{1.44}$$

In the following we shall assume that the chosen closure $\mathcal{U}(\mathcal{H}, \bar{\mathbf{u}})$ satisfy the last transformation rule.

Finally, let us discuss the invariance of the base model (1.40), (1.41). Basically, this property follows from the transformation rule (1.44), from the fact that $\mathbf{C} =$ const and the following observation:[10]

$$\mathscr{D}\mathcal{H} \equiv \mathscr{D}'\mathcal{H}', \qquad \mathscr{D}\bar{\mathbf{u}} \equiv \mathscr{D}'\bar{\mathbf{u}}'.$$

The pressure variables $\mathscr{P}$ and $\check{p}$ remain invariant as well, since they depend on velocity through $\mathscr{R}_1$ and $\mathscr{R}_2$, which depend in their term only on the full derivative and divergence of the velocity $\bar{\mathbf{u}}$. Thus, the base model (1.40), (1.41) is GALILEAN invariant under not very restrictive assumptions made above.

Remark 1.3 Many BOUSSINESQ-type equations derived and published in the literature are not GALILEAN invariant. As such a classical example we can mention Peregrine's (1967) system [264]. In [133] it was shown how to derive a weakly nonlinear model from the fully nonlinear one in such a way that the reduced BOUSSINESQ-type model has the GALILEAN invariance and energy conservation properties.

1.1.5 Serre–Green–Naghdi Equations

The celebrated SERRE–GREEN–NAGHDI (SGN) equations can be obtained by choosing the simplest possible closure, i.e.

$$\mathcal{U} \equiv \mathbf{0}.$$

This closure follows from the fact that the velocity variable $\bar{\mathbf{u}}$ chosen in SGN equations is precisely the depth-averaged velocity. Let us discuss this point more thoroughly. For this, we separate the discussion in two branches. Indeed, during the derivation of the GREEN–Naghdi (GN) equations it is assumed from the beginning that the horizontal velocity $\mathbf{u}$ does not depend on the vertical coordinate y. The

[10]Let us prove, for example, the first identity:

$$\mathscr{D}\mathcal{H} \equiv \mathcal{H}_t + \bar{\mathbf{u}}\cdot\nabla\mathcal{H} = \mathcal{H}'_{t'} + \mathbf{C}\nabla\mathcal{H}' + (\bar{\mathbf{u}}' - \mathbf{C})\cdot\nabla\mathcal{H}' = \\ \mathcal{H}'_{t'} + \bar{\mathbf{u}}'\cdot\nabla\mathcal{H}' \equiv \mathscr{D}'\mathcal{H}'.$$

velocity $\bar{\mathbf{u}}$ in the GN model coincides with $\mathbf{u}$ thanks to this property.[11] That is why, from the representation (1.30) we conclude that $\tilde{\mathbf{u}}(\mathbf{x}, y, t) \equiv \mathbf{0}$ and by Definition (1.32), we have that $\mathcal{U}(\mathbf{x}, t) \equiv \mathbf{0}$ in GN equations. Now, let us consider the SERRE equations. In this model the variable $\bar{\mathbf{u}}$ is taken to be precisely the depth-averaged velocity $\mathbf{u}$ according to Definition (1.29). In this derivation, $\mathbf{u}$ may depend on y, in other words, in representation (1.30) $\tilde{\mathbf{u}} \neq \mathbf{0}$. Hence, SERRE equations and GN equations are derived under slightly different assumptions. Nevertheless, in SERRE equations, one obtains that $\mathcal{U}(\mathbf{x}, t) \equiv \mathbf{0}$ as well. To verify this statement, one may integrate Equation (1.30) over the water depth and divide by $\mathcal{H}$. After the integration, on both sides one obtains the depth-averaged velocity $\bar{\mathbf{u}}$. Hence, by Definition (1.32) one invariably obtains that $\mathcal{U}(\mathbf{x}, t) \equiv \mathbf{0}$. To make a conclusion, the quantity $\tilde{\mathbf{u}}$ is different in SERRE and GREEN–NAGHDI equations. However, the quantity $\mathcal{U}$ is the same and it assumes the zero value. This observation allows us to unify both models and to speak of the so-called SERRE–GREEN–NAGHDI (SGN) equations.

By substituting the proposed closure into Eqs. (1.40), (1.41), we obtain the SGN equations:

$$\mathcal{H}_t + \nabla \cdot [\mathcal{H}\bar{\mathbf{u}}] = 0, \tag{1.45}$$

$$(\mathcal{H}\bar{\mathbf{u}})_t + \nabla \cdot \Big[\mathcal{H}\bar{\mathbf{u}} \otimes \bar{\mathbf{u}} + \mathscr{P}(\mathcal{H}, \bar{\mathbf{u}}) \cdot \mathbb{I}\Big] = \check{p}\,\nabla h, \tag{1.46}$$

where $\mathscr{P}(\mathcal{H}, \bar{\mathbf{u}})$ was defined in (1.39). The last equation can be written in a non-conservative form as well:

$$\bar{\mathbf{u}}_t + (\bar{\mathbf{u}} \cdot \nabla)\bar{\mathbf{u}} + \frac{\nabla \mathscr{P}}{\mathcal{H}} = \frac{\check{p}\,\nabla h}{\mathcal{H}}. \tag{1.47}$$

The SGN equations have been rediscovered independently by a number of authors. The steady version of these equations can be already found in the study of Rayleigh [221]. Then, this model in 1D was derived by Serre [282, 283] and by Su and Gardner [297]. A modern derivation was done by Green et al. [148]. Later, in Soviet Union this system was derived also by Pelinovsky and Zheleznyak [344] (see also Appendix A for a more detailed discussion). More recently, modern derivations of these equations based on variational principles have been proposed. Namely, Miles and Salmon [239] gave a derivation in LAGRANGIAN (e.g. particle) description. The variational derivation in EULERIAN description was given by Fedotova and Karepova [127] and later by Kim et al. [196] and Clamond and Dutykh [72]. Recently the multi-symplectic structure for SGN equations was proposed in [64].

[11] We have in mind here the property of being independent of y.

Energy Balance

In this section we shall illustrate the connection between the energy balance law (1.42) in SGN equations and the corresponding conservation law (1.11) in the full EULER equations in the case of the moving bottom. In other words, we are going to demonstrate the consistency of the energy equation in the SGN model [179].

Indeed, the energy density in the full EULER equations (1.1)–(1.3) in dimensionless variables becomes

$$\mathcal{E} = \tfrac{1}{2}\,(|\,\mathbf{u}\,|^2 + \mu^2 v^2) + y\,.$$

Let us investigate how the depth-averaging operator acts on the quantity $\mathcal{E}$. To achieve this goal, we substitute in the dimensionless expression above, the asymptotic expansions of $\mathbf{u}$ and v in terms of the small parameter μ and we apply this operator:

$$\frac{1}{\mathcal{H}}\int_{-h}^{\varepsilon\eta} \mathcal{E}\,\mathrm{d}y = \frac{1}{\mathcal{H}}\int_{-h}^{\varepsilon\eta} \Bigl\{\frac{1}{2}\Bigl[\bar{\mathbf{u}}\cdot\bar{\mathbf{u}} + 2\mu^2\,\bar{\mathbf{u}}\cdot\tilde{\mathbf{u}} + \mu^2\bigl(\mathscr{D}h + (y+h)\,\boldsymbol{\nabla}\cdot\bar{\mathbf{u}}\bigr)^2\Bigr] + y\Bigr\}\,\mathrm{d}y + \mathcal{O}(\mu^4)\,.$$

After performing the integration over y and taking into account that for SGN equations $\mathcal{U} \equiv \mathbf{0}$, we obtain

$$\frac{1}{\mathcal{H}}\int_{-h}^{\varepsilon\eta} \mathcal{E}\,\mathrm{d}y = \mathscr{E} + \mathcal{O}(\mu^4)\,, \tag{1.48}$$

where

$$\mathscr{E} = \tfrac{1}{2}\,|\,\bar{\mathbf{u}}\,|^2 + \mu^2\Bigl[\tfrac{1}{6}\,\mathcal{H}^2\,(\boldsymbol{\nabla}\cdot\bar{\mathbf{u}})^2 + \tfrac{1}{2}\,\mathcal{H}\,(\mathscr{D}h)\,(\boldsymbol{\nabla}\cdot\bar{\mathbf{u}}) + \tfrac{1}{2}\,(\mathscr{D}h)^2\Bigr] + \frac{\mathcal{H} - 2h}{2}\,.$$

Taking into account the asymptotic relation (1.48) between the depth-averaged energy of the 3D flow and the function $\mathscr{E}$ defined above, it is natural to identify this quantity with the (total) *energy* of the SGN system (1.45), (1.46).

Direct Derivation of the Energy Equation

For the sake of completeness, in this section we shall provide a direct derivation of the energy balance equation from the SGN system. Let us multiply Eq. (1.47) by $\bar{\mathbf{u}}$ and take into account the identity $\bar{\mathbf{u}}\cdot(\bar{\mathbf{u}}\cdot\boldsymbol{\nabla})\,\bar{\mathbf{u}} \equiv \tfrac{1}{2}\,\bar{\mathbf{u}}\cdot\boldsymbol{\nabla}\,|\,\bar{\mathbf{u}}\,|^2$, we then obtain:

$$\mathscr{D}\bigl(\tfrac{1}{2}\,|\,\bar{\mathbf{u}}\,|^2\bigr) + \frac{1}{\mathcal{H}}\,\nabla\cdot(\mathscr{P}\,\bar{\mathbf{u}}) - \underbrace{\Bigl(\frac{\mathscr{P}}{\mathcal{H}}\,\nabla\cdot\bar{\mathbf{u}} + \frac{\check{p}}{\mathcal{H}}\,\mathscr{D}h\Bigr)}_{(\mathcal{S})} = -\frac{\check{p}}{\mathcal{H}}\,h_t\,. \tag{1.49}$$

Two last terms $(\mathcal{S})$ on the left-hand side can be drastically simplified by using the short-hand notation for $\mathscr{R}_{1,\,2}$ along with the definitions of $\check{p}$ and $\mathscr{P}$:

$$\begin{aligned}
(\mathcal{S}) \equiv \frac{\mathscr{P}}{\mathcal{H}}\,\nabla\cdot\bar{\mathbf{u}} + \frac{\check{p}}{\mathcal{H}}\,\mathscr{D}h &= \frac{\mathcal{H}}{2}\,\nabla\cdot\bar{\mathbf{u}} + \mathscr{D}h \\
-\mu^2\Bigl[\Bigl(\frac{\mathcal{H}^2}{3}\,\mathscr{R}_1 + \frac{\mathcal{H}}{2}\,\mathscr{R}_2\Bigr)\nabla\cdot\bar{\mathbf{u}} + \mathscr{D}h\,\Bigl(\frac{\mathcal{H}}{2}\,\mathscr{R}_1 + \mathscr{R}_2\Bigr)\Bigr] &= -\mathscr{D}\Bigl(\frac{\mathcal{H} - 2h}{2}\Bigr) \\
-\mu^2\Bigl[\mathscr{D}\Bigl(\frac{\mathcal{H}^2}{2}\,(\nabla\cdot\bar{\mathbf{u}})^2\Bigr) + \mathscr{D}\Bigl(\frac{\mathcal{H}}{2}\,(\nabla\cdot\bar{\mathbf{u}})\,\mathscr{D}h\Bigr) + \mathscr{D}\Bigl(\frac{(\mathscr{D}h)^2}{2}\Bigr)\Bigr] &= \\
&- \mathscr{D}\bigl(\mathscr{E} - \tfrac{1}{2}\,|\,\bar{\mathbf{u}}\,|^2\bigr)\,.
\end{aligned}$$

The last sequence of equalities suggests that Eq. (1.49) can be rewritten in an equivalent form:

$$\mathscr{E}_t + \bar{\mathbf{u}}\cdot\nabla\,\mathscr{E} + \frac{1}{\mathcal{H}}\,\nabla\cdot(\mathscr{P}\,\bar{\mathbf{u}}) = -\frac{\check{p}}{\mathcal{H}}\,h_t\,.$$

The last equation can be multiplied by the total water depth $\mathcal{H}$ and using the continuity equation (1.45) multiplied by $\mathscr{E}$, we obtain the same energy balance equation, but in the conservative form:

$$(\mathcal{H}\,\mathscr{E})_t + \nabla\cdot\Bigl[\mathcal{H}\Bigl(\mathscr{E} + \frac{\mathscr{P}}{\mathcal{H}}\Bigr)\bar{\mathbf{u}}\Bigr] = -\,\check{p}\,h_t\,. \tag{1.50}$$

We shall offer another view on Eq. (1.50). In fact, this Equation can be derived from the depth-integrated energy balance equation (1.12) of the 3D EULER equations. Indeed, by substituting the asymptotic expansions for the horizontal velocity vector $\mathbf{u}$, fluid pressure p and taking into account the closure relation $\mathcal{U} \equiv \mathbf{0}$, we obtain:

$$\begin{aligned}
0 = \partial_t \int_{-h}^{\varepsilon\eta} \mathcal{E}\,\mathrm{d}y + \nabla\cdot\int_{-h}^{\varepsilon\eta} (\mathcal{E} + p)\,\mathbf{u}\,\mathrm{d}y + p\,|_{y=-h}\,h_t &= \\
(\mathcal{H}\,\mathscr{E})_t + \nabla\cdot\Bigl[\mathcal{H}\Bigl(\mathscr{E} + \frac{\mathscr{P}}{\mathcal{H}}\Bigr)\bar{\mathbf{u}}\Bigr] + \check{p}\,h_t + \mathcal{O}(\mu^4)\,.
\end{aligned}$$

The last result shows that Eq. (1.50) approximates the depth-integrated energy equation of the initial 3D hydrodynamic system to the order $\mathcal{O}(\mu^4)$. This observation gives us a rational basis to speak about the (asymptotic) agreement (or consistency) between the energy balance equations of the full and approximate hydrodynamic models.

Remark 1.4 Let us describe one more time what we call by the (asymptotic) *agreement* (or *consistency*) property of the energy balance equation in an approximate model. Namely, we take the energy conservation equation of the 3D complete model. We substitute the corresponding asymptotic expansions of all variables entering this equation up to the same asymptotic order, followed eventually by other simplifications or substitutions. If the result of this operation coincides with a differential consequence of the mass and momentum balance equations of the approximate model, then this differential consequence will be called *consistent*[12] with the energy conservation of the parent hydrodynamic model. In this section we showed that SGN model possesses this important property.

1.1.6 Other Particular Cases

The scope of the present section is slightly broader than its title may suggest. More precisely, we consider the whole class of models where the velocity variable is defined on a certain surface inside the fluid, see Eq. (1.28) for the definition. We show in this section that the base model (1.40), (1.41) can be closed using the partial irrotationality condition. Namely, we assume that only two horizontal components of vorticity vanish, i.e.

$$\mathbf{u}_{y} \;=\; \mu^{2}\,\boldsymbol{\nabla}\,v\,. \tag{1.51}$$

Integration of this identity over y and using representations (1.30), (1.33) leads

$$\tilde{\mathbf{u}}\,(\mathbf{x},\,y,\,t) \;=\; -\,(y\,+\,h)\,\bigl[\,\boldsymbol{\nabla}\,(\mathscr{D} h)\,+\,\boldsymbol{\nabla}\,h\,(\boldsymbol{\nabla}\boldsymbol{\cdot}\bar{\mathbf{u}})\,\bigr] \\ -\;\frac{(y\,+\,h)^{2}}{2}\;\boldsymbol{\nabla}\,(\boldsymbol{\nabla}\boldsymbol{\cdot}\bar{\mathbf{u}})\;+\;\tilde{\mathbf{u}}\big|_{\,y\,=\,-h}\;+\;\mathcal{O}\,(\mu^{2})\,.$$

Consequently, from (1.30) we obtain

$$\mathbf{u}\,(\mathbf{x},\;y,\;t)\;=\;\bar{\mathbf{u}}\,+\,\mu^{2}\Bigl[(y+h)\,\mathscr{A}\,+\,\frac{1}{2}\,(y+h)^{2}\,\mathscr{B}\,+\,\mathscr{C}\Bigr]\,+\,\mathcal{O}\,(\mu^{4})\,, \tag{1.52}$$

where we introduced for simplicity the following notation:

[12] Or in other words, we may say that the energy conservation equation of the approximate model *agrees* with the energy conservation of the parent model.

$$\mathscr{A}(\mathbf{x},\,t) \stackrel{\text{def}}{:=} -\nabla(\mathscr{D}h) - \nabla h\,(\nabla \cdot \bar{\mathbf{u}}),$$

$$\mathscr{B}(\mathbf{x},\,t) \stackrel{\text{def}}{:=} -\nabla(\nabla \cdot \bar{\mathbf{u}}),$$

$$\mathscr{C}(\mathbf{x},\,t) \stackrel{\text{def}}{:=} \tilde{\mathbf{u}}\,\Big|_{y\,=\,-h}\,.$$

Let us evaluate both sides of equation (1.52) at $y_\sigma = \mathcal{Y}_\sigma(\mathbf{x},\,t)$. According to (1.28) we must have

$$\mathbf{u}\big(\mathbf{x},\,\mathcal{Y}_\sigma(\mathbf{x},\,t),\,t\big) \equiv \bar{\mathbf{u}}(\mathbf{x},\,t)\,.$$

Consequently, we have

$$\mathscr{C}(\mathbf{x},\,t) \equiv -(y_\sigma + h)\,\mathscr{A} - \frac{1}{2}\,(y_\sigma + h)^2\,\mathscr{B}\,.$$

Thus, coefficient $\mathscr{C}$ can be eliminated from (1.52) to give the following representation

$$\mathbf{u}(\mathbf{x},\,y,\,t) = \bar{\mathbf{u}} + \mu^2\Big[(y - y_\sigma)\,\mathscr{A} + \frac{1}{2}\,\big[(y+h)^2 - (y_\sigma + h)^2\big]\,\mathscr{B}\Big] + \mathcal{O}(\mu^4)\,.$$

Substituting the last result into equation (1.32) yields the required closure relation:

$$\mathcal{U}(\mathcal{H},\,\bar{\mathbf{u}}) = \Big[\frac{\mathcal{H}}{2} - (y_\sigma + h)\Big]\,\mathscr{A} + \Big[\frac{1}{6}\,\mathcal{H}^2 - \frac{1}{2}\,(y_\sigma + h)^2\Big]\,\mathscr{B} + \mathcal{O}(\mu^2)\,. \tag{1.53}$$

To summarize, under the assumption (1.51) that the first two components of the vorticity field vanish, we can propose a closure to the base model, after neglecting the terms of order $\mathcal{O}(\mu^2)$ in (1.53).

Depth-Averaged Velocity

It is interesting to obtain also the 3D velocity reconstruction formula in the case, where $\bar{\mathbf{u}}(\mathbf{x},\,t)$ is defined as the depth-averaged velocity (1.29). To do it, we average Eq. (1.52) over the depth:

$$\frac{1}{\mathcal{H}}\int_{-h}^{\varepsilon\eta}\mathbf{u}(\mathbf{x},\,y,\,t)\,\mathrm{d}y = \bar{\mathbf{u}}(\mathbf{x},\,t) + \mu^2\Big[\frac{\mathcal{H}}{2}\,\mathscr{A} + \frac{\mathcal{H}^2}{6}\,\mathscr{B} + \mathscr{C}\Big] + \mathcal{O}(\mu^4)\,.$$

Using the definition (1.29) of the depth-averaged velocity, we conclude that

$$\mathscr{C} = -\frac{\mathcal{H}}{2}\,\mathscr{A} - \frac{\mathcal{H}^2}{6}\,\mathscr{B} + \mathcal{O}(\mu^2)\,.$$

By substituting the last expression into (1.52) we obtain the desired representation:

$$\mathbf{u}(\mathbf{x}, y, t) = \bar{\mathbf{u}}(\mathbf{x}, t) + \mu^2\Big[\Big(\frac{\mathcal{H}}{2} - y - h\Big)\cdot\Big(\boldsymbol{\nabla}\mathscr{D}h + (\boldsymbol{\nabla}\cdot\bar{\mathbf{u}})\boldsymbol{\nabla}h\Big)$$
$$+ \Big(\frac{\mathcal{H}^2}{6} - \frac{(y+h)^2}{2}\Big)\boldsymbol{\nabla}(\boldsymbol{\nabla}\cdot\bar{\mathbf{u}})\Big] + \mathcal{O}(\mu^4). \qquad (1.54)$$

The last formula will be used in Chap. 2 in order to reconstruct the 3D field under a propagating wave, which undergoes some nonlinear transformations. Formula (1.54) shows also that in shallow water flows the velocity distribution in the vertical coordinate y is nearly quadratic.

Remark 1.5 We underline that formula (1.54) is obtained under the assumption that the flow is irrotational. Without this assumption, in the most general case we can only use formula (1.30) by neglecting terms of the order $\mathcal{O}(\mu^2)$. In other words, the velocity variable $\bar{\mathbf{u}}(\mathbf{x}, t)$ approximates the 3D velocity field $\mathbf{u}(\mathbf{x}, y, t)$ throughout the fluid to the order $\mathcal{O}(\mu^2)$. However, in many applications this accuracy is not enough.

Lynett–Liu's Model

It can be shown that the base model (1.40), (1.41) supplemented by the proposed closure (1.53) is asymptotically equivalent to the well-known Lynett–Liu (2002) model derived in [226] under an additional assumption that the initial 3D flow is *irrotational*. This claim is true only up to the approximation order $\mathcal{O}(\mu^4)$ and it can be checked by straightforward but tedious calculations.

Various choices of the level y_σ, where the horizontal velocity is defined, allow to obtain in a straightforward manner the fully nonlinear analogues of various existing models. Some of popular choices are discussed below.

Mei–Le Méhauté's Model

Consider the horizontal velocity variable defined at the bottom, i.e.

$$y_\sigma = -h(\mathbf{x}, t).$$

Substituting this value into (1.53) we obtain straightforwardly the following closure:

$$\mathcal{U}(\mathcal{H}, \bar{\mathbf{u}}) = \frac{1}{2}\mathcal{H}\mathscr{A} + \frac{1}{6}\mathcal{H}^2\mathscr{B} + \mathcal{O}(\mu^2). \qquad (1.55)$$

In this way, the base model (1.40), (1.41) with the last closure becomes the celebrated Mei–Le Méhauté (1966) model [238].

Aleshkov's Model vs. Mei–Le Méhauté's Model

In this section we assume the flow to be irrotational. Consider the fluid velocity potential expansion around the bottom:

$$\phi(\mathbf{x}, y, t) = \check{\phi} - \mu^2 (y+h)\big(h_t + \nabla\check{\phi}\cdot\nabla h\big) - \mu^2 \frac{(y+h)^2}{2}\nabla^2\check{\phi} + \mathcal{O}(\mu^4), \tag{1.56}$$

where $\check{\phi}$ is the velocity potential trace at the bottom, i.e.

$$\check{\phi}(\mathbf{x}, t) \stackrel{\text{def}}{:=} \phi(\mathbf{x}, y, t)|_{y=-h}.$$

A similar formula can be found in [344] for the stationary bottom and in [128] for moving bottoms. The horizontal fluid velocity can be readily obtained by differentiating equation (1.56):

$$\mathbf{u}(\mathbf{x}, y, t) \equiv \nabla\phi = \nabla\check{\phi} - \mu^2\big(h_t + \nabla\check{\phi}\cdot\nabla h\big)\nabla h - \mu^2 (y+h)\nabla\big(h_t + \nabla\check{\phi}\cdot\nabla h\big)$$
$$- \mu^2 (y+h)(\nabla^2\check{\phi})\nabla h - \mu^2\frac{(y+h)^2}{2}\nabla(\nabla^2\check{\phi}) + \mathcal{O}(\mu^4).$$

Then, the whole family of models can be obtained by choosing the velocity variable $\bar{\mathbf{u}}(\mathbf{x}, t)$ at different levels in the fluid. Here we take the velocity at solid bottom:

$$\bar{\mathbf{u}}(\mathbf{x}, t) \stackrel{\text{def}}{:=} \mathbf{u}(\mathbf{x}, y, t)|_{y=-h} = \nabla\check{\phi} - \mu^2\big(h_t + \nabla\check{\phi}\cdot\nabla h\big)\nabla h.$$

Hence, from definition (1.30) we can compute the expression for $\tilde{\mathbf{u}}$:

$$\tilde{\mathbf{u}}(\mathbf{x}, t) = -(y+h)\nabla\big(h_t + \nabla\check{\phi}\cdot\nabla h\big) - (y+h)(\nabla^2\check{\phi})\nabla h$$
$$- \frac{(y+h)^2}{2}\nabla(\nabla^2\check{\phi}) + \mathcal{O}(\mu^2),$$

and taking into account the fact that $\bar{\mathbf{u}} = \nabla\check{\phi} + \mathcal{O}(\mu^2)$ we have

$$\tilde{\mathbf{u}}(\mathbf{x}, t) = -(y+h)\nabla\mathscr{D}h - (y+h)(\nabla\cdot\bar{\mathbf{u}})\nabla h - \frac{(y+h)^2}{2}\nabla(\nabla\cdot\bar{\mathbf{u}}) + \mathcal{O}(\mu^2)$$
$$\equiv (y+h)\mathscr{A} + \frac{(y+h)^2}{2}\mathscr{B} + \mathcal{O}(\mu^2).$$

After applying the depth-averaging operator we obtain the corresponding closure variable:

$$\mathcal{U}(\mathbf{x}, t) \stackrel{\text{def}}{:=} \frac{1}{\mathcal{H}} \int_{-h}^{\varepsilon\eta} \tilde{\mathbf{u}}(\mathbf{x}, y, t)\, \mathrm{d}y = \frac{\mathcal{H}}{2} \mathscr{A} + \frac{\mathcal{H}^2}{6} \mathscr{B} + \mathcal{O}(\mu^2).$$

It coincides exactly with the closure relation (1.55) given above. This concludes our clarifications regarding Mei–Le Méhauté's model [238].

In ALESHKOV's model the velocity variable $\bar{\mathbf{u}}(\mathbf{x}, t)$ is defined in a different way:

$$\bar{\mathbf{u}}(\mathbf{x}, t) \stackrel{\text{def}}{:=} \nabla \check{\phi}(\mathbf{x}, t).$$

Then, the fluid horizontal velocity takes the form

$$\begin{aligned}
\mathbf{u}(\mathbf{x}, y, t) &= \bar{\mathbf{u}} - \mu^2 \bigl(h_t + \nabla \check{\phi} \cdot \nabla h\bigr) \nabla h - \mu^2 (y + h) \nabla \bigl(h_t + \nabla \check{\phi} \cdot \nabla h\bigr) \\
&\quad - \mu^2 (y + h) (\nabla^2 \check{\phi}) \nabla h - \mu^2 \frac{(y + h)^2}{2} \nabla (\nabla^2 \check{\phi}) + \mathcal{O}(\mu^4) \\
&= \bar{\mathbf{u}} - \mu^2 \bigl(h_t + \bar{\mathbf{u}} \cdot \nabla h\bigr) \nabla h - \mu^2 (y + h) \nabla \bigl(h_t + \bar{\mathbf{u}} \cdot \nabla h\bigr) \\
&\quad - \mu^2 (y + h) (\nabla \cdot \bar{\mathbf{u}}) \nabla h - \mu^2 \frac{(y + h)^2}{2} \nabla (\nabla \cdot \bar{\mathbf{u}}) + \mathcal{O}(\mu^4) \\
&= \bar{\mathbf{u}} + \mu^2 \Bigl[- \mathscr{D} h (\nabla h) + (y + h) \mathscr{A} + \frac{(y + h)^2}{2} \mathscr{B} \Bigr] + \mathcal{O}(\mu^4).
\end{aligned}$$

From the last formula it is straightforward to obtain the closure relation (1.58) which yields ALESHKOV's model [4]. It explains also the differences between ALESHKOV's and MEI–LE MÉHAUTÉ's models.

Peregrine's Model and Its Generalizations

In 1967 Peregrine [264] considered a weakly nonlinear model with $y_\sigma = 0$. The fully nonlinear analogue of PEREGRINE's model can be obtained if we take

$$y_\sigma = \varepsilon \eta(\mathbf{x}, t).$$

Closure relation (1.53) then becomes:

$$\mathcal{U}(\mathcal{H}, \bar{\mathbf{u}}) = -\frac{1}{2} \mathcal{H} \mathscr{A} - \frac{1}{3} \mathcal{H}^2 \mathscr{B} + \mathcal{O}(\mu^2),$$

and base model (1.40), (1.41) becomes the fully nonlinear PEREGRINE's system. The momentum balance equation of this model takes a very simple form, when the

BOUSSINESQ regime is considered:

$$\bar{\mathbf{u}}_t + (\bar{\mathbf{u}} \cdot \nabla)\bar{\mathbf{u}} + \varepsilon \nabla \eta = \mathbf{0}.$$

In other words, if initially the vertical component of vorticity is zero, then it is so for all times, i.e.

$$\bar{u}_{2,x_1} - \bar{u}_{1,x_2} = 0, \qquad \forall t \geqslant 0. \tag{1.57}$$

The last assertion is true only in BOUSSINESQ approximation in for the CAUCHY problem. The irrotationality can break when boundary conditions are applied on finite (i.e. bounded) domains [94].

Nwogu's Model and Its Generalizations

In 1993 Nwogu proposed the following choice [254]:

$$y_\sigma \approx -\beta \cdot h(\mathbf{x}, t), \qquad \beta \approx 0.531.$$

This choice was motivated by linear dispersion relation considerations (optimization of dispersive characteristics). The nonlinearity of NWOGU's model was improved in, e.g., [182, 285]. The idea consists in finding surface between the bottom $y = -h(\mathbf{x}, t)$ and free surface $y = \varepsilon\eta(\mathbf{x}, t)$ (instead of the bottom and $y = 0$ in weakly nonlinear considerations). In this way, a free parameter $\beta \in [0, 1]$ at our disposal:

$$y_\sigma(\mathbf{x}, t) = -\beta h(\mathbf{x}, t) + (1 - \beta)\varepsilon\eta(\mathbf{x}, t).$$

In this case the closure relation becomes:

$$\mathcal{U}(\mathcal{H}, \bar{\mathbf{u}}) = \Big(\beta - \frac{1}{2}\Big)\mathcal{H}\mathscr{A} - \frac{\mathcal{H}^2}{6}\big(3\beta^2 - 6\beta + 2\big)\mathscr{B} + \mathcal{O}(\mu^2).$$

The '*optimal*' value of β will coincide with that given by Nwogu [254] since linearizations of both models coincide.

Aleshkov's Model

As the last example, we show here how to obtain Aleshkov's (1996) model [4], which was generalized later to include moving bottom effects in [128]. ALESHKOV's model (with moving bottom) can be obtained from the base model (1.40), (1.41) if we adopt the following closure:

$$\mathcal{U}(\mathcal{H}, \bar{\mathbf{u}}) = -(\mathscr{D}h)\nabla h + \frac{1}{2}\mathcal{H}\mathscr{A} + \frac{1}{6}\mathcal{H}^2\mathscr{B} + \mathcal{O}(\mu^2). \tag{1.58}$$

This closure is similar to MEI–LE MÉHAUTÉ closure (1.55) except for the first term. The horizontal velocity in ALESHKOV's model does not coincide with the horizontal fluid velocity at any surface inside fluid bulk. Instead, ALESHKOV's velocity variable is given by the gradient of the velocity potential evaluated at solid bottom. For non-flat bottoms it does not coincide with $\mathbf{u}|_{y=-h}$. These subtle differences are discussed in some detail in Sect. 1.1.6. Since this model is not widely known, we give here the governing equations:

$$\mathcal{H}_t + \nabla\cdot\bigl[\mathcal{H}\bar{\mathbf{u}}\bigr] = \mu^2\nabla\cdot\Bigl[\mathcal{H}(\nabla h)\mathscr{D}h + \frac{\mathcal{H}^2}{2}\bigl[\nabla(\mathscr{D}h) + (\nabla\cdot\bar{\mathbf{u}})\nabla h\bigr] + \frac{\mathcal{H}^3}{6}\nabla(\nabla\cdot\bar{\mathbf{u}})\Bigr], \tag{1.59}$$

$$\bar{\mathbf{u}}_t + (\bar{\mathbf{u}}\cdot\nabla)\bar{\mathbf{u}} + \varepsilon\nabla\eta = \mu^2\nabla\Bigl[\mathcal{H}\mathscr{R}_2 + \frac{\mathcal{H}^2}{2}\mathscr{R}_1 + \frac{1}{2}(\mathscr{D}h)^2\Bigr] + \mathcal{O}(\mu^4). \tag{1.60}$$

One big advantage of equations above is that the irrotational flow is preserved by its dynamics of equations (1.59), (1.60) in the sense of definition given in equation (1.57). The proof of this fact is given in the following Sect. 1.1.6.

Vorticity in Aleshkov's Model

In this section we study how the vertical component of vorticity evolves under the dynamics of ALESHKOV's model (1.59), (1.60). Consequently, we rewrite equations (1.60) in the following equivalent form:

$$\bar{u}_{1,t} + \bar{u}_1\bar{u}_{1,x_1} + \bar{u}_2\bar{u}_{1,x_2} + \mathcal{R}_{x_1} = 0,$$
$$\bar{u}_{2,t} + \bar{u}_1\bar{u}_{2,x_1} + \bar{u}_2\bar{u}_{2,x_2} + \mathcal{R}_{x_2} = 0,$$

where $\mathcal{R}$ is a scalar function defined as

$$\mathcal{R} \stackrel{\text{def}}{:=} \varepsilon\eta - \mu^2\Bigl[\mathcal{H}\mathscr{R}_2 + \tfrac{1}{2}\mathcal{H}^2\mathscr{R}_1 + \tfrac{1}{2}(\mathscr{D}h)^2\Bigr].$$

The same Eqs. (1.60) can be rewritten also as

$$\bar{u}_{1t} - \bar{u}_2\omega + \Bigl[\mathcal{R} + \frac{\bar{u}_1^2 + \bar{u}_2^2}{2}\Bigr]_{x_1} = 0,$$

$$\bar{u}_{2t} + \bar{u}_2\omega + \Bigl[\mathcal{R} + \frac{\bar{u}_1^2 + \bar{u}_2^2}{2}\Bigr]_{x_2} = 0,$$

where we introduced the vertical vorticity function $\omega \stackrel{\text{def}}{:=} \bar{u}_{2,x_1} - \bar{u}_{1,x_2}$. Making a cross differentiation of two last equations and subtracting them yields the following vorticity equation:

$$\omega_t + [\omega \bar{u}_1]_{x_1} + [\omega \bar{u}_2]_{x_2} = 0. \tag{1.61}$$

Let us assume that initially we have $\omega(\mathbf{x}, 0) \equiv 0$ and Eq. (1.61) admits a unique solution. By noticing that $\omega(\mathbf{x}, t) \equiv 0$ solves Eq. (1.61) and satisfies the initial condition, we obtain the required result.

Remark 1.6 There is a much shorter (but less insightful) proof of the same result. Namely, by definition of the velocity variable $\bar{\mathbf{u}}$ in ALESHKOV's model we have:

$$\bar{u}_1 = \check{\phi}_{x_1}, \qquad \bar{u}_2 = \check{\phi}_{x_2}.$$

Then straightforwardly we have

$$\omega = \bar{u}_{2,x_1} - \bar{u}_{1,x_2} = (\check{\phi}_{x_2})_{x_1} - (\check{\phi}_{x_1})_{x_2} \equiv 0,$$

provided that the trace of the velocity potential at the bottom $\check{\phi}$ is a continuously differentiable function.

1.2 Weakly Nonlinear Models

We considered the fully nonlinear version of the base model (1.40), (1.41) previously since the small amplitude assumption was never used (even if we introduced formally the nonlinearity parameter ε). The only constitutive assumption employed was the long wave hypothesis or, in other words, the waves are only weakly dispersive. In the present section we derive a weakly nonlinear variant of the base model (1.40), (1.41). In this way we achieve a further simplification of governing equations. Moreover, we shall work in the so-called BOUSSINESQ regime:

$$\varepsilon = \mathcal{O}(\mu^2) \quad \Longleftrightarrow \quad \mathsf{S_U} = \mathcal{O}(1), \tag{1.62}$$

where $\mathsf{S_U} \stackrel{\text{def}}{:=} \dfrac{\varepsilon}{\mu^2} \equiv \dfrac{\alpha \ell^2}{d^3}$ is the so-called STOKES–URSELL number [314]. In other words, we assume that the nonlinearity parameter and the squared dispersion parameter have *approximatively* the same order of magnitude. It is under this assumption that one can obtain numerous BOUSSINESQ-type models [37, 93]. Sometimes the simplifying BOUSSINESQ assumption (1.62) is accompanied also by explicitly (or implicitly) stated assumptions on the bottom variations, e.g. $\|\nabla h\| \sim \mathcal{O}(\varepsilon) \simeq \mathcal{O}(\mu^2)$, as it is the case for the base model.

The most difficult task here is to keep as many good properties of the base model as possible, while simplifying the governing equations. It is not always possible and some illustrations will be given below.

1.2.1 Weakly Nonlinear Base Model

In the present section we derive the Weakly Nonlinear Base Model (WNBM) starting from the base model equations (1.40), (1.41). The first goal here is to preserve at least the conservative form of the equations when simplifying the base model.

First of all, we notice that the vector $\mathcal{U}$ always enters into governing equations with coefficient μ^2, i.e. $\mu^2 \mathcal{U}$. Consequently, under the assumption (1.62), the vector $\mathcal{U}$ can be formally split as

$$\mathcal{U} = \underbrace{\mathcal{U}_0 + \mathcal{U}_1}_{\mathcal{O}(1)} + \mathcal{O}(\mu^2),$$

where $\mathcal{U}_0$ contains all the terms independent of the system solution and $\mathcal{U}_1$ contains everything else involving η, $\bar{\mathbf{u}}$. For instance, to illustrate this idea for the closure relation (1.53), which gives the LYNETT–LIU's model, we have the following decomposition:

$$\mathcal{U}_0 = \Big(\frac{h}{2} + y_\sigma\Big) \nabla h_t, \tag{1.63}$$

$$\mathcal{U}_1 = \Big(\frac{h}{2} + y_\sigma\Big) \cdot \big(\nabla (\bar{\mathbf{u}} \cdot \nabla h) + (\nabla \cdot \bar{\mathbf{u}}) \nabla h\big) - \Big(\frac{h^2}{6} - \frac{(y_\sigma + h)^2}{2}\Big) \nabla (\nabla \cdot \bar{\mathbf{u}}). \tag{1.64}$$

From now on we use the following notation for the main part of vector $\mathcal{U}$:

$$\mathcal{U}^\flat \stackrel{\text{def}}{:=} \mathcal{U}_0 + \mathcal{U}_1.$$

The mass conservation equation for WNBM model is directly obtained from (1.31):

$$\mathcal{H}_t + \nabla \cdot (\mathcal{H} \bar{\mathbf{u}}) = -\mu^2 \nabla \cdot (h\, \mathcal{U}^\flat). \tag{1.65}$$

We notice that the last equation is in the conservative form as well. In a similar way, we obtain the weakly nonlinear analogue of the momentum conservation equation:

$$(\mathcal{H} \bar{\mathbf{u}})_t + \nabla \cdot (\mathcal{H} \bar{\mathbf{u}} \otimes \bar{\mathbf{u}}) + \nabla \mathscr{P}^\flat = \check{p}^\flat \nabla h - \mu^2 \Big[(h\, \mathcal{U}^\flat)_t + \nabla \cdot \big(h\, \mathcal{U}_0 \otimes \bar{\mathbf{u}} + h\, \bar{\mathbf{u}} \otimes \mathcal{U}_0\big) \Big]. \tag{1.66}$$

In some cases, it is useful to have also a non-conservative form of the momentum conservation equation (1.66), which can be obtained using the weakly nonlinear form of the mass conservation (1.65):

$$\bar{\mathbf{u}}_t + (\bar{\mathbf{u}} \cdot \nabla)\bar{\mathbf{u}} + \frac{\nabla \mathscr{P}^\flat}{\mathcal{H}} = \frac{\check{p}^\flat \nabla h}{\mathcal{H}}$$
$$- \frac{\mu^2}{\mathcal{H}} \Big[(h\,\mathcal{U}^\flat)_t - \bar{\mathbf{u}}\nabla \cdot (h\,\mathcal{U}^\flat) + \nabla \cdot \big(h\,\mathcal{U}_0 \otimes \bar{\mathbf{u}} + h\,\bar{\mathbf{u}} \otimes \mathcal{U}_0 \big) \Big]. \qquad (1.67)$$

To complete the description of the WNBM, we have to explain how to compute the non-hydrostatic pressure in this model:

$$\mathscr{P}^\flat \stackrel{\text{def}}{:=} \frac{\mathcal{H}^2}{2} - \mu^2 \Big[\frac{h^3}{3}\,\mathscr{R}_1^\flat + \frac{h^2}{2}\,\mathscr{R}_2^\flat \Big],$$
$$\check{p}^\flat \stackrel{\text{def}}{:=} \mathcal{H} - \mu^2 \Big[\frac{h^2}{2}\,\mathscr{R}_1^\flat + h\,\mathscr{R}_2^\flat \Big],$$

where

$$\mathscr{R}_1^\flat \stackrel{\text{def}}{:=} (\nabla \cdot \bar{\mathbf{u}})_t, \qquad \mathscr{R}_2^\flat \stackrel{\text{def}}{:=} h_{tt} + 2\bar{\mathbf{u}} \cdot \nabla h_t + \bar{\mathbf{u}}_t \cdot \nabla h.$$

We underline that the non-conservative form (1.67) contains one nonlinear dispersive term $\bar{\mathbf{u}}\,\nabla \cdot (h\,\mathcal{U}_1)$, while in the conservative form (1.66) all dispersive terms are *linear*. Equations (1.65), (1.66) constitute the WNBM. Below we derive some important particular cases of WNBM.

1.2.2 Depth-Averaged WNBM

Consider a particular case of the WNBM when the velocity variable is chosen to be depth-averaged. In this case we showed above that $\mathcal{U} \equiv \mathbf{0}$. Consequently, $\mathcal{U}^\flat \equiv \mathbf{0}$ as well. WNBM equations (1.65), (1.66) take the simplest form in this particular case:

$$\mathcal{H}_t + \nabla \cdot (\mathcal{H}\,\bar{\mathbf{u}}) = 0,$$
$$(\mathcal{H}\,\bar{\mathbf{u}})_t + \nabla \cdot (\mathcal{H}\,\bar{\mathbf{u}} \otimes \bar{\mathbf{u}}) + \nabla \mathscr{P}^\flat = \check{p}^\flat \nabla h.$$

On the flat bottom the last equation becomes even simpler:

$$(\mathcal{H}\,\bar{\mathbf{u}})_t + \nabla \cdot (\mathcal{H}\,\bar{\mathbf{u}} \otimes \bar{\mathbf{u}}) + \nabla \mathscr{P}^\flat = \mathbf{0}.$$

The equivalent non-conservative form of the momentum conservation equation (on uneven bottoms) is

$$\bar{\mathbf{u}}_t + (\bar{\mathbf{u}} \cdot \nabla)\bar{\mathbf{u}} + \alpha \nabla \eta = \underbrace{\frac{\mu^2}{\mathcal{H}} \left\{ \nabla \left[\frac{h^3}{3} \mathscr{R}_1^{\mathrm{b}} + \frac{h^2}{2} \mathscr{R}_2^{\mathrm{b}} \right] - \left[\frac{h^2}{2} \mathscr{R}_1^{\mathrm{b}} + h \mathscr{R}_2^{\mathrm{b}} \right] \nabla h \right\}}_{(\checkmark)} . \tag{1.68}$$

Peregrine's System

In the pioneering work [264] Peregrine derived a weakly nonlinear model over a *stationary bottom*, i.e. $h_t \equiv 0$. The mass conservation in PEREGRINE's system coincides exactly with the mass conservation equation from the previous Sect. 1.2.2. We show below that the PEREGRINE's momentum conservation can be obtained from the non-conservative equation (1.68) under the BOUSSINESQ assumption (1.62). The right-hand side $(\checkmark)$ can be rewritten as

$$\begin{aligned} \frac{(\checkmark)}{\mu^2} &= \frac{1}{h} \nabla \left[\frac{h^3}{3} (\nabla \cdot \bar{\mathbf{u}})_t + \frac{h^2}{2} (\bar{\mathbf{u}}_t \cdot \nabla h) \right] - \left[\frac{h}{2} (\nabla \cdot \bar{\mathbf{u}})_t + \bar{\mathbf{u}}_t \cdot \nabla h \right] \nabla h \\ &= \left[\frac{h}{2} (\nabla \cdot \bar{\mathbf{u}}) \nabla h + \frac{h^2}{3} \nabla (\nabla \cdot \bar{\mathbf{u}}) + \frac{h}{2} \nabla (\bar{\mathbf{u}} \cdot \nabla h) \right]_t . \end{aligned}$$

Then, we use the relation

$$\bar{\mathbf{u}} \cdot \nabla h = \nabla \cdot (h \bar{\mathbf{u}}) - h \nabla \cdot \bar{\mathbf{u}} . \tag{1.69}$$

Finally, we obtain the right-hand side of PEREGRINE's model:

$$\frac{(\checkmark)}{\mu^2} = \left[\frac{h}{2} \nabla \big(\nabla \cdot (h \bar{\mathbf{u}}) \big) - \frac{h^2}{6} \nabla (\nabla \cdot \bar{\mathbf{u}}) \right]_t .$$

Hence, the non-conservative momentum equation reads

$$\bar{\mathbf{u}}_t + (\bar{\mathbf{u}} \cdot \nabla)\bar{\mathbf{u}} + \alpha \nabla \eta = \underbrace{\mu^2 \left[\frac{h}{2} \nabla \big(\nabla \cdot (h \bar{\mathbf{u}}) \big) - \frac{h^2}{6} \nabla (\nabla \cdot \bar{\mathbf{u}}) \right]_t}_{\simeq (\checkmark)} .$$

However, the simplifications we made above were drastic in some sense. For instance, the PEREGRINE's model cannot be recast in a conservative form even on a flat bottom. It goes without saying that the energy equation cannot be established for this model either.[13] These are the main drawbacks of the weakly nonlinear PEREGRINE's system. Moreover, the numerical schemes based on non-conservative equations may be divergent [215]. Despite all this critics, the PEREGRINE's system supplemented with moving bottom effects (i.e. $h_t \neq 0$) was successfully used to model wave generation in closed basins [109, 252].

It is interesting to note that the depth-averaged WNBM and PEREGRINE's system give the same linearization over the flat bottom $h(x) \equiv d$:

$$\eta_t + d\,\nabla\cdot\bar{\mathbf{u}} = 0, \tag{1.70}$$

$$\bar{\mathbf{u}}_t + \alpha\,\nabla\eta = \mu^2\frac{d^2}{3}\,\nabla(\nabla\cdot\bar{\mathbf{u}}_t). \tag{1.71}$$

In particular, it implies that dispersive properties are the same.

1.2.3 WNBM with the Velocity Given on a Surface

When the velocity variable is defined on a surface in the fluid bulk as in (1.28), WNBM equations are (1.65), (1.66) and the closure relation for variable $\mathcal{U}^{\flat}$ is given by formulas (1.63), (1.64). Consequently, the dispersive terms are present in both mass and momentum conservation equations. Moreover, in the case of the stationary bottom ($h_t \equiv 0$) we have automatically that $\mathcal{U}_0 \equiv \mathbf{0}$. Consequently, the WNBM equations with this choice of the velocity variable read:

$$\mathcal{H}_t + \nabla\cdot(\mathcal{H}\bar{\mathbf{u}}) = -\mu^2\,\nabla\cdot(h\,\mathcal{U}_1),$$
$$(\mathcal{H}\bar{\mathbf{u}})_t + \nabla\cdot(\mathcal{H}\bar{\mathbf{u}}\otimes\bar{\mathbf{u}}) + \nabla\mathscr{P}^{\flat} = \check{p}^{\flat}\,\nabla h - \mu^2\,(h\,\mathcal{U}_1)_t.$$

The last equation can be recast in the non-conservative form:

$$\bar{\mathbf{u}}_t + (\bar{\mathbf{u}}\cdot\nabla)\,\bar{\mathbf{u}} + \alpha\,\nabla\eta$$
$$= \underbrace{\frac{\mu^2}{\mathcal{H}}\Big\{\nabla\Big[\frac{h^3}{3}\,\mathscr{R}_1^{\flat} + \frac{h^2}{2}\,\mathscr{R}_2^{\flat}\Big] - \Big[\frac{h^2}{2}\,\mathscr{R}_1^{\flat} + h\,\mathscr{R}_2^{\flat}\Big]\nabla h - (h\,\mathcal{U}_1)_t + \bar{\mathbf{u}}\,\nabla\cdot(h\,\mathcal{U}_1)\Big\}}_{(\mathfrak{d})}.$$

---

[13] We have to make this statement more precise: an analogue of the energy conservation equation for the classical PEREGRINE system has been found in [97] for the flat bottom case. However, the method used in [97] does not give anything on general bottoms.

Below we show an important application of this variant of the WNBM.

Nwogu's System

NWOGU's model was derived in [254] under the assumption of the stationary bottom ($h_t \equiv 0$) that we adopt here as well. First of all, the expression (1.64) can be transformed using the relation (1.69):

$$\mathcal{U}_1 = \Big(\frac{h}{2} + y_\sigma\Big)\nabla\big(\nabla\cdot(h\,\bar{\mathbf{u}})\big) + \Big(\frac{y_\sigma^2}{2} - \frac{h^2}{6}\Big)\nabla(\nabla\cdot\bar{\mathbf{u}}).$$

In this way we obtain straightforwardly the mass conservation equation of NWOGU's system [254]. In order to obtain the momentum equation of NWOGU's system, first we neglect in ($\eth$) the nonlinear dispersive term $\bar{\mathbf{u}}\,\nabla\cdot(h\,\mathcal{U}_1)$. Then, the non-hydrostatic pressure terms are transformed similarly to PEREGRINE's system case studied above in Sect. 1.2.2. So, the right-hand side ($\eth$) of WNBM becomes:

$$\begin{aligned}\frac{(\eth)}{\mu^2} &= \frac{h}{2}\,\nabla\big(\nabla\cdot(h\,\bar{\mathbf{u}})_t\big) - \frac{h^2}{6}\,\nabla(\nabla\cdot\bar{\mathbf{u}})_t - \Big(\frac{h}{2} + y_\sigma\Big)\nabla\big(\nabla\cdot(h\,\bar{\mathbf{u}})_t\big)\\ &- \Big(\frac{y_\sigma^2}{2} - \frac{h^2}{6}\Big)\nabla(\nabla\cdot\bar{\mathbf{u}})_t \equiv -\Big[y_\sigma\,\nabla\big(\nabla\cdot(h\,\bar{\mathbf{u}})_t\big) + \frac{y_\sigma^2}{2}\,\nabla(\nabla\cdot\bar{\mathbf{u}})_t\Big].\end{aligned}$$

As a result, we obtain the momentum equation of Nwogu's system [254]:

$$\bar{\mathbf{u}}_t + (\bar{\mathbf{u}}\cdot\nabla)\,\bar{\mathbf{u}} + \alpha\,\nabla\eta = \underbrace{-\mu^2\Big[y_\sigma\,\nabla\big(\nabla\cdot(h\,\bar{\mathbf{u}})_t\big) + \frac{y_\sigma^2}{2}\,\nabla(\nabla\cdot\bar{\mathbf{u}})_t\Big]}_{\simeq\,(\eth)}.$$

Using the low-order linear terms in the dispersive terms again, other asymptotic equivalent models can also be derived [243].

The WNBM equations and NWOGU's system linearize on the flat bottom $h(\mathbf{x}) \equiv d$ to the same equations:

$$\eta_t + d\,\nabla\cdot\bar{\mathbf{u}} = -\mu^2 d^3\Big(\beta + \frac{1}{3}\Big)\nabla\cdot\big(\nabla(\nabla\cdot\bar{\mathbf{u}})\big), \tag{1.72}$$

$$\bar{\mathbf{u}}_t + \alpha\,\nabla\eta = -\mu^2 d^2\,\beta\,\nabla(\nabla\cdot\bar{\mathbf{u}}_t), \tag{1.73}$$

where we introduced the following parameter:

$$\beta \stackrel{\text{def}}{:=} \frac{y_\sigma}{d} + \frac{y_\sigma^2}{2d^2}. \tag{1.74}$$

However, in the nonlinear case the WNBM system has the advantage of admitting the conservative form on general (unsteady and uneven) bottoms. This fact can be used to develop efficient numerical algorithms to solve nonlinear dispersive equations numerically. For instance, this conservative property will be exploited in Chap. 2 in order to construct adaptive and efficient numerical discretizations.

1.2.4 Dispersive Properties of Long Wave Models

The study of dispersive properties of hydrodynamic shallow water models is performed with linearized governing equations around the still water level. If we consider additionally the flat bottom case, then we obtain Eqs. (1.70), (1.71) after the linearization operation. This linear form of equations is common to all the approximate hydrodynamic models mentioned above which use the depth-averaged velocity[14] as the second evolution variable. On the other hand, approximate models employing the horizontal velocity at certain level will look after linearization as the linearized NWOGU's system (1.72), (1.73).

In order to assess the quality of dispersive wave models, we may repeat for each model the dispersion relation analysis and compute its phase velocity $c_p(\mathbf{k})$. Then, this prediction can be compared with the reference value given in Eq. (1.22). The closeness of phase velocities with the full water wave problem prediction (1.22) will characterize the quality[15] of the proposed approximation. For approximate models based on the depth-averaged velocity, one obtains the following expression of the phase speed:

$$c_p^{\mathrm{SGN}}(\mathbf{k}) = \sqrt{g\,d}\cdot\sqrt{\frac{1}{1+\frac{(|\mathbf{k}|d)^2}{3}}}\,.$$

For the models with the horizontal velocity defined on the 'optimal' surface $y_\sigma = -0.531\,d$, the phase speed is given by the following formula [254]:

$$c_p^{\mathrm{Nw}}(\mathbf{k}) = \sqrt{g\,d}\cdot\sqrt{1-\frac{(|\mathbf{k}|d)^2}{3\,(1-\beta\,(|\mathbf{k}|d)^2)}}\,,$$

where the parameter $\beta \approx -0.390$ was defined in Eq. (1.74).

[14] There is another possibility leading to this system is to assume from the beginning the independence of the 3D horizontal velocity $\mathbf{u}$ of the flow from the vertical coordinate y.

[15] Here by quality we mean only the ability to approximate the linear dispersion relation of the full water wave problem. Unfortunately, the nonlinear properties are not captured by this analysis.

From these formulas for $c_p(\mathbf{k})$ we can see that the value of the phase speed does not depend on the direction of the wavenumber vector $\mathbf{k}$. It means that approximate models inherit also the *isotropy property* of the wave propagation. The *monotonicity property* of the phase speed is also preserved (see Sect. 1.1.2). This can be seen also on Fig. 1.3a, which shows the dependence of three phase velocities on the wavelength $\lambda \equiv \ell = \dfrac{2\pi}{|\mathbf{k}|}$. However, the third property of the phase speed *vanishing* for infinitely short waves is present in the SGN model, but it is not verified in NWOGU's model. This may have undesirable effects during numerical simulations with NWOGU's equations on very fine grids. Such NYQUIST waves[16] will propagate with unphysically high speeds. On the other hand, we must admit that for longer waves, NWOGU's system approximates better the reference phase celerity than the SGN model (see Fig. 1.3a).

The phase speed dependence on the dispersion parameter $\mu \stackrel{\text{def}}{:=} \dfrac{d}{\ell} = \dfrac{d\,|\mathbf{k}|}{2\pi}$ for all three models is depicted in Fig. 1.3b. It is not difficult to estimate that the approximation error is smaller than 5% in the SGN model for $\mu \leqslant 0.22$ and for $\mu \leqslant 0.5$ in NWOGU's system. We can conclude that in the linear approximation, the last model reproduces the dispersion relation of the full problem in the larger interval of the dispersion parameter μ. However, we remind that today we run mostly nonlinear (and even *fully* nonlinear) simulations and in the nonlinear case other properties such as the GALILEAN invariance, conservation of energy and momentum, finally, the consistency of the energy with the full water wave

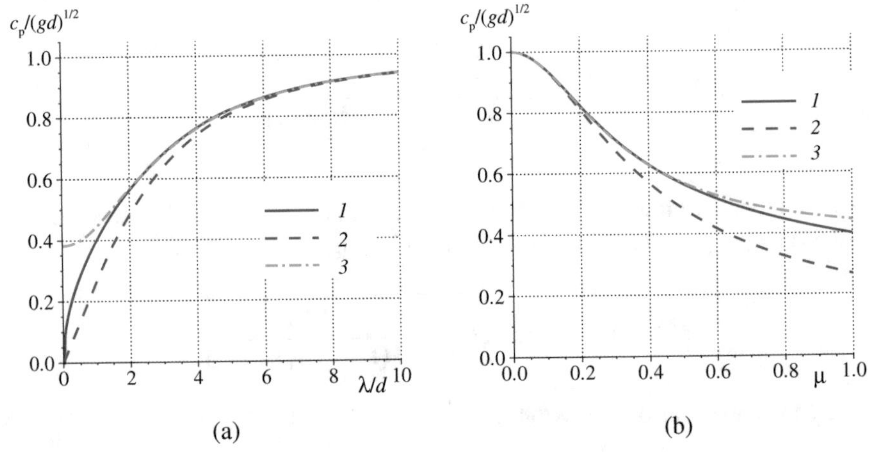

Fig. 1.3 Dispersive properties of hydrodynamic models: (**a**) dependence of the phase velocity on the wavelength and (**b**) dependence of the phase velocity on the dispersion parameter μ. Legend: (1) full water wave problem, (2) SGN model and (3) NWOGU's model

[16]The shortest waves which can be supported by the given grid.

problem [179] become equally important. For the SGN model all these properties are verified.

1.3 Green–Naghdi Equations

In Sect. 1.1.5 we briefly introduced the SGN equations. However, this model along with its cousins (see e.g. [74, 120]) is becoming so popular nowadays that we would like to devote a dedicated section to this model.

1.3.1 Historical Background

The first systems of nonlinear dispersive equations have appeared in the works of J. Boussinesq already in [41] to explain the phenomenon of the solitary wave observed some 30 years before by J.S. Russell [269]. The magnitude of nonlinear effects in long wave models is measured by the dimensionless parameter $\varepsilon = \dfrac{\alpha}{d}$, where α is the characteristic wave amplitude and d is the typical water depth. The magnitude of the dispersive effects is measured by another dimensionless number $\mu = \dfrac{d}{\ell}$, where ℓ is the characteristic wave length. Ursell [314] proposed to introduce the third dimensionless parameter that we call today the STOKES–URSELL number $\mathsf{S_U} = \dfrac{\varepsilon}{\mu^2}$, which measures the relative importance of nonlinear and dispersive effects (see also our discussion in Sect. 1.2). The applicability of various nonlinear dispersive wave models can be decided based on the magnitude of $\mathsf{S_U}$. For instance, if in some problem $\mathsf{S_U} \gg 1$, i.e. the nonlinear effects prevail over the dispersive ones, then it is enough to use just nonlinear shallow water equations. In the contrary case, the dispersive terms cannot be neglected. When both effects are balanced, i.e. $\mathsf{S_U} = \mathcal{O}(1)$, then we are in the so-called BOUSSINESQ regime.

One-dimensional BOUSSINESQ-type systems of equations for uneven bottoms have been derived for the first time in 'modern era' in [229, 238, 263]. In the pioneering work of Peregrine [264], a two-dimensional version of BOUSSINESQ-type equations has been derived. This step gave an important impulse towards the development of various BOUSSINESQ-type systems in 2D. Very quickly, a huge family of BOUSSINESQ-type equations with various properties and serving various purposes has appeared in the following years. It would be an impossible task to overview all these models nowadays. We would like just to mention that some of these models are being used for the solution of various practical problems in coastal engineering.

The original BOUSSINESQ-type equations have been derived under the assumptions of weak nonlinearity and $\mathsf{S_U} = \mathcal{O}(1)$, which limits somehow their range of applicability. Subsequent studies have shown that they are valid for frankly long waves, i.e. with the increase of the parameter μ substantial discrepancies with the reality (and also with Direct Numerical Simulations (DNS)) become apparent. This observation has generated an important activity of extending BOUSSINESQ-type systems towards deeper waters. For example, in [231] some additional terms have been added to the BOUSSINESQ equations, which after the linearization improve the dispersive properties of the system. Later this method has been extended to slowly varying bathymetries as well [230]. In [27] a formal dispersion-improved generalization of the classical PEREGRINE system has been obtained. We would like to mention that in the early developments the question of the GALILEAN invariance [96] and energy conservation of resulting equations has not been asked. Almost all of the proposed models at that time violated these fundamental principles. Moreover, the divergences among models over non-flat bathymetries have generated discussions in the literature [28, 275]. Fortunately, it was established later in [276] that all these systems are equivalent and can be obtained for various values of a parameter introduced in [274]. Another direction has been initiated by Bona and Smith [35] and Nwogu [254], who proposed to use the velocity variable at a certain level of the water depth to introduce a free parameter into governing equations. This parameter can be also used further to tune the dispersion relation properties.

In traditional derivations of the shallow water equations, the asymptotic methods are used to simplify EULER equations by expanding solution variables in powers of the shallowness parameter[17] μ. By substituting these expansions into the complete water wave problem and after neglecting higher order terms, one obtains the dispersionless equations to the first approximation and nonlinear dispersive terms at the second approximation. This approach allows to obtain various weakly nonlinear, but also Fully Nonlinear Weakly Dispersive (FNWD) wave equations. Fully nonlinear models do not assume any limitation on the wave amplitude. The famous SERRE–GREEN–NAGHDI (SGN) equations [123, 147, 282, 283] belong to the class of fully nonlinear models. The classical BOUSSINESQ equations can be recovered as a weakly nonlinear limit of SGN equations. Below we provide a slightly different derivation based on depth-averaging of the full EULER equations. The application of this derivation procedure is possible only if we assume a specific structure of the three-dimensional (3D) flow.

1.3.2 Derivation of the Governing Equations

Shallow water equations are usually formulated in terms of the total water depth $\mathcal{H} = h + \eta$ and depth-averaged velocity $\bar{\mathbf{u}}$ (or another horizontal velocity variable related to the corresponding velocity $\mathbf{u}$ of the 3D flow).

[17] This operation can be seen also as an expansion in terms of the powers of $(y + h)$.

In the derivation of the SGN equations it is assumed that the horizontal velocity vector $\mathbf{u}$ does not depend on the vertical coordinate y and the vertical component v depends on y only linearly [123, 147]. Thus, $\mathbf{u}$ is approximated by $\bar{\mathbf{u}}$ and

$$v(\mathbf{x}, y, t) = v_0(\mathbf{x}, t) + \bigl(y + h(\mathbf{x}, t)\bigr)\, v_1(\mathbf{x}, t). \tag{1.75}$$

This will be called the shallow water ansatz. From integral relation (1.7) written with $\zeta = -h$ follows the following equality:

$$\mathcal{H}(\boldsymbol{\nabla} \cdot \mathbf{u}) + v\,|^{y=\eta} - v\,|_{y=-h} = 0.$$

By taking into account kinematic boundary conditions (1.4) and (1.6), one obtains the continuity equation of the SGN model:

$$\mathcal{H}_t + \boldsymbol{\nabla} \cdot (\mathcal{H}\bar{\mathbf{u}}) = 0. \tag{1.76}$$

From the ansatz (1.75) it readily follows that $v_0 = -\mathscr{D}h$, where $\mathscr{D}$ is the total derivative operator defined in (1.34). From the free surface kinematic boundary condition (1.4) it follows that $v_1 = \dfrac{\mathscr{D}\mathcal{H}}{\mathcal{H}}$. After putting these elements together, we obtain a more precise form of ansatz (1.75):

$$v = -\mathscr{D}h + \frac{y + h}{\mathcal{H}}\,\mathscr{D}\mathcal{H}. \tag{1.77}$$

After some algebra, we obtain two following identities:

$$\mathscr{D}v = -\mathscr{D}^2 h + \mathscr{D}h\,\frac{\mathscr{D}\mathcal{H}}{\mathcal{H}} + (y + h)\left(\frac{\mathscr{D}^2\mathcal{H}}{\mathcal{H}} - \left(\frac{\mathscr{D}\mathcal{H}}{\mathcal{H}}\right)^2\right), \tag{1.78}$$

$$v\,v_y = -\mathscr{D}h\,\frac{\mathscr{D}\mathcal{H}}{\mathcal{H}} + (y + h)\left(\frac{\mathscr{D}\mathcal{H}}{\mathcal{H}}\right)^2. \tag{1.79}$$

Then, from the integral relation (1.9), using the dynamic free surface boundary condition (1.5), we obtain that the flow pressure p is a quadratic function of the vertical coordinate y, defined in the segment $-h \leqslant y \leqslant \eta$:

$$p = \frac{\mathcal{H}}{2}\,\mathscr{D}^2\mathcal{H} + \bigl(g - \mathscr{D}^2 h\bigr)(\eta - y) - \frac{(y + h)^2}{2}\,\frac{\mathscr{D}^2\mathcal{H}}{\mathcal{H}}. \tag{1.80}$$

From the shallow water ansatz it follows that $\mathbf{u}_y \equiv 0$ and $\mathbf{u} \equiv \bar{\mathbf{u}}$ and we can see also that

$$\frac{1}{\mathcal{H}}\int_{-h}^{\eta}\bigl[\mathbf{u}_t + (\mathbf{u} \cdot \boldsymbol{\nabla})\mathbf{u} + v\,\mathbf{u}_y\bigr]\mathrm{d}y = \bar{\mathbf{u}}_t + (\bar{\mathbf{u}} \cdot \boldsymbol{\nabla})\bar{\mathbf{u}}. \tag{1.81}$$

Hence, the integral relation (1.8) with $\zeta = -h$ can be rewritten using the dynamic free surface boundary condition (1.5) as follows:

$$\mathcal{H}\left(\bar{\mathbf{u}}_t + (\bar{\mathbf{u}} \cdot \nabla)\,\bar{\mathbf{u}}\right) + \nabla \int_{-h}^{\eta} p\,\mathrm{d}y - p\,|_{y=-h}\,\nabla h = 0. \tag{1.82}$$

Now we can substitute just found quadratic expression (1.80) for the pressure p to obtain the second equation of motion in the SGN model:

$$\bar{\mathbf{u}}_t + (\bar{\mathbf{u}} \cdot \nabla)\,\bar{\mathbf{u}} + g\,\nabla\,\eta =$$
$$\frac{1}{\mathcal{H}}\nabla\left(-\frac{\mathcal{H}^2}{3}\,\mathcal{D}^2\,\mathcal{H} + \frac{\mathcal{H}^2}{2}\,\mathcal{D}^2\,h\right) - \nabla\,h\left(-\frac{\mathcal{D}^2\,\mathcal{H}}{2} + \mathcal{D}^2\,h\right). \tag{1.83}$$

After some elementary transformations of the right-hand side, we may rewrite the last equation in the form given for the first time in [123]:

$$\bar{\mathbf{u}}_t + (\bar{\mathbf{u}} \cdot \nabla)\,\bar{\mathbf{u}} + g\,\nabla\,\eta = \frac{1}{6}\Big(-\mathcal{D}^2\,\eta\,\nabla\,(4\,\eta + h) +$$
$$\mathcal{D}^2\,h\,\nabla\,(2\,\eta - h) - (\eta + h)\,\nabla\left(2\,\mathcal{D}^2\,\eta - \mathcal{D}^2\,h\right)\Big). \tag{1.84}$$

Equations (1.76), (1.83) are equivalent to well-known SGN equations [147].

1.3.3 Energy Balance in the SGN Model

One of the most important characteristics of the ideal fluid dynamics is the energy balance property [22, 22, 67, 104, 127, 133, 292, 331]. In the case of an ideal homogeneous and incompressible fluid flow, this balance law can be obtained as a differential consequence of the continuity and momentum balance equations (*cf.* Sect. 1.1.1). When one derives approximate models, the loss of several important properties is possible. In particular, the energy conservation property can be lost on the way. For example, there are several models which do not conserve the mechanical energy even over the flat bottom. This remark concerns the famous NWOGU system [254] and during long time it concerned also the classical PEREGRINE model [264]. However, recently it was shown [97] that PEREGRINE's system admits a kind of energy functional over flat bottoms, even if its physical interpretation remains unclear.

The energy conservation property is extremely important for the physical soundness of the approximation. However, under mild coercivity conditions, it can be used also in the mathematical analysis of approximate models (see e.g.

[22, 38, 39]) to provide a rigorous justification of these models. Namely, a well-defined positive energy allows to introduce the appropriate energy norm in the space where the convergence being studied. For instance, it is well-known [319] that the conservation of energy plays a crucial rôle in the justification of the non-dispersive (NSWE) first order approximation model.

In order to establish the energy conservation property in the SGN model, we shall use the integral identity (1.12) and substitute the shallow water ansatz into it to obtain

$$\left(\mathcal{H}\mathscr{E}\right)_t + \nabla\cdot\left[\mathcal{H}\mathscr{E}\,\mathbf{u} + \mathbf{u}\int_{-h}^{\eta} p\,\mathrm{d}y\right] + p\,|_{y=-h}\,h_t = 0, \qquad (1.85)$$

where we used the fact that $\mathbf{u}$ does not depend on y in our approximation and $\mathscr{E}$ is the depth-averaged total energy defined as

$$\mathscr{E} \stackrel{\text{def}}{:=} \frac{1}{\mathcal{H}}\int_{-h}^{\eta} \mathcal{E}\,\mathrm{d}y\,.$$

By using the continuity equation rewritten in the equivalent form

$$\mathscr{D}\mathcal{H} + \mathcal{H}\nabla\cdot\bar{\mathbf{u}} = 0,$$

the vertical velocity v ansatz (1.77) can be rewritten accordingly as

$$v = -\mathscr{D}h - (y + h)\,\nabla\cdot\bar{\mathbf{u}} \qquad (1.86)$$

and the pressure p representation becomes

$$p = \bigl(\mathcal{H} - (y + h)\bigr)(g - \mathscr{R}_2) - \left(\frac{\mathcal{H}^2}{2} - \frac{(y + h)^2}{2}\right)\mathscr{R}_1,$$

where we used already familiar definitions (see the base model derivation above):

$$\mathscr{R}_1 \stackrel{\text{def}}{:=} \mathscr{D}(\nabla\cdot\bar{\mathbf{u}}) - (\nabla\cdot\bar{\mathbf{u}})^2, \qquad \mathscr{R}_2 \stackrel{\text{def}}{:=} \mathscr{D}^2 h\,. \qquad (1.87)$$

Then, the approximation of the energy density $\mathcal{E}$ with (1.86) becomes

$$\mathcal{E} = \frac{\bar{\mathbf{u}}\cdot\bar{\mathbf{u}}}{2} + \tfrac{1}{2}\bigl(\mathscr{D}h + (y + h)(\nabla\cdot\bar{\mathbf{u}})\bigr)^2 + g\,y\,.$$

To obtain a short-hand energy balance equation for the SGN model, we use the depth-integrated pressure variable:

$$\mathscr{P} \stackrel{\text{def}}{:=} \int_{-h}^{\eta} p\,\mathrm{d}y = g\,\frac{\mathcal{H}^2}{2} - \frac{\mathcal{H}^3}{3}\mathscr{R}_1 - \frac{\mathcal{H}^2}{2}\mathscr{R}_2\,.$$

Similarly, we have to employ also the pressure trace at the bottom:

$$\check{p} \stackrel{\text{def}}{:=} p\,|_{y=-h} = g\,\mathcal{H} - \left(\frac{\mathcal{H}^2}{2}\,\mathscr{R}_1 + \mathcal{H}\mathscr{R}_2\right).$$

With all these notations, Eq. (1.85) becomes

$$(\mathcal{H}\mathscr{E})_t + \nabla\cdot\left[\mathcal{H}\left(\mathscr{E} + \frac{\mathscr{P}}{\mathcal{H}}\right)\bar{\mathbf{u}}\right] = -\,\check{p}\,h_t\,. \tag{1.88}$$

The last equation is precisely the energy balance equation of the SGN model. Moreover, the depth-averaged energy density $\mathscr{E}$ can be expressed through other variables $(\mathcal{H},\ \bar{\mathbf{u}},\ h)$ as follows:

$$\mathscr{E} = \frac{\bar{\mathbf{u}}\cdot\bar{\mathbf{u}}}{2} + \frac{\mathcal{H}^2}{6}\,(\nabla\cdot\bar{\mathbf{u}})^2 + \frac{\mathcal{H}}{2}\,(\nabla\cdot\bar{\mathbf{u}})\,\mathscr{D}h + \frac{(\mathscr{D}h)^2}{2} + \frac{g\,(\mathcal{H} - 2\,h)}{2}\,.$$

Remark 1.7 It is interesting to note that the energy $\mathscr{E}$ balance equation for the classical nonlinear shallow water (NSWE) equations has precisely the same form as (1.88). The only difference is that the depth-integrated pressure $\mathscr{P}$, the pressure trace at the bottom $\check{p}$ and the depth-averaged energy density assume correspondingly simpler expressions (*cf.* [104]):

$$\mathscr{P} = g\,\frac{\mathcal{H}^2}{2}\,, \qquad \check{p} = g\,\mathcal{H}\,, \qquad \mathscr{E} = \frac{\bar{\mathbf{u}}\cdot\bar{\mathbf{u}}}{2} + \frac{g\,(\mathcal{H} - 2\,h)}{2}\,.$$

1.3.4 Section Conclusions

The described above SGN model turns out to be the first fully nonlinear two-dimensional (2D) nonlinear dispersive wave model. In other words, no assumption on the wave amplitude has been made during the derivation procedure. In the one-dimensional (1D) case this model was derived[18] presumably for the first time by F. Serre [282, 283]. For general, but stationary, bottoms, this model was generalized in [278]. The assumption of the independence of the horizontal velocity field **u** on y was not explicitly stated there. The extension of these equations to the 2D case was done by P. Naghdi and his collaborators [147, 148]. This justifies the name of SERRE–GREEN–NAGHDI (SGN) equations.

[18]The steady version of 1D SGN equations can be even found in the works by Lord Rayleigh [221].

1.4 Discussion

We presented a certain number of developments going from the derivation of the base model (1.40), (1.41) to obtaining some particular models as particular cases. The main conclusions and perspectives of this chapter are outlined below.

1.4.1 Conclusions

In this chapter we attempted to meet two main goals. First of all, we tried to make a review of the continuously growing field of long wave modelling. In particular, we focused on nonlinear dispersive wave models such as some improved BOUSSINESQ-type and SERRE–GREEN–NAGHDI (SGN) equations [147, 148, 283], which were not covered in previously published review papers. We apologize in advance if we forgot to mention somebody's contribution to this field. The topic being so broad that it is practically impossible to referred to all the published literature.

Then, we attempted to present a unified approach which incorporates some well-known and some less known models in the same modelling framework. The derivation procedure is based on the minimal set of assumptions. Various models can be obtained as particular cases of the so-called *base model* presented in our book. In the same time, the base model allows to obtain fully nonlinear analogues of previously derived weakly nonlinear models. The linearizations of old and new models will coincide exactly, hence leaving dispersive characteristics unchanged. Moreover, the resulting models admit an elegant conservative form by construction. The improvement of dispersive characteristics can be achieved by a judicious choice of the closure relation $\mathcal{U}\,(\mathcal{H},\ \bar{\mathbf{u}})$ as it was illustrated, for example, in Sect. 1.1.6.

1.4.2 Perspectives

In the present chapter we discussed modelling and derivation of models for shallow water waves flowing over uneven bottoms, but the whole system was defined on a flat domain Ω of the EUCLIDEAN space $\mathbb{R}^n$, with dimension $n = 1, 2$. The bottom represents only a deformation (not necessarily small) of the mean water depth. Among the main perspectives of this chapter we would like to mention the derivation of fully nonlinear shallow water models defined on more general geometries. In particular, the spherical geometry represents a lot of interest in view of applications to atmospheric sciences. The first steps in this direction have already been made in [130, 131]. The derivation of shallow water equations on a sphere will be discussed in Chap. 3. The numerical discretization of the derived above equations on moving adaptive grids will be considered in details in the Chap. 2, while the numerical simulation of shallow water waves on a sphere will be considered in Chap. 4 of this book.

Chapter 2
Numerical Simulation on a Globally Flat Space

In this chapter we describe a numerical method to solve numerically the weakly dispersive fully nonlinear SERRE–GREEN–NAGHDI (SGN) celebrated model. Namely, our scheme is based on reliable finite volume methods, proven to be very efficient for the hyperbolic part of equations. The particularity of our study is that we develop an adaptive numerical model using moving grids [32, 339]. Moreover, we use a special form of the SGN equations where non-hydrostatic part of pressure is found by solving a nonlinear elliptic equation. Moreover, this form of governing equations allows to determine the natural form of boundary conditions to obtain a well-posed (numerical) problem.

As we saw in the previous Chap. 1, in 1967 D. PEREGRINE derived the first two-dimensional BOUSSINESQ-type system of equations [264]. This model described the propagation of long weakly nonlinear waves over a general non-flat bottom. From this landmark study the modern era of long wave modelling started. However, researchers focused on the development of new models and in parallel the numerical algorithms have been developed. We refer to [46] for a recent '*reasoned*' review of this topic.

In Chap. 1 we derived the so-called base model, which encompasses a number of previously known models (but, of course, not all of nonlinear dispersive systems). The governing equations of the base model are

$$\mathcal{H}_t \;+\; \nabla\cdot[\,\mathcal{H}\,\mathbf{U}\,] \;=\; 0\,, \tag{2.1}$$

$$\begin{aligned}\bar{\mathbf{u}}_t \;+\; (\bar{\mathbf{u}}\cdot\nabla)\,\bar{\mathbf{u}} \;+\; \frac{\nabla\mathscr{P}}{\mathcal{H}} \;=\; \frac{\check{p}}{\mathcal{H}}\,\nabla h \;-\; \frac{1}{\mathcal{H}}\Big[\,(\mathcal{H}\,\mathcal{U})_t \;+\; (\bar{\mathbf{u}}\cdot\nabla)(\mathcal{H}\,\mathcal{U}) \\ +\; \mathcal{H}\,(\mathcal{U}\cdot\nabla)\,\bar{\mathbf{u}} \;+\; \mathcal{H}\,\mathcal{U}\,\nabla\cdot\bar{\mathbf{u}}\,\Big]\,,\end{aligned} \tag{2.2}$$

where $\mathbf{U} \stackrel{\text{def}}{:=} \bar{\mathbf{u}} + \mathcal{U}$ is the modified horizontal velocity and $\mathcal{U} = \mathcal{U}\,(\mathcal{H},\,\bar{\mathbf{u}})$ is the closure relation to be specified later. Depending on the choice of this variable various

G. Khakimzyanov et al., *Dispersive Shallow Water Waves*, Lecture Notes in Geosystems Mathematics and Computing, https://doi.org/10.1007/978-3-030-46267-3_2

models can be obtained (see Sect. 1.1.6). Variables $\mathscr{P}$ and $\check{p}$ are related to the fluid pressure. The physical meaning of these variables is reminded below in Sect. 2.1. In the present chapter we propose an adaptive numerical discretization for a particular, but very popular nowadays model which can be obtained from the base model (2.1), (2.2). Namely, if we choose $\mathcal{U} \equiv \mathbf{0}$ (thus, $\mathbf{U}$ becomes the depth-averaged velocity $\mathbf{u}$) then we obtain equations equivalent to the celebrated SERRE–GREEN–NAGHDI (SGN) equations [148, 282, 283] (rediscovered later independently by many other researchers). This system will be the main topic of our numerical study. Most often, adaptive techniques for dispersive wave equations involve the so-called Adaptive Mesh Refinement (AMR) [271] (see also [30] for nonlinear shallow water equations). The particularity of our study is that we conserve the total number of grid points and the adaptivity is achieved by judiciously redistributing them in space [172, 173]. The ideas of redistributing grid nodes is stemming from the works of BAKHVALOV [16], IL'IN [174] and others [3, 305].

The base model (2.1), (2.2) admits an elegant conservative form (see Sect. 1.1.4):

$$\mathcal{H}_t + \nabla \cdot [\mathcal{H}\mathbf{U}] = 0, \qquad (2.3)$$

$$(\mathcal{H}\mathbf{U})_t + \nabla \cdot \Big[\mathcal{H}\bar{\mathbf{u}} \otimes \mathbf{U} + \mathscr{P}(\mathcal{H}, \bar{\mathbf{u}}) \cdot \mathbb{I} + \mathcal{H}\mathcal{U} \otimes \bar{\mathbf{u}}\Big] = \check{p}\,\nabla h, \qquad (2.4)$$

where $\mathbb{I} \in \mathrm{Mat}_{2\times 2}(\mathbb{R})$ is the identity matrix and the operator $\otimes$ denotes the tensorial product. We note that the pressure function $\mathscr{P}(\mathcal{H}, \bar{\mathbf{u}})$ incorporates the familiar hydrostatic pressure part $\dfrac{g\,\mathcal{H}^2}{2}$ well-known from the Nonlinear Shallow Water Equations (NSWE) [20, 84]. By setting $\mathcal{U} \equiv \mathbf{0}$ we obtain readily from (2.3), (2.4) the conservative form of the SGN equations (one can notice that the mass conservation equation (2.1) was already in conservative form).

Nonlinear dispersive wave equations represent certain numerical difficulties since they involve mixed derivatives (usually of the horizontal velocity variable, but sometimes of the total water depth as well) in space and time. These derivatives have to be approximated numerically, thus leaving a lot of room for the creativity. Most often the so-called *Method Of Lines* (MOL) is employed [203, 268, 277, 284], where the spatial derivatives are discretized first and the resulting system of coupled Ordinary Differential Equations (ODEs) is then approached with more or less standard ODE techniques, see e.g. [159, 160]. The MOL separates the choice of time discretization from the procedure of discretization in space, even if the interplay between two schemes might be important. For example, it would be natural to choose the same order of accuracy for both schemes.

Let us review the available spatial discretization techniques employed in recent numerical studies. We focus essentially on fully nonlinear weakly dispersive models, even if some interesting works devoted to BOUSSINESQ-type and unidirectional equations will be mentioned. First of all, dispersive wave equations with the dispersion relation given by a rational function (*à la* BBM [29, 263]) usually involve the inversion of an elliptic operator. This gives the first idea of employing the splitting technique between the hyperbolic and elliptic operators. This idea was successfully realized in e.g. [17, 18, 40, 168]. Historically, perhaps, the finite difference techniques were applied first to dispersive (and more general non-hydrostatic) wave equations [50, 68, 69, 71, 230, 231, 327, 342]. Then, naturally we arrive to the development of continuous GALERKIN/Finite Element type discretizations [9, 36, 92, 94, 244, 295, 320]. See also a recent review [93] and the references therein. Pseudo-spectral FOURIER-type methods can also be successfully applied to the SGN equations [116]. See [125] for a pedagogical review of pseudo-spectral and radial basis function methods for some shallow water equations. More recently, the finite volume type methods were applied to dispersive equations [59, 112, 116, 117, 181, 212]. In the present study we also employ a predictor–corrector finite volume type scheme [287], which is described in detail below.

The present chapter is organized as follows. In Sect. 2.1 we present the governing equations in 2D and 1D spatial dimensions. The numerical method is described in Sect. 2.2. Several numerical illustrations are shown in Sect. 2.3 including the solitary wave/wall or bottom interactions and even a realistic underwater landslide simulation. Finally, in Sect. 2.4 we outline the main conclusions and perspectives of the present study.

2.1 Mathematical Model

In this study we consider the following system of the SERRE–GREEN–NAGHDI (SGN) equations, which describes the incompressible homogeneous fluid flow in a layer bounded from below by the impermeable bottom $y = -h(\mathbf{x}, t)$ and above by the free surface $y = \eta(\mathbf{x}, t), \mathbf{x} = (x_1, x_2) \in \mathbb{R}^2$:

$$\mathcal{H}_t + \nabla \cdot \bigl[\mathcal{H}\mathbf{u}\bigr] = 0, \tag{2.5}$$

$$\mathbf{u}_t + (\mathbf{u} \cdot \nabla)\mathbf{u} + \frac{\nabla \mathscr{P}}{\mathcal{H}} = \frac{\check{p}}{\mathcal{H}} \nabla h, \tag{2.6}$$

where for simplicity we drop the bars over the horizontal velocity variable $\mathbf{u}(\mathbf{x}, t) = \bigl(u_1(\mathbf{x}, t), u_2(\mathbf{x}, t)\bigr)$. Function $\mathcal{H}(\mathbf{x}, t) \stackrel{\text{def}}{:=} h(\mathbf{x}, t) + \eta(\mathbf{x}, t)$ being the total water depth. The sketch of the fluid domain is schematically depicted in Fig. 2.1. For the derivation of Eqs. (2.5), (2.6) we refer to Chap. 1. The depth-integrated pressure $\mathscr{P}(\mathbf{u}, \mathcal{H})$ is defined as

$$\mathscr{P}(\mathbf{u}, \mathcal{H}) \stackrel{\text{def}}{:=} \frac{g\,\mathcal{H}^2}{2} - \wp(\mathbf{x}, t),$$

where $\wp(\mathbf{x}, t)$ is the non-hydrostatic part of the pressure:

$$\wp(\mathbf{x}, t) \stackrel{\text{def}}{:=} \frac{\mathcal{H}^3}{3}\mathscr{R}_1 + \frac{\mathcal{H}^2}{2}\mathscr{R}_2, \tag{2.7}$$

Fig. 2.1 Sketch of the fluid domain in 2D

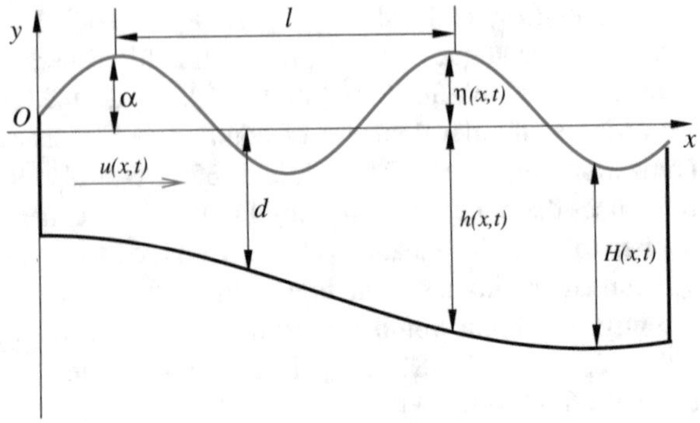

with

$$\mathscr{R}_1 \stackrel{\text{def}}{:=} \mathscr{D}(\nabla\cdot\mathbf{u}) - (\nabla\cdot\mathbf{u})^2, \qquad \mathscr{R}_2 \stackrel{\text{def}}{:=} \mathscr{D}^2 h, \qquad \mathscr{D} \stackrel{\text{def}}{:=} \partial_t + \mathbf{u}\cdot\nabla.$$

Above, $\mathscr{D}$ is the total or material derivative operator. On the right-hand side of Eq. (2.6) we have the pressure trace at the bottom $\check{p} \stackrel{\text{def}}{:=} p|_{y=-h}$, which can be written as

$$\check{p}(\mathbf{x}, t) = g\mathcal{H} - \check{\wp}(\mathbf{x}, t),$$

where $\check{\wp}(\mathbf{x}, t)$ is again the non-hydrostatic pressure contribution:

$$\check{\wp}(\mathbf{x}, t) \stackrel{\text{def}}{:=} \frac{\mathcal{H}^2}{2}\mathscr{R}_1 + \mathcal{H}\mathscr{R}_2. \tag{2.8}$$

Equations above are much more complex comparing to the classical NSWE (or Saint-Venant equations) [84], since they contain mixed derivatives up to the third order. From the numerical perspective these derivatives have to be approximated. However, the problem can be simplified if we 'extract' a second order sub-problem for the non-hydrostatic component of the pressure. Indeed, it can be shown (see Sect. 2.1.1) that function $\wp(\mathbf{x}, t)$ satisfies the following second order linear elliptic equation with variable coefficients (by analogy with incompressible Navier–Stokes equations, where the pressure is found numerically by solving a Poisson-type problem [66, 163]):

$$\nabla\cdot\left[\frac{\nabla\wp}{\mathcal{H}} - \frac{(\nabla\wp\cdot\nabla h)\,\nabla h}{\mathcal{H}\Upsilon}\right] - 6\underbrace{\left[\frac{2}{\mathcal{H}^3}\cdot\frac{\Upsilon-3}{\Upsilon} + \nabla\cdot\left(\frac{\nabla h}{\mathcal{H}^2\Upsilon}\right)\right]}_{(\star)}\wp = \mathcal{F}, \tag{2.9}$$

where $\Upsilon \stackrel{\text{def}}{:=} 4 + |\nabla h|^2$ and $\mathcal{F}$, $\mathscr{R}$ are defined as

$$\mathcal{F} \stackrel{\text{def}}{:=} \nabla \cdot \left[g \, \nabla \eta \, + \, \frac{\mathscr{R} \, \nabla h}{\Upsilon} \right] \, - \, \frac{6 \, \mathscr{R}}{\mathcal{H} \, \Upsilon} \, + \, 2 \, (\nabla \cdot \mathbf{u})^2 \, - \, 2 \begin{vmatrix} u_{1x_1} & u_{1x_2} \\ u_{2x_1} & u_{2x_2} \end{vmatrix}, \tag{2.10}$$

$$\mathscr{R} \stackrel{\text{def}}{:=} -g \, \nabla \eta \cdot \nabla h \, + \, \big[(\mathbf{u} \cdot \nabla) \, \nabla h \big] \cdot \mathbf{u} \, + \, h_{tt} \, + \, 2 \, \mathbf{u} \cdot \nabla h_t \, . \tag{2.11}$$

Symbol $| \cdot |$ in (2.10) denotes the determinant of a 2×2 matrix.

Equation (2.9) is uniformly elliptic and it does not contain time derivatives of the fluid velocity $\mathbf{u}$. If the coefficient ($\bigstar$) is positive (for instance, it is the case for a flat bottom $h(\mathbf{x}, t) \equiv$ const), we deal with a positive operator and stable robust discretizations can be proposed. Taking into account the fact that Eq. (2.9) is linear with respect to the variable $\wp(\mathbf{x}, t)$, its discrete counterpart can be solved by direct or iterative methods[1]. Well-posedness of this equation is discussed below (see Sect. 2.1.2). The boundary conditions for equation (2.9) will be discussed below in Sect. 2.1.4 (in 1D case only, the generalization to 2D is done by projecting on the normal direction to the boundary).

Introduction of the variable $\wp(\mathbf{x}, t)$ allows to rewrite Eq. (2.6) in the following equivalent form:

$$\mathbf{u}_t \, + \, (\mathbf{u} \cdot \nabla) \, \mathbf{u} \, + \, g \, \nabla \mathcal{H} \, = \, g \, \nabla h \, + \, \frac{\nabla \wp \, - \, \check{\wp} \, \nabla h}{\mathcal{H}} \, . \tag{2.12}$$

The non-hydrostatic pressure at the bottom $\check{\wp}(\mathbf{x}, t)$ can be expressed through $\wp$ in the following way:

$$\check{\wp}(\mathbf{x}, t) \, = \, \frac{1}{\Upsilon} \left[\frac{6 \, \wp}{\mathcal{H}} \, + \, \mathcal{H} \, \mathscr{R} \, + \, \nabla \wp \cdot \nabla h \right] . \tag{2.13}$$

The derivation of this Eq. (2.13) is given in Sect. 2.1.1 as well. So, thanks to this relation (2.13), the usage of Eq. (2.8) is not necessary anymore. Once we found the function $\wp(\mathbf{x}, t)$, we can compute the bottom component from (2.13).

Remark 2.1 It can be easily seen that taking formally the limit $\wp \to 0$ and $\check{\wp} \to 0$ of vanishing non-hydrostatic pressures allows us to recover the classical NSWE (or SAINT–VENANT equations) [84]. Thus, the governing equations verify the BOHR correspondence principle [33].

[1] In our implementation we use the direct THOMAS algorithm, since in 1D the resulting linear system of equations is tridiagonal with the dominant diagonal.

2.1.1 Non-hydrostatic Pressure Equation

In this section we give some hints for the derivation of the non-hydrostatic pressure equation (2.9) and relation (2.13). Let us start with the latter. For this purpose we rewrite equation (2.12) in a more compact form using the total derivative operator:

$$\mathscr{D}\mathbf{u} = -g\,\nabla\eta + \frac{\nabla\wp - \check{\wp}\,\nabla h}{\mathcal{H}}, \tag{2.14}$$

By definition of non-hydrostatic quantities $\wp$ and $\check{\wp}$ (see Eqs. (2.7) and (2.8) correspondingly) we obtain:

$$\check{\wp} = \frac{3\,\wp}{2\,\mathcal{H}} + \frac{\mathcal{H}}{4}\,\mathscr{R}_2\,.$$

We have to substitute into the last relation the expression for $\mathscr{R}_2$:

$$\mathscr{R}_2 = (\mathscr{D}\mathbf{u})\cdot\nabla h + \mathbf{u}\cdot\bigl((\mathbf{u}\cdot\nabla)\nabla h\bigr) + h_{tt} + 2\,\mathbf{u}\cdot\nabla h_t\,,$$

along with the expression (2.14) for the horizontal acceleration $\mathscr{D}\mathbf{u}$ of fluid particles. After simple algebraic computations one obtains (2.13).

The derivation of Eq. (2.9) is somehow similar. First, from definitions (2.7), (2.8) we obtain another relation between non-hydrostatic pressures:

$$\wp = \frac{\mathcal{H}^3}{12}\,\mathscr{R}_1 + \frac{\mathcal{H}}{2}\,\check{\wp}\,, \tag{2.15}$$

with $\mathscr{R}_1$ rewritten in the following form:

$$\mathscr{R}_1 = \nabla\cdot(\mathscr{D}\mathbf{u}) - 2\,(\nabla\cdot\mathbf{u})^2 + 2\begin{vmatrix} u_{1x_1} & u_{1x_2} \\ u_{2x_1} & u_{2x_2} \end{vmatrix}.$$

Substituting into Eq. (2.15) the just shown relation (2.13) with the last expression for $\mathscr{R}_1$ yields the required Eq. (2.9).

2.1.2 Well-Posedness Conditions

In order to obtain a well-posed elliptic problem (2.9), one has to ensure that coefficient ($\bigstar$) is positive. This coefficient involves the bathymetry function $h\,(\mathbf{x},\,t)$ and the total water depth $\mathcal{H}\,(\mathbf{x},\,t)$. In other words, the answer depends on local depth and wave elevation. It is not excluded that for some wave conditions the coefficient ($\bigstar$) may become negative. In the most general case the positivity condition is trivial and, thus, not very helpful, i.e.

$$(\bigstar) \equiv \frac{2}{\mathcal{H}^3} \cdot \frac{\Upsilon - 3}{\Upsilon} + \nabla \cdot \left(\frac{\nabla h}{\mathcal{H}^2 \Upsilon}\right) \geqslant 0. \tag{2.16}$$

On the flat bottom $h(\mathbf{x}, t) \to d = \text{const}$ we know that the above condition is satisfied since $\Upsilon \to 4$ and $(\bigstar) \to \dfrac{1}{2\mathcal{H}^3} > 0$. Consequently, by continuity of the coefficient $(\bigstar)$ we conclude that the same property will hold for some (sufficiently small) variations of the depth $h(\mathbf{x}, t)$, i.e. $|\nabla h| \ll 1$. In practice it can be verified that bathymetry variations can be even finite so that condition (2.16) still holds.

Remark 2.2 It may appear that restrictions on the bathymetry variations are inherent to our formulation only. However, it is the case of all long wave models, even if this assumption does not appear explicitly in the derivation. For instance, bottom irregularities will inevitably generate short waves (i.e. higher frequencies) during the wave propagation process. A priori, this part of the spectrum is not modelled correctly by approximate equations, unless some special care is taken (see e.g. [99, 101]).

Linear Waves

Let us take the limit of linear waves $\eta \to 0$ in expression $(\bigstar)$. It will become then

$$(\bigstar) \to (\text{☆}) \stackrel{\text{def}}{:=} \frac{2}{h^3} \cdot \frac{\Upsilon - 3}{\Upsilon} + \nabla \cdot \left(\frac{\nabla h}{h^2 \Upsilon}\right).$$

The positivity[2] condition of (☆) then takes the following form:

$$2\Upsilon + h\left\{h_{x_1x_1}\left(1 - h_{x_1}^2 + h_{x_2}^2\right) + h_{x_2x_2}\left(1 + h_{x_1}^2 - h_{x_2}^2\right) - 4\,h_{x_1} h_{x_2} h_{x_1x_2}\right\} \geqslant 0.$$

If we restrict our attention to the one-dimensional bathymetries (i.e. $h_{x_2} \to 0$), then we obtain an even simpler condition:

$$h_{xx} \geqslant -\frac{2}{h} \cdot \frac{1 + h_x^2}{1 - h_x^2},$$

where by x we denote x_1 for simplicity. The last condition can be easily checked at the problem outset. A further simplification is possible if we additionally assume that

$$|\nabla h| \equiv |h_x| < 1, \tag{2.17}$$

[2]Non-negativity, to be more precise.

then we have the following elegant condition:

$$h_{xx} > -\frac{2}{h}. \tag{2.18}$$

2.1.3 Conservative Form of Equations

Equations (2.5), (2.6) admit an elegant conservative form, which is suitable for numerical simulations:

$$\mathcal{H}_t + \nabla \cdot \bigl[\mathcal{H}\mathbf{u}\bigr] = 0, \tag{2.19}$$

$$(\mathcal{H}\mathbf{u})_t + \nabla \cdot \mathscr{F} = g\,\mathcal{H}\nabla h + \nabla\wp - \check{\wp}\,\nabla h, \tag{2.20}$$

where the flux matrix $\mathscr{F}(\mathcal{H},\,\mathbf{u})$ is the same as in NSWE (or SAINT–VENANT equations):

$$\mathscr{F}(H,\,\mathbf{u}) \stackrel{\text{def}}{:=} \begin{bmatrix} \mathcal{H}u_1^2 + \dfrac{g\,\mathcal{H}^2}{2} & \mathcal{H}u_1 \cdot u_2 \\ \mathcal{H}u_1 \cdot u_2 & \mathcal{H}u_2^2 + \dfrac{g\,\mathcal{H}^2}{2} \end{bmatrix}.$$

Notice that it is slightly different from the (fully-)conservative form given in Chap. 1. Conservative equations[3] (2.19), (2.20) can be supplemented by the energy conservation equation which can be used to check the accuracy of simulation (in conservative case, i.e. $h_t \equiv 0$) and/or to estimate the energy of generated waves [104]:

$$(\mathcal{H}\mathscr{E})_t + \nabla \cdot \Bigl[\mathcal{H}\mathbf{u}\Bigl(\mathscr{E} + \frac{\mathscr{P}}{\mathcal{H}}\Bigr)\Bigr] = -\check{p}\,h_t, \tag{2.21}$$

where the total energy $\mathscr{E}$ is defined as

$$\mathscr{E} \stackrel{\text{def}}{:=} \tfrac{1}{2}\,|\mathbf{u}|^2 + \tfrac{1}{6}\,\mathcal{H}^2(\nabla\cdot\mathbf{u})^2 + \tfrac{1}{2}\,\mathcal{H}(\mathscr{D}h)(\nabla\cdot\mathbf{u}) + \tfrac{1}{2}\,(\mathscr{D}h)^2 + \frac{g}{2}\,(\mathcal{H} - 2h).$$

Notice that Eq. (2.21) is not independent. It is a differential consequence of the mass and momentum conservations (2.19), (2.20) (as it is the case for incompressible flows in general).

[3] It is not difficult to see that the mass conservation equation (2.5) is already in a conservative form in the SGN model. Thus, Eqs. (2.5) and (2.19) are obviously identical.

Intermediate Conclusions

As a result, the system of nonlinear dispersive equations (2.5), (2.6) was split in two main parts:

1. Governing equations (2.19), (2.20) in the form of (hyperbolic) balance laws with source terms;
2. A scalar nonlinear elliptic equation to determine the non-hydrostatic part of the pressure $\wp(\mathbf{x}, t)$ (and consequently $\check{\wp}(\mathbf{x}, t)$ as well).

This splitting idea will be exploited below in the numerical algorithm in order to apply the most suitable and robust algorithm for each part of the solution process [18].

2.1.4 One-Dimensional Case

In this chapter, for the sake of simplicity, we focus on the two-dimensional physical problem, i.e. one horizontal and one vertical dimensions. The vertical flow structure being resolved using the asymptotic expansion (see Chap. 1), thus we deal with PDEs involving one spatial (horizontal) dimension ($x \stackrel{\text{def}}{:=} x_1$) and one temporal variable $t \in \mathbb{R}^+$. The horizontal velocity variable $u(x, t)$ becomes a scalar function in this case. Below we provide the full set of governing equations (which follow directly from (2.19), (2.20) and (2.9)):

$$\mathcal{H}_t + [\mathcal{H} u]_x = 0, \tag{2.22}$$

$$(\mathcal{H} u)_t + \Bigl[\mathcal{H} u^2 + \frac{g\,\mathcal{H}^2}{2}\Bigr]_x = g\,\mathcal{H} h_x + \wp_x - \check{\wp}\, h_x, \tag{2.23}$$

$$4\Bigl[\frac{\wp_x}{\mathcal{H}\Upsilon}\Bigr]_x - 6\Bigl[\frac{2}{\mathcal{H}^3}\cdot\frac{\Upsilon - 3}{\Upsilon} + \Bigl[\frac{h_x}{\mathcal{H}^2\Upsilon}\Bigr]_x\Bigr]\wp = \mathcal{F}, \tag{2.24}$$

where $\Upsilon \stackrel{\text{def}}{:=} 4 + h_x^2$ and

$$\mathcal{F} \stackrel{\text{def}}{:=} \Bigl[g\,\eta_x + \frac{\mathcal{R} h_x}{\Upsilon}\Bigr]_x - \frac{6\mathcal{R}}{\mathcal{H}\Upsilon} + 2u_x^2,$$

$$\mathcal{R} \stackrel{\text{def}}{:=} -g\,\eta_x h_x + u^2 h_{xx} + h_{tt} + 2u\,h_{xt}.$$

The last equations can be trivially obtained from corresponding two-dimensional versions given in (2.10), (2.11). This set of equations will be solved numerically below (see Sect. 2.2).

Boundary Conditions on the Elliptic Part

First, we rewrite elliptic equation (2.24) in the following equivalent form:

$$[\mathcal{K}\wp_x]_x - \mathcal{K}_0\wp = \mathcal{F}, \tag{2.25}$$

where

$$\mathcal{K} \stackrel{\text{def}}{:=} \frac{4}{\mathcal{H}\Upsilon}, \qquad \mathcal{K}_0 \stackrel{\text{def}}{:=} 6\left[\frac{2}{\mathcal{H}^3}\frac{\Upsilon - 3}{\Upsilon} + \left(\frac{h_x}{\mathcal{H}^2\Upsilon}\right)_x\right].$$

We assume that we have to solve an initial-boundary value problem for the system (2.22)–(2.24). If we have a closed numerical wave tank[4] (as it is always the case in laboratory experiments), then on vertical walls the horizontal velocity satisfies:

$$u(x,t)\,|_{x=0} = u(x,t)\,|_{x=\ell} \equiv 0, \qquad \forall t \in \mathbb{R}^+.$$

For the situation where the same boundary condition holds on both boundaries, we introduce a short-hand notation:

$$u(x,t)\,|_{x=0}^{x=\ell} \equiv 0, \qquad \forall t \in \mathbb{R}^+.$$

Assuming that Eq. (2.12) is valid up to the boundaries, we obtain the following boundary conditions for the elliptic equation (2.24):

$$\left(\frac{\wp_x - \check{\wp}\,h_x}{\mathcal{H}} - g\,\eta_x\right)\Bigg|_{x=0}^{x=\ell} = 0, \qquad \forall t \in \mathbb{R}^+.$$

Or in terms of Eq. (2.25) we equivalently have:

$$\left(\mathcal{K}\wp_x - \frac{6\,h_x}{\mathcal{H}^2\Upsilon}\,\wp\right)\Bigg|_{x=0}^{x=\ell} = \left(g\,\eta_x + \frac{\mathscr{R}\,h_x}{\Upsilon}\right)\Bigg|_{x=0}^{x=\ell}, \qquad \forall t \in \mathbb{R}^+. \tag{2.26}$$

The boundary conditions for the non-hydrostatic pressure component $\wp$ are of the 3rd kind (sometimes they are referred to as of ROBIN-type). For the case where locally at the boundaries the bottom is flat (to the first order), i.e. $h_x|_{x=0}^{x=\ell} \equiv 0$, then we have the (non-homogeneous) NEUMANN boundary condition of the 2nd kind:

$$\mathcal{K}\wp_x\,|_{x=0}^{x=\ell} = g\,\eta_x\,|_{x=0}^{x=\ell}, \qquad \forall t \in \mathbb{R}^+.$$

[4]Other possibilities have to be discussed separately.

For a classical POISSON-type equation this condition would not be enough to have a well-posed problem. However, we deal rather with a HELMHOLTZ-type equation (if $\mathcal{K}_0 > 0$). So, the flat bottom does not represent any additional difficulty for us and the unicity of the solution can be shown in this case as well.

Unicity of the Elliptic Equation Solution

The mathematical structure of Eq. (2.25) is very advantageous since it allows to show the following

Theorem 2.3 *Suppose that the Boundary Value Problem (BVP)* (2.26) *for equation* (2.25) *admits a solution and the following conditions are satisfied:*

$$\mathcal{K}_0 > 0, \quad h_x|_{x=0} \geqslant 0, \quad h_x|_{x=\ell} \leqslant 0, \tag{2.27}$$

then this solution is unique.

Proof Assume that there are two such solutions $\wp_1$ and $\wp_2$. Then, their difference $\wp \stackrel{\text{def}}{:=} \wp_1 - \wp_2$ satisfies the following homogeneous BVP:

$$[\mathcal{K}\wp_x]_x - \mathcal{K}_0\wp = 0, \tag{2.28}$$

$$\left(\mathcal{K}\wp_x - \frac{6h_x}{\mathcal{H}^2\Upsilon}\wp\right)\Bigg|_{x=0}^{x=\ell} = 0. \tag{2.29}$$

Let us multiply the first equation (2.28) by $\wp$ and integrate over the computational domain:

$$\underbrace{\int_0^\ell [\mathcal{K}\wp_x]_x\wp\,\mathrm{d}x}_{(\blacklozenge)} - \int_0^\ell \mathcal{K}_0\wp^2\,\mathrm{d}x = 0.$$

Integration by parts of the first integral (♦) yields:

$$\mathcal{K}\wp_x\,\wp|^{x=\ell} - \mathcal{K}\wp_x\,\wp|_{x=0} - \int_0^\ell \mathcal{K}\wp_x^2\,\mathrm{d}x - \int_0^\ell \mathcal{K}_0\wp^2\,\mathrm{d}x = 0.$$

And using boundary conditions (2.29) we finally obtain:

$$\frac{6h_x}{\mathcal{H}^2\Upsilon}\wp^2\Big|^{x=\ell} - \frac{6h_x}{\mathcal{H}^2\Upsilon}\wp^2\Big|_{x=0} - \int_0^\ell \mathcal{K}\wp_x^2\,\mathrm{d}x - \int_0^\ell \mathcal{K}_0\wp^2\,\mathrm{d}x = 0.$$

Taking into account this theorem assumptions (2.27) and the fact that $\mathcal{K} > 0$, the last identity leads to a contradiction, since the left-hand side is strictly negative.

Consequently, the solution to equation (2.25) with boundary condition (2.26) is unique.

Remark 2.4 Conditions in Theorem 2.3 are quite natural. The non-negativity of coefficient $\mathcal{K}_0$ has already been discussed in Sect. 2.1.2. Two other conditions mean that the water depth is increasing in the offshore direction ($h_x|_{x=0} \geqslant 0$) and again it is decreasing ($h_x|_{x=\ell} \leqslant 0$) when we approach the opposite shore.

2.1.5 Vector Short-Hand Notation

For the sake of convenience we shall rewrite governing equations (2.22), (2.23) in the following vectorial form:

$$\mathbf{v}_t + \left[\mathscr{F}(\mathbf{v})\right]_x = \mathscr{G}(\mathbf{v}, \wp_x, \check{\wp}, h), \tag{2.30}$$

where we introduced the following vector-valued functions:

$$\mathbf{v} \stackrel{\text{def}}{:=} \begin{pmatrix} \mathcal{H} \\ \mathcal{H}u \end{pmatrix}, \qquad \mathscr{F}(\mathbf{v}) \stackrel{\text{def}}{:=} \begin{pmatrix} \mathcal{H}u \\ \mathcal{H}u^2 + \dfrac{g\mathcal{H}^2}{2} \end{pmatrix},$$

and the source term is defined as

$$\mathscr{G}(\mathbf{v}, \wp_x, \check{\wp}, h) \stackrel{\text{def}}{:=} \begin{pmatrix} 0 \\ g\mathcal{H}h_x + \wp_x - \check{\wp}h_x \end{pmatrix}.$$

The point of view that we adopt in this study is to view the SGN equations as a system of hyperbolic equations (2.30) with source terms $\mathscr{G}(\mathbf{v}, \wp_x, \check{\wp}, h)$. Obviously, one has to solve also the elliptic equation (2.24) in order to compute the source term $\mathscr{G}$.

The JACOBIAN matrix of the advection operator coincides with that of classical NSWE equations:

$$\mathscr{A}(\mathbf{v}) \stackrel{\text{def}}{:=} \frac{\mathrm{d}\mathscr{F}(\mathbf{v})}{\mathrm{d}\mathbf{v}} = \begin{pmatrix} 0 & 1 \\ -u^2 + g\mathcal{H} & 2u \end{pmatrix}.$$

Eigenvalues of the JACOBIAN matrix $\mathscr{A}(\mathbf{v})$ can be readily computed:

$$\lambda^- = u - s, \qquad \lambda^+ = u + s, \qquad s \stackrel{\text{def}}{:=} \sqrt{g\mathcal{H}}. \tag{2.31}$$

The JACOBIAN matrix appears naturally in the non-divergent form of equations (2.30):

$$\mathbf{v}_t + \mathscr{A}(\mathbf{v}) \cdot \mathbf{v}_x = \mathscr{G}, \tag{2.32}$$

By multiplying both sides of the last equation by $\mathscr{A}(\mathbf{v})$ we obtain the equations for the advection flux function $\mathscr{F}(\mathbf{v})$:

$$\mathscr{F}_t + \mathscr{A}(\mathbf{v}) \cdot \mathscr{F}_x = \mathscr{A} \cdot \mathscr{G}. \tag{2.33}$$

In order to study the characteristic form of equations one needs also to know the matrix of left and right eigenvectors correspondingly:

$$L \stackrel{\mathrm{def}}{:=} \frac{1}{s^2} \begin{pmatrix} -\lambda^+ & 1 \\ -\lambda^- & 1 \end{pmatrix}, \qquad R \stackrel{\mathrm{def}}{:=} \frac{s}{2} \begin{pmatrix} -1 & 1 \\ -\lambda^- & \lambda^+ \end{pmatrix}. \tag{2.34}$$

If we introduce also the diagonal matrix of eigenvalues

$$\Lambda \stackrel{\mathrm{def}}{:=} \begin{pmatrix} \lambda^- & 0 \\ 0 & \lambda^+ \end{pmatrix},$$

the following relations can be easily checked:

$$R \cdot \Lambda \cdot L \equiv \mathscr{A}, \qquad R \cdot L = L \cdot R \equiv \mathbb{I},$$

where $\mathbb{I}$ is the identity 2×2 matrix.

Flat Bottom

Equations above become particularly simple on the flat bottom. In this case the bathymetry functions are constant, i.e.

$$h(x, t) \equiv d = \mathrm{const} > 0.$$

Substituting it into governing equations above, we straightforwardly obtain:

$$\mathcal{H}_t + [\mathcal{H}u]_x = 0, \tag{2.35}$$

$$(\mathcal{H}u)_t + \left[\mathcal{H}u^2 + \frac{g\mathcal{H}^2}{2}\right]_x = \wp_x, \tag{2.36}$$

$$\left[\frac{\wp_x}{\mathcal{H}}\right]_x - \frac{3}{\mathcal{H}^3}\wp = g\eta_{xx} + 2u_x^2. \tag{2.37}$$

Solitary Wave Solution

Equations above admit an elegant analytical solution known as the *solitary wave*[5]. It is given by the following expressions:

$$\eta(x,t) = \alpha \cdot \mathrm{sech}^2\left[\frac{\sqrt{3\alpha g}}{2d\upsilon}\left(x - x_0 - \upsilon t\right)\right],$$

$$u(x,t) = \frac{\upsilon \cdot \eta(x,t)}{d + \eta(x,t)}, \tag{2.38}$$

where α is the wave amplitude, $x_0 \in \mathbb{R}$ is the initial wave crest position and υ is the velocity defined as

$$\upsilon \stackrel{\mathrm{def}}{:=} \sqrt{g(d + \alpha)}.$$

The non-hydrostatic pressure under the solitary wave can be readily computed as well:

$$\wp(x,t) = \frac{g}{2}\left[\mathcal{H}^2(x,t) - d^2\right] - d\upsilon u(x,t).$$

One can derive also periodic travelling waves known as *cnoidal waves*. For their expressions we refer to e.g. [108, 116].

2.1.6 Solitary Wave Solution to the Modified Boussinesq Equations

On flat bottoms, the continuity and momentum conservation equations of the modified BOUSSINESQ system (see Appendix B for the presentation of the model) take the same form as Eqs. (2.35), (2.36) given above. The difference can be seen only in the third elliptic Eq. (2.37) which, in the case of the modified BOUSSINESQ system, reads:

$$\left[\frac{\wp_x}{\mathcal{H}}\right]_x - \frac{3}{\mathcal{H}d^2}\wp = g\,\eta_{xx} + u_x^2.$$

It turns out that the modified BOUSSINESQ equations do also possess a family of solitary wave solutions. The analytical formulas for the horizontal velocity u

[5] Solitary waves are to be distinguished from the so-called *solitons* which interact elastically [95]. Since the SGN equations are not integrable (for the notion of integrability, we refer to e.g. [337]), the interaction of solitary waves is inelastic [244].

and depth-integrated pressure $\wp$ remain the same as in the fully nonlinear case. However, the free surface elevation η is computed differently. Its shape is given by the initial position:

$$\eta(x,t) = \eta_0(x - x_0 - \upsilon t), \qquad \eta(x,0) = \eta_0(x - x_0),$$

where $x_0 \in \mathbb{R}$ is the solitary wave crest position. Thus, the wave amplitude $\alpha = \eta(x_0, 0) = \eta_0(0)$. The solitary wave shape can be found from the formula:

$$\eta_0(x) = \alpha - q^2(x),$$

where $q(x)$ denotes the root of the following equation (for each fixed value of x):

$$q \longmapsto \left(\frac{\sqrt{\alpha + d} - q}{\sqrt{\alpha + d} + q}\right)^{\gamma} - \frac{\sqrt{\alpha} - q}{\sqrt{\alpha} + q}\, \mathrm{e}^{\beta|x - x_0|} = 0, \tag{2.39}$$

with

$$\beta \stackrel{\text{def}}{:=} \frac{\sqrt{3\alpha g}}{\upsilon d}, \qquad \gamma \stackrel{\text{def}}{:=} \sqrt{\frac{\alpha}{\alpha + d}}, \qquad 0 < \gamma < 1.$$

So that Eq. (2.39) admits a unique solution, it is enough that the following inequality is verified:

$$P(\gamma) \stackrel{\text{def}}{:=} \left(\frac{2\gamma}{1+\gamma}\right)^2 \cdot \left(\frac{1+\gamma}{1-\gamma}\right)^{1-\gamma} < 1.$$

To give an example, let us assume that $\frac{\alpha}{d} \leqslant \frac{9}{16} = 0.5625$. Then, $\gamma \leqslant 0.6$ and, consequently we have

$$P(\gamma) \leqslant P(0.6) = \frac{9\sqrt[5]{16}}{16} \approx 0.98 < 1, \qquad \forall \gamma \in [0, 0.6].$$

Thus, under the condition $\frac{\alpha}{d} \leqslant 0.5625$ Eq. (2.39) will admit a unique solution for every value of the abscissa x.

Several solitary wave profiles to the modified BOUSSINESQ system are depicted in Fig. 2.2 (see the dashed line). For the sake of comparison, we plot also the solitary wave solutions to the SGN system for the same wave amplitudes (with solid lines). One can see, in particular, that for small amplitude solitary waves (i.e. $\frac{\alpha}{d} \sim 0.1$) the differences between corresponding profiles are barely visible to the graphical resolution. However, when the amplitude increases, the differences become clearly visible, which is consistent with the weakly nonlinear approximation. It is interesting to note that the modified BOUSSINESQ equations

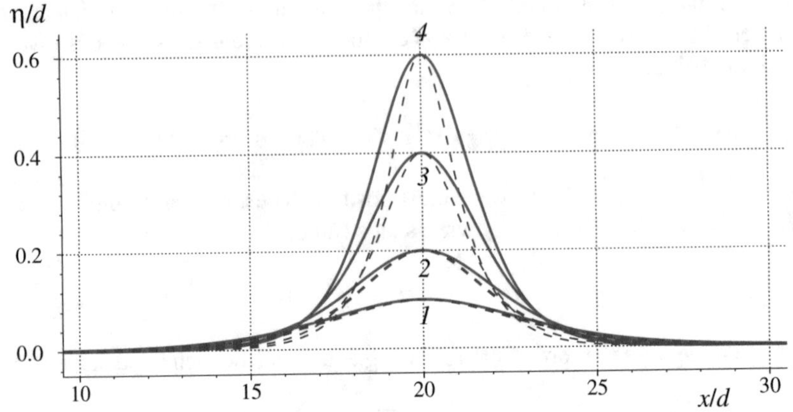

Fig. 2.2 Solitary wave profiles to the SGN model (solid lines) and modified BOUSSINESQ equations (dashed lines) for several values of the wave amplitudes: (1): $\dfrac{\alpha}{d} = 0.1$, (2): $\dfrac{\alpha}{d} = 0.2$, (3): $\dfrac{\alpha}{d} = 0.4$, (4): $\dfrac{\alpha}{d} = 0.6$

predict steeper solitary waves comparing to the SGN (fully nonlinear) model. Thus, the *effective* wavelength of the weakly nonlinear solitary waves is smaller than their fully nonlinear counterpart.

2.1.7 *Linear Dispersion Relation*

The governing equations (2.22)–(2.24) after linearizations take the following form:

$$\eta_t + d\, u_x = 0,$$

$$u_t + g\, \eta_x = \frac{\wp_x}{d},$$

$$\wp_{xx} - \frac{3}{d^2}\, \wp = c^2\, \eta_{xx},$$

where $c \stackrel{\text{def}}{:=} \sqrt{g\, d}$ is the linear gravity wave speed. By looking for plane wave solutions of form

$$\eta(x,\, t) = \alpha\, \mathrm{e}^{\mathrm{i}(k x - \omega t)}, \quad u(x,\, t) = \upsilon\, \mathrm{e}^{\mathrm{i}(k x - \omega t)}, \quad \wp(x,\, t) = \rho\, \mathrm{e}^{\mathrm{i}(k x - \omega t)},$$

where k is the wave number, $\omega(k)$ is the wave frequency and $\{\alpha,\, \upsilon,\, \rho\} \in \mathbb{R}$ are some (constant) real amplitudes. The necessary condition for the existence of plane wave solutions reads

$$\omega(k) = \pm \frac{c\,k}{\sqrt{1 + \dfrac{(k\,d)^2}{3}}}. \tag{2.40}$$

By substituting the definition of $k = \dfrac{2\pi}{\lambda}$ into the last formula and dividing both sides by k we obtain the relation between the phase speed c_p and the wavelength λ:

$$c_p(\lambda) \stackrel{\text{def}}{:=} \frac{\omega\big(k(\lambda)\big)}{k(\lambda)} = \frac{c}{\sqrt{1 + \dfrac{4\pi^2 d^2}{3\lambda^2}}}.$$

This dispersion relation is accurate to 2nd order at the limit of long waves $k\,d \to 0$. There are many other nonlinear dispersive wave models which share the same linear dispersion relation, see e.g. [123, 128, 264, 344]. However, their nonlinear properties might be very different.

2.2 Numerical Method

The construction of numerical schemes for hyperbolic conservation laws on moving grids was described in our previous work [192]. In the present manuscript we make an extension of this technology to dispersive PDEs illustrated on the example of the SGN equations (2.22)–(2.24). The main difficulty which arises in the dispersive case is handling of high order (possibly mixed) derivatives. The SGN system is an archetype of such systems with sufficient degree of nonlinearity and practically important applications in Coastal Engineering [194].

2.2.1 Adaptive Mesh Construction

In the present work we employ the method of moving grids initially proposed in early 1960s by Tikhonov and Samarskii [306, 307] and developed later by Bakhvalov (1969) [16] and Il'in (1969) [174]. Later, it was successfully employed to various problems by P. Zegeling and his co-authors [32, 315, 338, 339]. This technology was recently described by the authors for steady (e.g. boundary layers) [184] and unsteady [192] wave propagation problems. For more details, we refer to our recent publications [184, 192]. An alternative recent approach can be found in e.g. [11, 12]. In the present section we just recall briefly the main steps of the method.

The main idea consists in assuming that there exists a (time-dependent) diffeomorphism from the reference domain $\mathcal{Q} \stackrel{\text{def}}{:=} [0,\ 1]$ to the computational domain $\mathcal{I} = [0,\ \ell]$:

$$x\,(q,\ t) : \mathcal{Q} \longmapsto \mathcal{I}.$$

It is natural to assume that boundaries of the domains correspond to each other, i.e.

$$x\,(0,\ t) = 0, \qquad x\,(1,\ t) = \ell, \qquad \forall t \geqslant 0.$$

We shall additionally assume that the JACOBIAN of this map is bounded from below and above

$$0 < m \leqslant \mathcal{J}\,(q,\ t) \stackrel{\text{def}}{:=} \frac{\partial x}{\partial q} \leqslant M < +\infty \tag{2.41}$$

by some real constants m and M.

The construction of this diffeomorphism $x\,(q,\ t)$ is the heart of the matter in the moving grid method. We employ the so-called equidistribution method. The required non-uniform grid $\mathcal{I}_h$ of the computational domain $\mathcal{I}$ is then obtained as the image of the uniformly distributed nodes $\mathcal{Q}_h$ under the mapping $x\,(q,\ t)$:

$$x_j = x\,(q_j,\ t), \qquad q_j = j\,\Delta q, \qquad \Delta q = \frac{1}{N},$$

where N is the total number of grid points. Notice that, strictly speaking, we do not even need to know the mapping $x\,(q,\ t)$ in other points besides $\{q_j\}_{j=0}^{N}$. Under condition (2.41) it easily follows that the maximal discretization step in the physical space vanishes when we refine the mesh in the reference domain $\mathcal{Q}_h$:

$$\max_{j=0\ldots,\,N-1} |x_{j+1} - x_j| \leqslant M\,\Delta q \to 0, \quad \text{as} \quad \Delta q \to 0.$$

Initial Grid Generation

Initially, the desired mapping $x\,(q,\ 0)$ is obtained as a solution to the following nonlinear elliptic problem

$$\frac{\mathrm{d}}{\mathrm{d}q}\left[\varpi\,(x)\,\frac{\mathrm{d}x}{\mathrm{d}q}\right] = 0, \qquad x\,(0) = 0, \quad x\,(1) = \ell, \tag{2.42}$$

where we drop in this section the 2nd constant argument 0. The function $\varpi\,(x)$ is the so-called *monitor function*. Its choice will be specified below, but we can say that this functions has to be positive defined and bounded from below, i.e.

$$\varpi(x) \geqslant C > 0, \qquad \forall x \in \mathbb{R}.$$

In practice the lower bound C is taken for simplicity to be equal to 1. A popular choice of the monitor function is, for example,

$$\varpi[\eta](x) = 1 + \vartheta_0 |\eta|, \qquad \vartheta_0 \in \mathbb{R}^+,$$

where η is the free surface elevation. Another possibility consists in taking into account the free surface gradient:

$$\varpi[\eta](x) = 1 + \vartheta_1 |\eta_x|, \qquad \vartheta_1 \in \mathbb{R}^+,$$

or even both effects:

$$\varpi[\eta](x) = 1 + \vartheta_0 |\eta| + \vartheta_1 |\eta_x|, \qquad \vartheta_{0,1} \in \mathbb{R}^+.$$

In some simple cases equation (2.42) can be solved analytically (see e.g. [184]). However, in most cases we have to solve the nonlinear elliptic problem (2.42) numerically. For this purpose we use an iterative scheme, where at every stage we have a linear three-diagonal problem to solve:

$$\frac{1}{\Delta q}\Biggl[\varpi(x^{(n)}_{j+1/2})\,\frac{x^{(n+1)}_{j+1} - x^{(n+1)}_{j}}{\Delta q} - \varpi(x^{(n)}_{j-1/2})\,\frac{x^{(n+1)}_{j} - x^{(n+1)}_{j-1}}{\Delta q}\Biggr] = 0, \qquad n \in \mathbb{N}_0. \tag{2.43}$$

The iterations are continued until the convergence is achieved to the prescribed tolerance parameter (typically $\propto 10^{-10}$).

Grid Motion

In unsteady computations the grid motion is given by the following nonlinear parabolic equation:

$$\frac{\partial}{\partial q}\left[\varpi(x, t)\,\frac{\partial x}{\partial q}\right] = \beta\,\frac{\partial x}{\partial t}, \qquad \beta \in \mathbb{R}^+. \tag{2.44}$$

The parameter β plays the rôle of the diffusion coefficient here. It is used to control the smoothness of nodes trajectories. Equation (2.44) is discretized using an implicit scheme:

$$\frac{1}{\Delta q}\left\{\varpi^{\,n}_{\,j+1/2}\,\frac{x^{\,n+1}_{\,j+1} - x^{\,n+1}_{\,j}}{\Delta q} - \varpi^{\,n}_{\,j-1/2}\,\frac{x^{\,n+1}_{\,j} - x^{\,n+1}_{\,j-1}}{\Delta q}\right\}$$
$$= \beta\,\frac{x^{\,n+1}_{\,j} - x^{\,n}_{\,j}}{\tau}, \tag{2.45}$$

with boundary conditions $x^{\,n+1}_{\,0} = 0$, $x^{\,n+1}_{\,N} = \ell$ as above. We would like to reiterate that at every time step we solve only one additional (tridiagonal) linear system. Nonlinear iterative computations are performed only once when we project the initial condition on the ad-hoc non-uniform grid. So, the additional overhead due to the mesh motion is linear in complexity, i.e. $\mathcal{O}(N)$.

Similarly to the elliptic case (2.42), Eq. (2.44) admits smooth solutions provided that the monitor function $\varpi(x,\,t)$ is bounded from below by a positive constant. In numerical examples shown below we always take monitor functions which satisfy the condition $\varpi(x,\,t) \geqslant 1, \forall x \in \mathcal{I}, \forall t \geqslant 0$. Thus, for any $t > 0$ Eq. (2.44) provides us the required diffeomorphism between the reference domain $\mathcal{Q}$ and the computational domain $\mathcal{I}$.

2.2.2 The SGN Equations on a Moving Grid

Before discretizing the SGN equations (2.22)–(2.24), we have to pose them on the reference domain $\mathcal{Q}$. The composed functions will be denoted as:

$$\mathring{u}(q,\,t) \stackrel{\text{def}}{:=} (u \circ x)(q,\,t) \equiv u\bigl(x(q,\,t),\,t\bigr).$$

And we introduce similar notations for all other variables, e.g. $\mathring{\mathcal{H}}(q,\,t) \stackrel{\text{def}}{:=} \mathcal{H}\bigl(x(q,\,t),\,t\bigr)$, etc. The conservative (2.30) and non-conservative (2.32), (2.33) forms of hyperbolic equations read:

$$(\mathfrak{J}\,\mathring{\mathbf{v}})_t + \bigl[\mathring{\mathscr{F}} - x_t\,\mathring{\mathbf{v}}\bigr]_q = \mathring{\mathscr{G}}, \tag{2.46}$$

$$\mathring{\mathbf{v}}_t + \frac{1}{\mathfrak{J}}\bigl[\mathring{\mathscr{F}}_q - x_t\,\mathring{\mathbf{v}}_q\bigr] = \frac{1}{\mathfrak{J}}\,\mathring{\mathscr{G}}, \tag{2.47}$$

$$\mathring{\mathscr{F}}_t + \frac{1}{\mathfrak{J}}\,\mathring{\mathscr{A}}\cdot\bigl[\mathring{\mathscr{F}}_q - x_t\,\mathring{\mathbf{v}}_q\bigr] = \frac{1}{\mathfrak{J}}\,\mathring{\mathscr{A}}\cdot\mathring{\mathscr{G}}, \tag{2.48}$$

where the terms on the right-hand sides are defined similarly as above:

$$\mathring{\mathscr{G}} \stackrel{\text{def}}{:=} \begin{pmatrix} 0 \\ g\,\mathring{\mathcal{H}}\,\mathring{h}_q + \mathring{\wp}_q - \mathring{\breve{\wp}}\,\mathring{h}_q \end{pmatrix}.$$

The non-hydrostatic pressure on the bottom is computed in $\mathcal{Q}$ space as:

$$\check{\mathring{\wp}} \stackrel{\text{def}}{:=} \frac{1}{\mathring{\Upsilon}}\left[\frac{6\mathring{\wp}}{\mathring{\mathcal{H}}} + \mathring{\mathcal{H}}\mathring{\mathscr{R}} + \frac{\mathring{\wp}_q\mathring{h}_q}{\mathcal{J}^2}\right], \qquad \mathring{\Upsilon} \stackrel{\text{def}}{:=} 4 + \frac{\mathring{h}_q^2}{\mathcal{J}^2} \tag{2.49}$$

Finally, we just have to specify the expression for $\mathring{\mathscr{R}}$:

$$\mathring{\mathscr{R}} \stackrel{\text{def}}{:=} -g\frac{\mathring{\eta}_q\mathring{h}_q}{\mathcal{J}^2} + \frac{\mathring{u}^2}{\mathcal{J}}\left[\frac{\mathring{h}_q}{\mathcal{J}}\right]_q + \left(\mathring{h}_t - \frac{x_t}{\mathcal{J}}\mathring{h}_q\right)_t + \frac{2\mathring{u} - x_t}{\mathcal{J}}\cdot\left[\mathring{h}_t - \frac{x_t}{\mathcal{J}}\mathring{h}_q\right]_q.$$

We have to specify also the equations which allow us to find the non-hydrostatic part of the pressure field $\mathring{\wp}$. Equation (2.25) posed on the reference domain $\mathcal{Q}$ reads:

$$\left[\mathring{\mathcal{K}}\mathring{\wp}_q\right]_q - \mathring{\mathcal{K}}_0\mathring{\wp} = \mathring{\mathcal{F}}, \tag{2.50}$$

where the coefficients and the right-hand side are defined as

$$\mathring{\mathcal{K}} \stackrel{\text{def}}{:=} \frac{4}{\mathcal{J}\mathring{\mathcal{H}}\mathring{\Upsilon}}, \qquad \mathring{\mathcal{K}}_0 \stackrel{\text{def}}{:=} 6\left[\frac{2\mathcal{J}}{\mathring{\mathcal{H}}^3}\cdot\frac{\mathring{\Upsilon} - 3}{\mathring{\Upsilon}} + \left(\frac{\mathring{h}_q}{\mathcal{J}\mathring{\mathcal{H}}^2\mathring{\Upsilon}}\right)_q\right],$$

$$\mathring{\mathcal{F}} \stackrel{\text{def}}{:=} \left[g\frac{\mathring{\eta}_q}{\mathcal{J}} + \frac{\mathring{\mathscr{R}}\mathring{h}_q}{\mathcal{J}\mathring{\Upsilon}}\right]_q - \frac{6\mathring{\mathscr{R}}\mathcal{J}}{\mathring{\mathcal{H}}\mathring{\Upsilon}} + 2\frac{\mathring{u}_q^2}{\mathcal{J}}.$$

Finally, the boundary conditions are specified now at $q = 0$ and $q = 1$. For the hyperbolic part of the equations they are

$$\mathring{u}(0, t) = 0 \qquad \mathring{u}(1, t) = 0 \qquad \forall t \geqslant 0.$$

For the elliptic part we have the following mixed-type boundary conditions:

$$\left[\frac{4}{\mathcal{J}\mathring{\mathcal{H}}\mathring{\Upsilon}}\mathring{\wp}_q - \frac{6\mathring{h}_q}{\mathcal{J}\mathring{\mathcal{H}}^2\mathring{\Upsilon}}\mathring{\wp}\right]\Bigg|_{q=0}^{q=1} = \frac{1}{\mathcal{J}}\left[g\mathring{\eta}_q + \frac{\mathring{\mathscr{R}}}{\mathring{\Upsilon}}\mathring{h}_q\right]\Bigg|_{q=0}^{q=1}. \tag{2.51}$$

2.2.3 *Predictor–Corrector Scheme on Moving Grids*

In this section we describe the numerical finite volume discretization of the SGN equations on a moving grid. We assume that the reference domain $\mathcal{Q}$ is discretized with a uniform grid $\mathcal{Q}_h \stackrel{\text{def}}{:=} \{q_j = j\,\Delta q\}_{j=0}^N$, with the uniform spacing $\Delta q = \frac{1}{N}$. Then, the grid $\mathcal{I}_h^n$ in the physical domain $\mathcal{I}$ at every time instance $t = t^n \geqslant 0$ is given by the image of the uniform grid $\mathcal{Q}_h$ under the mapping $x(q, t)$, i.e.

$x_j^n = x(q_j, t^n)$, $j = 0, 1, \ldots, N$ or simply $\mathfrak{J}_h^n = x(\mathfrak{Q}_h, t^n)$. We assume that we know the discrete solution[6] $\mathring{\mathbf{v}}_\sharp^n \stackrel{\text{def}}{:=} \{\mathring{\mathbf{v}}_j^n\}_{j=0}^N$, $\mathring{\wp}_\sharp^n \stackrel{\text{def}}{:=} \{\mathring{\wp}_j^n\}_{j=0}^N$ at the current time $t = t^n$ and we already constructed the non-uniform grid $x_\sharp^{n+1} \stackrel{\text{def}}{:=} \{x_j^{n+1}\}_{j=0}^N$ at the following time layer t^{n+1} using the equidistribution method described above. We remind that the non-uniform grid at the following layer is constructed based only on the knowledge of $\mathring{\mathbf{v}}_\sharp^n$.

Predictor Step

In the nonlinear case, during the predictor step the hyperbolic part of equations is solved two times:

- First, using Eq. (2.47) we compute the discrete solution values $\mathring{\mathbf{v}}_{\sharp,c}^* \stackrel{\text{def}}{:=} \{\mathring{\mathbf{v}}_{j+1/2}^*\}_{j=0}^{N-1}$ in the cell centres $\mathfrak{Q}_{h,c} \stackrel{\text{def}}{:=} \{q_{j+1/2} = q_j + \frac{\Delta q}{2}\}_{j=0}^{N-1}$.
- Then, using Eq. (2.48) we compute the values of the flux vector equally in the cell centres $\mathring{\mathscr{F}}_{\sharp,c}^* \stackrel{\text{def}}{:=} \{\mathring{\mathscr{F}}_{j+1/2}^*\}_{j=0}^{N-1}$.

We rewrite Eqs. (2.47), (2.48) in the characteristic form by multiplying them on the left by the matrix $\mathring{L}$ (of left eigenvectors of the JACOBIAN $\mathring{\mathscr{A}}$):

$$\mathring{L} \cdot \mathring{\mathbf{v}}_t + \frac{1}{\mathfrak{J}} \mathring{L} \cdot \left[\mathring{\mathscr{F}}_q - x_t \mathring{\mathbf{v}}_q\right] = \frac{1}{\mathfrak{J}} \mathring{L} \cdot \mathring{\mathscr{G}},$$

$$\mathring{L} \cdot \mathring{\mathscr{F}}_t + \frac{1}{\mathfrak{J}} \mathring{\Lambda} \cdot \mathring{L} \cdot \left[\mathring{\mathscr{F}}_q - x_t \mathring{\mathbf{v}}_q\right] = \frac{1}{\mathfrak{J}} \mathring{\Lambda} \cdot \mathring{L} \cdot \mathring{\mathscr{G}},$$

The discretization of last equations reads:

$$\left(\mathfrak{D}^{-1} \cdot \mathring{L}\right)_{j+1/2}^n \cdot \frac{\mathring{\mathbf{v}}_{j+1/2}^* - \mathring{\mathbf{v}}_{j+1/2}^n}{\tau/2} + \left(\frac{1}{\mathfrak{J}} \mathring{L} \cdot \left[\mathring{\mathscr{F}}_q - x_t \mathring{\mathbf{v}}_q\right]\right)_{j+1/2}^n = \left(\frac{1}{\mathfrak{J}} \mathring{L} \cdot \mathring{\mathscr{G}}\right)_{j+1/2}^n, \tag{2.52}$$

$$\left(\mathfrak{D}^{-1} \cdot \mathring{L}\right)_{j+1/2}^n \cdot \frac{\mathring{\mathscr{F}}_{j+1/2}^* - \mathring{\mathscr{F}}_{j+1/2}^n}{\tau/2} + \left(\frac{1}{\mathfrak{J}} \mathring{\Lambda} \cdot \mathring{L} \cdot \left[\mathring{\mathscr{F}}_q - x_t \mathring{\mathbf{v}}_q\right]\right)_{j+1/2}^n = \left(\frac{1}{\mathfrak{J}} \mathring{\Lambda} \cdot \mathring{L} \cdot \mathring{\mathscr{G}}\right)_{j+1/2}^n, \tag{2.53}$$

[6]With symbol $\sharp$ we denote the set of solution values at discrete spatial grid nodes.

where τ is the time step, $\mathring{L}^n_{j+1/2}$ is an approximation of matrix $\mathring{L}$ in the cell centres $\mathcal{Q}_{h,c}$ (it will be specified below). The matrix $\mathcal{D}$ is composed of cell parameters for each equation:

$$\mathcal{D}^n_{j+1/2} \stackrel{\text{def}}{:=} \begin{pmatrix} 1 + \theta^{1,n}_{j+1/2} & 0 \\ 0 & 1 + \theta^{2,n}_{j+1/2} \end{pmatrix},$$

$$\mathring{\Lambda}^n_{j+1/2} \stackrel{\text{def}}{:=} \begin{pmatrix} 1 + \lambda^{-,n}_{j+1/2} & 0 \\ 0 & 1 + \lambda^{+,n}_{j+1/2} \end{pmatrix},$$

with $\lambda^{\pm,n}_{j+1/2}$ being the approximations of eigenvalues (2.31) in the cell centres $\mathcal{Q}_{h,c}$ (it will be specified below). On the right-hand side the source term is

$$\mathring{\mathcal{G}}^n_{j+1/2} \stackrel{\text{def}}{:=} \begin{pmatrix} 0 \\ \left(g\,\mathcal{H}\,\mathring{h}_q + \mathring{\wp}_q - \mathring{\check{\wp}}\,\mathring{h}_q\right)^n_{j+1/2} \end{pmatrix},$$

where derivatives with respect to q are computed using central differences:

$$\mathring{\wp}^n_{q,\,j+1/2} \stackrel{\text{def}}{:=} \frac{\mathring{\wp}^n_{j+1} - \mathring{\wp}^n_j}{\Delta q}, \qquad \mathring{h}^n_{q,\,j+1/2} \stackrel{\text{def}}{:=} \frac{\mathring{h}^n_{j+1} - \mathring{h}^n_j}{\Delta q}.$$

The value of the non-hydrostatic pressure trace at the bottom $\mathring{\check{\wp}}^n_{j+1/2}$ is computed according to formula (2.49). Solution vector $\mathring{\mathbf{v}}^n_{\sharp,c}$ and the fluxes $\mathring{\mathcal{F}}^n_{\sharp,c}$ in cell centres are computed as:

$$\mathring{\mathbf{v}}^n_{j+1/2} \stackrel{\text{def}}{:=} \frac{\mathring{\mathbf{v}}^n_{j+1} + \mathring{\mathbf{v}}^n_j}{2}, \qquad \mathring{\mathcal{F}}^n_{j+1/2} \stackrel{\text{def}}{:=} \frac{\mathring{\mathcal{F}}^n_{j+1} + \mathring{\mathcal{F}}^n_j}{2}.$$

The derivatives of these quantities are estimated using simple finite differences:

$$\mathring{\mathbf{v}}^n_{q,\,j+1/2} \stackrel{\text{def}}{:=} \frac{\mathring{\mathbf{v}}^n_{j+1} - \mathring{\mathbf{v}}^n_j}{\Delta q}, \qquad \mathring{\mathcal{F}}^n_{q,\,j+1/2} \stackrel{\text{def}}{:=} \frac{\mathring{\mathcal{F}}^n_{j+1} - \mathring{\mathcal{F}}^n_j}{\Delta q}.$$

Finally, we have to specify the computation of some mesh-related quantities:

$$x^n_{t,\,j} \stackrel{\text{def}}{:=} \frac{x^{n+1}_j - x^n_j}{\tau}, \qquad x^n_{t,\,j+1/2} \stackrel{\text{def}}{:=} \frac{x^n_{t,\,j+1} + x^n_{t,\,j}}{2},$$

$$\mathcal{J}^n_{j+1/2} \equiv x^n_{q,\,j+1/2} \stackrel{\text{def}}{:=} \frac{x^n_{j+1} - x^n_j}{\Delta q}.$$

The approximation of the matrix of left eigenvectors $\mathring{L}^{\,n}_{j+1/2}$ and eigenvalues $\lambda^{\pm,\,n}_{j+1/2}$ depends on the specification of the JACOBIAN matrix $\mathring{\mathscr{A}}^{\,n}_{j+1/2}$. Our approach consists in choosing the discrete approximation in order to have at discrete level

$$\mathring{\mathscr{F}}^{\,n}_{q,\,j+1/2} \equiv \left(\mathring{\mathscr{A}} \cdot \mathring{\mathbf{v}}_q\right)^n_{j+1/2}, \tag{2.54}$$

which is the discrete analogue of the continuous identity $\mathring{\mathfrak{F}}_q \equiv \mathring{\mathscr{A}} \cdot \mathring{\mathbf{v}}_q$. Basically, our philosophy consists in preserving as many as possible continuous properties at the discrete level. For example, the following matrix satisfies the condition (2.54):

$$\mathring{\mathscr{A}}^{\,n}_{j+1/2} = \begin{pmatrix} 0 & 1 \\ -u^n_j\, u^n_{j+1} + g\,\mathcal{H}^n_{j+1/2} & 2\,u^n_{j+1/2} \end{pmatrix} = \left(\mathring{R} \cdot \mathring{\Lambda} \cdot \mathring{L}\right)^n_{j+1/2}$$

The matrices $L^n_{j+1/2}$ and $R^n_{j+1/2} = (L^n_{j+1/2})^{-1}$ are computed by formulas (2.34). The JACOBIAN matrix $\mathring{\mathscr{A}}^{\,n}_{j+1/2}$ eigenvalues can be readily computed:

$$\lambda^{\pm,\,n}_{j+1/2} \overset{\text{def}}{:=} (u \pm s)^n_{j+1/2},$$

$$s^n_{j+1/2} \overset{\text{def}}{:=} \sqrt{(u^n_{j+1/2})^2 - u^n_j\, u^n_{j+1} + g\,\mathcal{H}^n_{j+1/2}} \geqslant \sqrt{g\,\mathcal{H}^n_{j+1/2}} > 0\,.$$

Thanks to the discrete differentiation rule (2.54), we can derive elegant formulas for the predicted values $\mathring{\mathbf{v}}^{\,*}_{\sharp,\,c}$, $\mathring{\mathscr{F}}^{\,*}_{\sharp,\,c}$ by drastically simplifying the scheme (2.52), (2.53):

$$\mathring{\mathbf{v}}^{\,*}_{j+1/2} = \left[\mathring{\mathbf{v}} - \frac{\tau}{2\,\mathfrak{J}}\,\mathring{R} \cdot \mathfrak{D} \cdot \left(\bar{\mathring{\Lambda}} \cdot \mathring{\mathbb{P}} - \mathring{L} \cdot \mathring{\mathscr{G}}\right)\right]^n_{j+1/2}, \tag{2.55}$$

$$\mathring{\mathscr{F}}^{\,*}_{j+1/2} = \left[\mathring{\mathscr{F}} - \frac{\tau}{2\,\mathfrak{J}}\,\mathring{R} \cdot \mathfrak{D} \cdot \mathring{\Lambda} \cdot \left(\bar{\mathring{\Lambda}} \cdot \mathring{\mathbb{P}} - \mathring{L} \cdot \mathring{\mathscr{G}}\right)\right]^n_{j+1/2}, \tag{2.56}$$

where we introduced two matrices:

$$\bar{\mathring{\Lambda}}^{\,n}_{j+1/2} \overset{\text{def}}{:=} \mathring{\Lambda}^{\,n}_{j+1/2} - x^n_{t,\,j+1/2} \cdot \mathbb{I}, \qquad \mathring{\mathbb{P}}^{\,n}_{j+1/2} \overset{\text{def}}{:=} \left(\mathring{L} \cdot \mathring{\mathbf{v}}_q\right)^n_{j+1/2}.$$

Finally, the scheme parameters $\theta^{1,2}_{j+1/2}$ are chosen as it was explained in our works [192, 287] for the case of Nonlinear Shallow Water Equations. This choice guarantees the TVD property of the resulting scheme.

Non-hydrostatic Pressure Computation

Once we determined the predicted values $\mathring{\mathbf{v}}^*_{\sharp,c}$, $\mathring{\mathscr{F}}^*_{\sharp,c}$, we have to determine also the predicted value for the non-hydrostatic pressure components $\mathring{\wp}^*_{\sharp,c}$ located in cell centres $\mathcal{Q}_{h,c}$. In order to discretize the elliptic equation (2.50) we apply the same finite volume philosophy. Namely, we integrate equation (2.50) over one cell $[q_j, q_{j+1}]$. Right now for simplicity we consider an interior element. The approximation near boundaries will be discussed below. The integral form of Eq. (2.50) reads

$$\int_{q_j}^{q_{j+1}} [\mathring{\mathcal{K}}\mathring{\wp}^*_q]_q \, \mathrm{d}q - \int_{q_j}^{q_{j+1}} \mathring{\mathcal{K}}_0 \mathring{\wp}^* \, \mathrm{d}q = \int_{q_j}^{q_{j+1}} \mathring{\mathcal{F}} \, \mathrm{d}q. \tag{2.57}$$

The coefficients $\mathring{\mathcal{K}}$, $\mathring{\mathcal{K}}_0$ are evaluated using the predicted value of the total water depth $\mathring{\mathcal{H}}^*_{\sharp,c}$. If the scheme parameter $\theta^n_{j+1/2} \equiv 0, \forall j = 0, \ldots, N-1$, then the predictor value would lie completely on the middle layer $t = t^n + \frac{\tau}{2}$. However, this simple choice of $\{\theta^n_{j+1/2}\}_{j=0}^{N-1}$ does not ensure the desired TVD property [19, 287].

The solution of this integral equation will give us the predictor value for the non-hydrostatic pressure $\mathring{\wp}^*_{\sharp,c}$. The finite difference scheme for Eq. (2.50) is obtained by applying the following quadrature formulas to all the terms in integral equation (2.57):

$$\int_{q_j}^{q_{j+1}} [\mathring{\mathcal{K}}\mathring{\wp}^*_q]_q \, \mathrm{d}q \simeq \frac{\mathring{\mathcal{K}}_{j+3/2} + \mathring{\mathcal{K}}_{j+1/2}}{2} \cdot \frac{\mathring{\wp}^*_{j+3/2} - \mathring{\wp}^*_{j+1/2}}{\Delta q} - \frac{\mathring{\mathcal{K}}_{j+1/2} + \mathring{\mathcal{K}}_{j-1/2}}{2} \cdot \frac{\mathring{\wp}^*_{j+1/2} - \mathring{\wp}^*_{j-1/2}}{\Delta q},$$

$$\int_{q_j}^{q_{j+1}} \mathring{\mathcal{K}}_0 \mathring{\wp}^* \, \mathrm{d}q \simeq \Bigg[\Delta q \cdot \Big[\frac{12\mathfrak{J}^n}{(\mathring{\mathcal{H}}^*)^3} \cdot \frac{\mathring{\Upsilon} - 3}{\mathring{\Upsilon}}\Big]_{j+1/2} + \Big[\frac{3\mathring{h}^n_q}{\mathring{\Upsilon}\mathfrak{J}^n(\mathring{\mathcal{H}}^*)^2}\Big]_{j+3/2} - \Big[\frac{3\mathring{h}^n_q}{\mathring{\Upsilon}\mathfrak{J}^n(\mathring{\mathcal{H}}^*)^2}\Big]_{j-1/2}\Bigg]\mathring{\wp}^*_{j+1/2},$$

$$\int_{q_j}^{q_{j+1}} \mathring{\mathcal{F}} \, \mathrm{d}q \simeq \Delta q \cdot \Big(2\frac{(\mathring{u}^*_q)^2}{\mathfrak{J}^n} - \frac{6\mathring{\mathscr{R}}\mathfrak{J}^n}{\mathring{\Upsilon}\mathring{\mathcal{H}}^*}\Big)_{j+1/2} + \Big(g\frac{\mathring{\eta}^*_q}{\mathfrak{J}^n} + \frac{\mathring{\mathscr{R}}\mathring{h}^n_q}{\mathring{\Upsilon}\mathfrak{J}^n}\Big)_{j+1} - \Big(g\frac{\mathring{\eta}^*_q}{\mathfrak{J}^n} + \frac{\mathring{\mathscr{R}}\mathring{h}^n_q}{\mathring{\Upsilon}\mathfrak{J}^n}\Big)_{j}.$$

In approximation formulas above we introduced the following notations:

$$\mathring{\mathcal{K}}_{j+1/2} \stackrel{\text{def}}{:=} \left[\frac{4}{\mathring{\Upsilon}\,\mathfrak{g}^n\,\mathring{\mathcal{H}}^*}\right]_{j+1/2}, \qquad \mathring{\Upsilon}_{j+1/2} \stackrel{\text{def}}{:=} 4 + \left(\frac{\mathring{h}^n_q}{\mathfrak{g}^n}\right)^2_{j+1/2},$$

$$\mathfrak{g}^n_j \stackrel{\text{def}}{:=} \frac{\mathfrak{g}^n_{j+1/2} + \mathfrak{g}^n_{j-1/2}}{2}.$$

In this way we obtain a three-point finite difference approximation of the elliptic equation (2.50) in interior of the domain, i.e. $j = 1, \ldots, N-2$:

$$\begin{aligned}
&\frac{\mathring{\mathcal{K}}_{j+3/2} + \mathring{\mathcal{K}}_{j+1/2}}{2}\cdot\frac{\mathring{\wp}^*_{j+3/2} - \mathring{\wp}^*_{j+1/2}}{\Delta q} \\
&\qquad - \frac{\mathring{\mathcal{K}}_{j+1/2} + \mathring{\mathcal{K}}_{j-1/2}}{2}\cdot\frac{\mathring{\wp}^*_{j+1/2} - \mathring{\wp}^*_{j-1/2}}{\Delta q} \\
&- \left[\Delta q\cdot\left[\frac{12\,\mathfrak{g}^n}{(\mathring{\mathcal{H}}^*)^3}\cdot\frac{\mathring{\Upsilon}-3}{\mathring{\Upsilon}}\right]_{j+1/2} + \left[\frac{3\,\mathring{h}^n_q}{\mathring{\Upsilon}\,\mathfrak{g}^n\,(\mathring{\mathcal{H}}^*)^2}\right]_{j+3/2} - \left[\frac{3\,\mathring{h}^n_q}{\mathring{\Upsilon}\,\mathfrak{g}^n\,(\mathring{\mathcal{H}}^*)^2}\right]_{j-1/2}\right] \\
&\qquad = \Delta q\cdot\left(2\,\frac{(\mathring{u}^*_q)^2}{\mathfrak{g}^n} - \frac{6\,\mathring{\mathcal{R}}\,\mathfrak{g}^n}{\mathring{\Upsilon}\,\mathring{\mathcal{H}}^*}\right)_{j+1/2} + \left(g\,\frac{\mathring{\eta}^*_q}{\mathfrak{g}^n} + \frac{\mathring{\mathcal{R}}\,\mathring{h}^n_q}{\mathring{\Upsilon}\,\mathfrak{g}^n}\right)_{j+1} \\
&\qquad\qquad - \left(g\,\frac{\mathring{\eta}^*_q}{\mathfrak{g}^n} + \frac{\mathring{\mathcal{R}}\,\mathring{h}^n_q}{\mathring{\Upsilon}\,\mathfrak{g}^n}\right)_j.
\end{aligned} \tag{2.58}$$

Two missing equations are obtained by approximating the integral equation (2.57) in intervals adjacent to the boundaries. As a result, we obtain a linear system of equations where unknowns are $\{\mathring{\wp}^*_{j+1/2}\}_{j=0}^{N-1}$. The approximation in boundary cells will be illustrated on the left boundary $[q_0 \equiv 0, q_1]$. The right-most cell $[q_{N-1}, q_N \equiv 1]$ can be treated similarly. Let us write down one-sided quadrature formulas for the first cell:

$$\begin{aligned}
&\frac{\mathring{\mathcal{K}}_{3/2}+\mathring{\mathcal{K}}_{1/2}}{2}\cdot\frac{\mathring{\wp}^*_{3/2}-\mathring{\wp}^*_{1/2}}{\Delta q} - \underbrace{\left.\frac{4\,\mathring{\wp}^*_q}{\mathfrak{g}\,\mathring{\mathcal{H}}^*\,\mathring{\Upsilon}}\right|_{q=0}}_{\otimes_1} - \mathring{\wp}^*_{1/2}\left[\Delta q\cdot\left[\frac{12\,\mathfrak{g}^n}{(\mathring{\mathcal{H}}^*)^3}\cdot\frac{\mathring{\Upsilon}-3}{\mathring{\Upsilon}}\right]_{1/2}\right. \\
&\quad \left. + \left[\frac{3\,\mathring{h}^n_q}{\mathring{\Upsilon}\,\mathfrak{g}^n\,(\mathring{\mathcal{H}}^*)^2}\right]_{3/2} + \left[\frac{3\,\mathring{h}^n_q}{\mathring{\Upsilon}\,\mathfrak{g}^n\,(\mathring{\mathcal{H}}^*)^2}\right]_{1/2}\right] + \underbrace{\left.\frac{6\,\mathring{h}^n_q\,\mathring{\wp}^*}{\mathfrak{g}\,(\mathring{\mathcal{H}}^*)^2\,\mathring{\Upsilon}}\right|_{q=0}}_{\otimes_2}
\end{aligned}$$

$$= \Delta q \cdot \Bigl(2\,\frac{(\mathring{u}_q^{\,*})^2}{\mathcal{J}^n} - \frac{6\,\mathscr{R}\,\mathcal{J}^n}{\mathring{\Upsilon}\,\mathring{\mathcal{H}}^*}\Bigr)_{1/2} + \Bigl(g\,\frac{\mathring{\eta}_q^{\,*}}{\mathcal{J}^n} + \frac{\mathscr{R}\,\mathring{h}_q^{\,n}}{\mathring{\Upsilon}\,\mathcal{J}^n}\Bigr)_1 - \underbrace{\Bigl(g\,\frac{\mathring{\eta}_q^{\,*}}{\mathcal{J}^n} + \frac{\mathscr{R}\,\mathring{h}_q^{\,n}}{\mathring{\Upsilon}\,\mathcal{J}^n}\Bigr)\Bigr|_{q\,=\,0}}_{\circledast_3}.$$

It can be readily noticed that terms $\circledast_1 + \circledast_2 + \circledast_3$ vanish, thanks to the boundary condition (2.51) (the part at $q = 0$). The same trick applies to the right-most cell $[q_{N-1}, q_N \equiv 1]$. We reiterate on the fact that in our scheme the boundary conditions are taken into account *exactly*. Consequently, in two boundary cells we obtain a two-point finite difference approximation to Eq. (2.50). The resulting linear system of equations can be solved using e.g. the direct THOMAS algorithm with linear complexity $\mathcal{O}(N)$. Under the conditions $\mathring{\mathcal{K}}_0 > 0$, $\mathring{h}_q\big|_{q\,=\,0} \geqslant 0$, $\mathring{h}_q\big|_{q\,=\,1} \leqslant 0$ the numerical solution exists, it is unique and stable [272].

Corrector Step

During the corrector step we solve again separately the hyperbolic and elliptic parts of the SGN equations. In order to determine the vector of conservative variables $\mathring{\mathbf{v}}_\sharp^{\,n+1}$ we use an explicit finite volume scheme based on the conservative equation (2.46):

$$\frac{(\mathcal{J}\,\mathring{\mathbf{v}})_j^{\,n+1} - (\mathcal{J}\,\mathring{\mathbf{v}})_j^{\,n}}{\tau} + \frac{\bigl(\mathring{\mathscr{F}}^* - x_t\cdot\mathring{\mathbf{v}}^*\bigr)_{j+1/2} - \bigl(\mathring{\mathscr{F}}^* - x_t\cdot\mathring{\mathbf{v}}^*\bigr)_{j-1/2}}{\Delta q} = \mathring{\mathscr{G}}_j^{\,*}, \qquad (2.59)$$

where

$$\mathring{\mathscr{G}}_j^{\,*} \stackrel{\text{def}}{:=} \left(\begin{array}{c} 0 \\ \Bigl(\bigl(g\,\mathring{\mathcal{H}}^{n+\flat} - \mathring{\wp}^*\bigr)\,\mathring{h}_q^{\,n+\flat} + \mathring{\wp}_q^{\,*}\Bigr)_j \end{array}\right), \qquad \mathring{\wp}_{q,\,j}^{\,*} \stackrel{\text{def}}{:=} \frac{\mathring{\wp}_{j+1/2}^{\,*} - \mathring{\wp}_{j-1/2}^{\,*}}{\Delta q},$$

and

$$\mathring{\mathcal{H}}_j^{\,n+\flat} \stackrel{\text{def}}{:=} \frac{\mathring{\mathcal{H}}_{j+1}^{\,n+1} + \mathring{\mathcal{H}}_{j-1}^{\,n+1} + 2\,\mathring{\mathcal{H}}_j^{\,n+1} + 2\,\mathring{\mathcal{H}}_j^{\,n} + \mathring{\mathcal{H}}_{j+1}^{\,n} + \mathring{\mathcal{H}}_{j-1}^{\,n}}{8}, \qquad (2.60)$$

$$\mathring{h}_q^{\,n+\flat} \stackrel{\text{def}}{:=} \frac{\mathring{h}_{j+1}^{\,n+1} - \mathring{h}_{j-1}^{\,n+1} + \mathring{h}_{j+1}^{\,n} - \mathring{h}_{j-1}^{\,n}}{4\,\Delta q}. \qquad (2.61)$$

The algorithm of the corrector scheme can be summarized as follows:

1. From the mass conservation equations (the first component in (2.59)) we find the total water depth $\mathring{\mathcal{H}}_{\sharp}^{n+1}$ in interior nodes of the grid.
2. Using the method of characteristics and the boundary conditions $\mathring{u}_0^{n+1} = \mathring{u}_N^{n+1} \equiv 0$ we determine the total water depth $\mathring{\mathcal{H}}_0^{n+1}$, $\mathring{\mathcal{H}}_N^{n+1}$ in boundary points $q_0 \equiv 0$ and $q_N \equiv 1$.
3. Then, using the momentum conservation equation (the second component in (2.59)) we find the momentum values $(\mathring{\mathcal{H}}\,\mathring{u})_{\sharp}^{n+1}$ on the next time layer.

In this way, we obtain an explicit scheme despite the fact that the right-hand side $\mathring{\mathcal{G}}_{\sharp}^{*}$ depends on the water depth $\mathring{\mathcal{H}}_{\sharp}^{n+1}$ at the new time layer $t = t^{n+1}$.

Non-hydrostatic Pressure Correction

The non-hydrostatic pressure correction $\mathring{\wp}_{\sharp}^{n+1}$ is computed by integrating locally the elliptic equation (2.50) around each grid point:

$$\int_{q_{j-1/2}}^{q_{j+1/2}} [\mathring{\mathcal{K}}\,\mathring{\wp}_q^{n+1}]_q \,\mathrm{d}q - \int_{q_{j-1/2}}^{q_{j+1/2}} \mathring{\mathcal{K}}_0\,\mathring{\wp}^{n+1} \,\mathrm{d}q = \int_{q_{j-1/2}}^{q_{j+1/2}} \mathring{\mathcal{F}}^{n+1} \,\mathrm{d}q, \qquad j = 1, \ldots, N-1,$$

The details of integrals approximations are similar to the predictor step described above. Consequently, we provide directly the difference scheme in interior nodes:

$$\begin{aligned}
&\mathcal{K}_{j+1/2}\,\frac{\mathring{\wp}_{j+1}^{n+1}-\mathring{\wp}_{j}^{n+1}}{\Delta q}-\mathcal{K}_{j-1/2}\,\frac{\mathring{\wp}_{j}^{n+1}-\mathring{\wp}_{j-1}^{n+1}}{\Delta q}\\
&-6\,\mathring{\wp}_{j}^{n+1}\Big[\Big(\Delta q\,\frac{(\mathring{\Upsilon}-3)\,\mathcal{J}}{\mathring{\Upsilon}\,\mathring{\mathcal{H}}^{3}}-\frac{\mathring{h}_q}{\mathring{\Upsilon}\,\mathcal{J}\,\mathring{\mathcal{H}}^{2}}\Big)_{j-1/2}^{n+1}+\Big(\Delta q\,\frac{(\mathring{\Upsilon}-3)\,\mathcal{J}}{\mathring{\Upsilon}\,\mathring{\mathcal{H}}^{3}}+\frac{\mathring{h}_q}{\mathring{\Upsilon}\,\mathcal{J}\,\mathring{\mathcal{H}}^{2}}\Big)_{j+1/2}^{n+1}\Big]\\
&\qquad=\Delta q\,\Big(2\,\frac{\mathring{u}_q^{2}}{\mathcal{J}}-\frac{6\,\mathring{\mathscr{R}}\,\mathcal{J}}{\mathring{\Upsilon}\,\mathring{\mathcal{H}}}\Big)_{j}^{n+1}+\Big(g\,\frac{\mathring{\eta}_q}{\mathcal{J}}+\frac{\mathring{\mathscr{R}}\,\mathring{h}_q}{\mathring{\Upsilon}\,\mathcal{J}}\Big)_{j+1/2}^{n+1}\\
&\qquad\qquad-\Big(g\,\frac{\mathring{\eta}_q}{\mathcal{J}}+\frac{\mathring{\mathscr{R}}\,\mathring{h}_q}{\mathring{\Upsilon}\,\mathcal{J}}\Big)_{j-1/2}^{n+1},
\end{aligned} \tag{2.62}$$

where

$$\mathcal{K}_{j+1/2} \stackrel{\text{def}}{:=} \frac{4}{(\mathring{\Upsilon}\,\mathcal{J}\,\mathcal{H})^{n+1}_{j+1/2}}\,, \qquad \mathring{\Upsilon}^{n+1}_{j+1/2} \stackrel{\text{def}}{:=} 4 + \Biggl[\frac{\mathring{h}^{n+1}_{j+1} - \mathring{h}^{n+1}_{j}}{x^{n+1}_{j+1} - x^{n+1}_{j}}\Biggr]^2,$$

$$\mathcal{J}^{n+1}_{j+1/2} \stackrel{\text{def}}{:=} \frac{x^{n+1}_{j+1} - x^{n+1}_{j}}{\Delta q}\,.$$

In order to complete the scheme description, we have to specify the discretization of the elliptic equation (2.50) in boundary cells. To be specific we take again the left-most cell $[\,q_0 \equiv 0,\ q_{1/2}\,]$. The integral equation in this cell reads:

$$\int_{q_0}^{q_{1/2}} \bigl[\mathring{\mathcal{K}}\,\mathring{\wp}^{\,n+1}_{q}\bigr]_q\,\mathrm{d}q - \int_{q_0}^{q_{1/2}} \mathring{\mathcal{K}}_0\,\mathring{\wp}^{\,n+1}\,\mathrm{d}q = \int_{q_0}^{q_{1/2}} \mathring{\mathcal{F}}^{\,n+1}\,\mathrm{d}q\,.$$

And the corresponding difference equation is

$$\mathcal{K}_{1/2}\,\frac{\mathring{\wp}^{\,n+1}_1 - \mathring{\wp}^{\,n+1}_0}{\Delta q} - \underbrace{\frac{4\,\mathring{\wp}_q}{\mathcal{J}\,\mathring{\mathcal{H}}\,\mathring{\Upsilon}}\Bigg|^{n+1}_{q\,=\,0}}_{\Diamond_1} - 6\,\mathring{\wp}^{\,n+1}_0\Biggl[\Delta q\,\frac{(\mathring{\Upsilon}-3)\,\mathcal{J}}{\mathring{\Upsilon}\,\mathring{\mathcal{H}}^3} + \frac{\mathring{h}_q}{\mathcal{J}\,\mathring{\mathcal{H}}^2\,\mathring{\Upsilon}}\Biggr]^{n+1}_{1/2}$$

$$+ \underbrace{\frac{6\,\mathring{h}_q\,\mathring{\wp}}{\mathcal{J}\,\mathring{\mathcal{H}}^2\,\mathring{\Upsilon}}\Bigg|^{n+1}_{q\,=\,0}}_{\Diamond_2} = \Biggl[g\,\frac{\mathring{\eta}_q}{\mathcal{J}} + \frac{\mathring{\mathscr{R}}\,\mathring{h}_q}{\mathring{\Upsilon}\,\mathcal{J}} + \Delta q\,\Bigl(\frac{\mathring{u}_q^2}{\mathcal{J}} - \frac{3\,\mathring{\mathscr{R}}\,\mathcal{J}}{\mathring{\Upsilon}\,\mathring{\mathcal{H}}}\Bigr)\Biggr]^{n+1}_{1/2}$$

$$- \underbrace{\Bigl(g\,\frac{\mathring{\eta}_q}{\mathcal{J}} + \frac{\mathring{\mathscr{R}}\,\mathring{h}_q}{\mathring{\Upsilon}\,\mathcal{J}}\Bigr)\Bigg|^{n+1}_{q\,=\,0}}_{\Diamond_3}\,.$$

By taking into account the boundary condition (2.51) we obtain that three underbraced terms vanish:

$$\Diamond_1 + \Diamond_2 + \Diamond_3 \equiv 0\,.$$

A similar two-point approximation can be obtained by integrating over the right-most cell $\bigl[\,q_{N-1/2},\ q_N\,\bigr] \equiv \Bigl[\,1 - \dfrac{\Delta q}{2},\ 1\,\Bigr]$. In this way we obtain again a three-diagonal system of linear equations which can be efficiently solved with the THOMAS algorithm [166].

Stability of the Scheme

In order to ensure the stability of (nonlinear) computations, we impose a slightly stricter restriction on the time step τ than the linear analysis given below predicts (see Sect. 2.2.4). Namely, at every time layer we apply the same restriction as for hyperbolic (non-dispersive) Nonlinear Shallow Water Equations [192]:

$$\max_j \{ \mathscr{C}^{n,\pm}_{j+1/2} \} \leqslant 1,$$

where $\mathscr{C}^{n,\pm}_{j+1/2}$ are local COURANT numbers [76] which are defined as follows:

$$\mathscr{C}^{n,\pm}_{j+1/2} \stackrel{\text{def}}{:=} \frac{\tau}{\Delta q} \left[\frac{|\lambda^{\pm} - x_t|}{\mathcal{J}} \right]^n_{j+1/2}.$$

Well-Balanced Property

It can be easily established that the predictor–corrector scheme presented above preserves exactly the so-called states 'lake-at-rest':

Lemma 2.5 *Assume that the bottom is stationary (i.e. $h_t \equiv 0$, but not necessary flat) and initially the fluid is at the 'lake-at-rest' state, i.e.*

$$\mathring{\eta}^0_j \equiv 0, \qquad \mathring{u}^0_j \equiv 0 \qquad j = 0, 1, 2, \dots, N. \tag{2.63}$$

Then, the predictor–corrector scheme will preserve this state at all time layers.

Proof In order to prove this lemma, we employ the mathematical induction [165]. First, we have to discuss the generation of the initial grid and how it will be transformed to the next time layer along with the discrete numerical solution:

$$x^0_\sharp \hookrightarrow x^1_\sharp, \qquad \mathring{\mathbf{v}}^0_\sharp \hookrightarrow \mathring{\mathbf{v}}^*_{c,\sharp} \hookrightarrow \mathring{\mathbf{v}}^1_\sharp.$$

Then, by assuming that our statement is true at the nth time layer, we will have to show that it is true on the upcoming $(n+1)^{\text{th}}$ layer. This will complete the proof [165].

If the monitoring function $\varpi(x, t)$ depends only on the free surface elevation $\eta(x, t)$ and fluid velocity $u(x, t)$, then the monitoring function $\varpi(x, t) \equiv 1$, thanks to Lemma assumption (2.63). And the equidistribution principle (2.42) will give us the uniform mesh. However, in most general situations one can envisage the grid adaptation upon the bathymetry profile[7] $h(x, t)$. Consequently, in general we

[7] In the present study we do not consider such example. However, the idea of grid adaptation upon the bathymetry function certainly deserves to be studied more carefully.

can expect that the mesh will be non-uniform even under condition (2.63), since $h_x \neq 0$. However, we know that the initial grid satisfies the fully converged discrete equidistribution principle (2.43). From now on we assume that the initial grid is generated and it is not necessarily uniform. In order to construct the grid at the next layer, we solve just one linear equation (2.45). Since, system (2.45) is diagonally dominant, its solution exists and it is unique [272]. It is not difficult to check that the set of values $\{x_j^1 \equiv x_j^0\}_{j=0}^{N}$ solves the system (2.45). It follows from two observations:

- The right-hand side of (2.45) vanishes when $x_j^1 \equiv x_j^0, \forall j = 0, \ldots, N$.
- The monitor function $\{\varpi_{j+1/2}^0\}_{j=0}^{N-1}$ is evaluated on the previous time layer $t = 0$.

Thus, we obtain that $x_\sharp^1 \equiv x_\sharp^0$. Consequently, we have $x_{t,j}^0 \equiv 0$ and $\mathfrak{z}_j^1 = \mathfrak{z}_j^0$, $\forall j = 0, \ldots, N$.

In order to complete the predictor step we need to determine the quantities $\mathring{\wp}_\sharp^0$ and $\check{\mathring{\wp}}_\sharp^0$ on which depends the source term $\mathring{\mathscr{G}}_{j+1/2}^0$. These quantities are uniquely determined by prescribed initial conditions. For instance, $\mathring{\wp}_\sharp^0$ are obtained by solving linear equations (2.62). We showed above also that the solution to this equation is unique. We notice also that the right-hand side in Eq. (2.62) vanishes under conditions of this lemma. Consequently, we obtain $\mathring{\wp}_\sharp^0 \equiv 0$. By applying a finite difference analogue of Eq. (2.49) we obtain also that $\check{\mathring{\wp}}_\sharp^0 \equiv 0$. As a result, at the 'lake-at-rest' state the right-hand side of predictor Eqs. (2.52), (2.53) reads

$$\mathring{\mathscr{G}}_{j+1/2}^0 = \begin{pmatrix} 0 \\ (g\,\mathring{h}\,\mathring{h}_q)_{j+1/2}^0 \end{pmatrix}.$$

Taking into account the fact that the mesh does not evolve $x_\sharp^0 \hookrightarrow x_\sharp^1 \equiv x_\sharp^0$, we obtain $x_{t,j}^0 \equiv 0$ and thus $\bar{\mathring{\Lambda}}_{j+1/2}^0 \equiv \mathring{\Lambda}_{j+1/2}^0$, $s_{j+1/2}^0 \equiv \sqrt{g\,\mathring{h}_{j+1/2}}$,

$$(\bar{\mathring{\Lambda}} \cdot \mathbb{P})_{j+1/2}^0 \equiv \begin{pmatrix} \mathring{h}_{q,\,j+1/2} \\ \mathring{h}_{q,\,j+1/2} \end{pmatrix}, \qquad (\mathring{L} \cdot \mathring{\mathscr{G}})_{j+1/2}^0 \equiv \begin{pmatrix} \mathring{h}_{q,\,j+1/2} \\ \mathring{h}_{q,\,j+1/2} \end{pmatrix}.$$

Consequently, the predictor step (2.55), (2.56) gives us the following values:

$$\mathring{\mathbf{v}}_{j+1/2}^* \equiv \mathring{\mathbf{v}}_{j+1/2}^0, \qquad \mathring{\mathscr{F}}_{j+1/2}^* \equiv \mathring{\mathscr{F}}_{j+1/2}^0.$$

For the sake of clarity, we rewrite the last predictions in component-wise form:

$$\mathring{\mathbf{v}}_{j+1/2}^* \equiv \begin{pmatrix} \mathring{h}_{j+1/2} \\ 0 \end{pmatrix}, \qquad \mathring{\mathscr{F}}_{j+1/2}^* \equiv \begin{pmatrix} 0 \\ \dfrac{g\,\mathring{h}_{j+1/2}^2}{2} \end{pmatrix}.$$

Thus, $\mathring{\mathcal{H}}^{*}_{j+1/2} \equiv \mathring{h}_{j+1/2}$. As an intermediate conclusion of the predictor step we have:

$$\eta^{*}_{j+1/2} \equiv 0, \qquad \mathring{u}^{*}_{j+1/2} \equiv 0,$$

and all dispersive corrections $\mathring{\wp}^{*}_{\sharp}$, $\mathring{\breve{\wp}}^{*}_{\sharp}$ vanish as well by applying similar arguments to Eq. (2.58).

The corrector step (2.59), written component-wise reads:

$$\frac{(\mathcal{J}\mathring{\mathcal{H}})^{1}_{j} - (\mathcal{J}\mathring{\mathcal{H}})^{0}_{j}}{\tau} = 0,$$

$$\frac{(\mathcal{J}\mathring{u}\mathring{\mathcal{H}})^{1}_{j} - (\mathcal{J}\mathring{u}\mathring{\mathcal{H}})^{0}_{j}}{\tau} + \frac{g\mathring{h}^{2}_{j+1/2} - g\mathring{h}^{2}_{j-1/2}}{2\,\Delta q} = g\,(\mathring{\mathcal{H}}\mathring{h}_{q})^{\flat}_{j}$$

From the first equation above taking into account that $\mathcal{J}^{1}_{j} \equiv \mathcal{J}^{0}_{j}$ and $\mathring{\mathcal{H}}^{0}_{j} = \mathring{h}_{j}$ we obtain $\mathring{\mathcal{H}}^{1}_{j} = \mathring{h}_{j}$. And thus, by the definition of the total water depth we obtain $\mathring{\eta}^{1}_{j} \equiv 0$. In the second equation above by condition (2.63) we have that $\mathring{u}^{0}_{j} \equiv 0$. Moreover, in the left-hand side:

$$\frac{g\mathring{h}^{2}_{j+1/2} - g\mathring{h}^{2}_{j-1/2}}{2\,\Delta q} = g\,\frac{(\mathring{h}_{j+1} - \mathring{h}_{j-1})\cdot(\mathring{h}_{j+1} + 2\mathring{h}_{j} + \mathring{h}_{j-1})}{8\,\Delta q}. \tag{2.64}$$

The right-hand side of the same corrector equation can be rewritten using definitions (2.60), (2.61) as

$$g\,(\mathring{\mathcal{H}}\mathring{h}_{q})^{\flat}_{j} = g\,\frac{2\mathring{h}_{j+1} + 4\mathring{h}_{j} + 2\mathring{h}_{j-1}}{8}\cdot\frac{2\mathring{h}_{j+1} - 2\mathring{h}_{j-1}}{4\,\Delta q}. \tag{2.65}$$

Comparing equation (2.64) with (2.65) yields the desired well-balanced property of the predictor–corrector scheme and thus $\mathring{u}^{1}_{j} \equiv 0$.

By assuming that (2.63) is verified at the time layer $t = t^{n}$ and repeating precisely the same reasoning as above (by substituting superscripts $0 \leftarrow n$ and $1 \leftarrow n+1$) we obtain that (2.63) is verified at the next time layer $t = t^{n+1}$. It completes the proof of this lemma.

We would like to mention that the well-balanced property of the proposed scheme was checked also in numerical experiments on various configurations of general uneven bottoms (not reported here for the sake of manuscript compactness)—in all cases we witnessed the preservation of the 'lake-at-rest' state up to the machine precision. This validates our numerical implementation of the proposed algorithm since the well-balanced property is absolutely crucial for qualitatively correct simulation of conservation laws [146].

2.2.4 Numerical Scheme for Linearized Equations

In order to study the numerical scheme stability and its dispersive properties, we consider the discretization of the linearized SGN equations on a uniform unbounded grid (for simplicity we consider an IVP without boundary conditions). The governing equations after linearization can be written as (we already gave these equations in Sect. 2.1.7)

$$\eta_t + d u_x = 0,$$
$$u_t + g\eta_x = \frac{1}{d}\wp_x,$$
$$\wp_{xx} - \frac{3}{d^2}\wp = c^2\eta_{xx},$$

where $c = \sqrt{g d}$ is the speed of linear gravity waves. We shall apply to these PDEs precisely the same scheme as described above. Since the grid is uniform, we can return to the original notation, i.e. $\mathring{\mathbf{v}} \equiv \mathbf{v}$, etc. Let Δx be the discretization step in the computational domain $\mathfrak{I}_h$ and τ is the local time step. We introduce the following finite difference operators (illustrated on the free surface elevation $\eta^n_\sharp$):

$$\eta^n_{t,j} \stackrel{\text{def}}{:=} \frac{\eta^{n+1}_j - \eta^n_j}{\tau}, \qquad \eta^n_{x,j} \stackrel{\text{def}}{:=} \frac{\eta^n_{j+1} - \eta^n_j}{\Delta x}, \qquad \eta^n_{(x),j} \stackrel{\text{def}}{:=} \frac{\eta^n_{j+1} - \eta^n_{j-1}}{2\,\Delta x},$$

$$\eta^n_{xx,j} \stackrel{\text{def}}{:=} \frac{\eta^n_{j+1} - 2\eta^n_j + \eta^n_{j-1}}{\Delta x^2}, \qquad \eta^n_{xx,j+1/2} \stackrel{\text{def}}{:=} \frac{\eta^n_{xx,j} + \eta^n_{xx,j+1}}{2},$$

$$\eta^n_{xxx,j} \stackrel{\text{def}}{:=} \frac{\eta^n_{xx,j+1} - \eta^n_{xx,j}}{\Delta x}.$$

Then, at the predictor step we compute several auxiliary quantities $\{\eta^*_{j+1/2}\}_{j=-\infty}^{+\infty}$, $\{u^*_{j+1/2}\}_{j=-\infty}^{+\infty}$ and $\{\wp^*_{j+1/2}\}_{j=-\infty}^{+\infty}$. First, we solve the hyperbolic part of the linearized SGN equations:

$$\frac{\eta^*_{j+1/2} - \frac{1}{2}\left(\eta^n_{j+1} + \eta^n_j\right)}{\tau^*_{j+1/2}} + d\,u^n_{x,j} = 0,$$

$$\frac{u^*_{j+1/2} - \frac{1}{2}\left(u^n_{j+1} + u^n_j\right)}{\tau^*_{j+1/2}} + g\,\eta^n_{x,j} = \frac{1}{d}\wp^n_{x,j},$$

and then we solve the elliptic equation to find $\{\wp^*_{j+1/2}\}_{j=-\infty}^{+\infty}$:

$$\frac{\wp^*_{j+3/2} - 2\wp^*_{j+1/2} + \wp^*_{j-1/2}}{\Delta x^2} - \frac{3}{d^2}\wp^*_{j+1/2} = c^2 \frac{\eta^*_{j+3/2} - 2\eta^*_{j+1/2} + \eta^*_{j-1/2}}{\Delta x^2},$$

where $\tau^*_{j+1/2} \stackrel{\text{def}}{:=} \frac{\tau}{2}(1 + \theta^n_{j+1/2})$ and $\theta^n_{j+1/2}$ is the numerical scheme parameter [192], whose choice guarantees the TVD property (strictly speaking the proof was done for scalar hyperbolic equations only).

Then, the predicted values are used on the second—corrector step, to compute all physical quantities $\{\eta^{n+1}_j\}^{+\infty}_{j=-\infty}$, $\{u^{n+1}_j\}^{+\infty}_{j=-\infty}$ and $\{\wp^{n+1}_j\}^{+\infty}_{j=-\infty}$ on the next time layer $t = t^{n+1}$:

$$\eta^n_{t,j} + d\,\frac{u^*_{j+1/2} - u^*_{j-1/2}}{\Delta x} = 0, \tag{2.66}$$

$$u^n_{t,j} + g\,\frac{\eta^*_{j+1/2} - \eta^*_{j-1/2}}{\Delta x} = \frac{1}{d}\,\frac{\wp^*_{j+1/2} - \wp^*_{j-1/2}}{\Delta x}, \tag{2.67}$$

$$\wp^{n+1}_{xx,j} - \frac{3}{d^2}\wp^{n+1}_j = c^2\eta^{n+1}_{xx,j}. \tag{2.68}$$

It can be easily checked that the scheme presented above has the first order accuracy if $\theta^n_{j+1/2} = \text{const}, \forall j$ and the second order if $\theta^n_{j+1/2} \equiv 0, \forall j$. However, the last condition can be somehow relaxed. There is an interesting case of quasi-constant values of the scheme parameter:

$$\theta^n_{j+1/2} = \mathcal{O}(\tau + \Delta x).$$

In this case the scheme is second order accurate as well. In the present section we perform a theoretical analysis of the scheme and we shall assume for simplicity that $\theta^n_{j+1/2} \equiv$ const. Consequently, from now on we shall drop the index $j + 1/2$ in the intermediate time step $\tau^*_{j+1/2}$.

Linear Stability of the Scheme

In this section we apply the so-called VON NEUMANN stability analysis to the predictor–corrector scheme described above [57]. In order to study the scheme stability, first we exclude algebraically the predicted values $\{\eta^*_{j+1/2}\}^{+\infty}_{j=-\infty}$ and $\{u^*_{j+1/2}\}^{+\infty}_{j=-\infty}$ from difference equations. The resulting system reads:

$$\eta^n_{t,j} + d\,u^n_{(x),j} = \tau^* c^2 \eta^n_{xx,j} - \tau^* \wp^n_{xx,j}, \tag{2.69}$$

$$u^{n}_{t,\,j} + g\,\eta^{n}_{(x),\,j} = \tau^{*} c^{2} u^{n}_{xx,\,j} + \frac{1}{d}\,\frac{\wp^{*}_{j+1/2} - \wp^{*}_{j-1/2}}{\Delta x}, \tag{2.70}$$

$$\frac{\wp^{*}_{j+3/2} - 2\wp^{*}_{j+1/2} + \wp^{*}_{j-1/2}}{\Delta x^{2}} - \frac{3}{d^{2}}\,\wp^{*}_{j+1/2} = c^{2}\left(\eta^{n}_{xx,\,j+1/2} - \tau^{*} d\, u^{n}_{xxx,\,j}\right). \tag{2.71}$$

We substitute in all difference relations above the following elementary harmonics

$$\eta^{n}_{j} = \Lambda_{0}\,\rho^{n}\,\mathrm{e}^{\mathrm{i}\,j\,\xi}, \quad u^{n}_{j} = \Psi_{0}\,\rho^{n}\,\mathrm{e}^{\mathrm{i}\,j\,\xi},$$
$$\wp^{n}_{j} = \Phi_{0}\,\rho^{n}\,\mathrm{e}^{\mathrm{i}\,j\,\xi}, \quad \wp^{*}_{j+1/2} = \Phi^{*}_{0}(\rho)\,\mathrm{e}^{\mathrm{i}\,(j+1/2)\,\xi}, \tag{2.72}$$

where $\xi \overset{\mathrm{def}}{:=} k \cdot \Delta x \in [0, \pi]$ is the *scaled* wavenumber and ρ is the transmission factor between the time layers t^{n} and t^{n+1}. As a result, from Eqs. (2.68) and (2.71) we obtain the following expressions for Φ_{0} and Φ^{*}_{0}:

$$\Phi_{0} = \frac{4c^{2}d^{2}}{3\hbar\,\Delta x^{2}}\,\gimel^{2}\,\Lambda_{0}, \quad \Phi^{*}_{0}(\rho) = \rho^{n}\,\frac{2c^{2}d^{2}}{3\hbar\,\Delta x^{2}}\,\gimel\left[\Lambda_{0}\,\sin(\xi) - \mathrm{i}\,\tau^{*}\,\frac{4d}{\Delta x}\,\gimel^{2}\,\Psi_{0}\right],$$

where we introduced some short-hand notations:

$$\aleph \overset{\mathrm{def}}{:=} \frac{\tau}{\Delta x}, \qquad \gimel \overset{\mathrm{def}}{:=} \sin\left(\frac{\xi}{2}\right), \qquad \daleth \overset{\mathrm{def}}{:=} 4c^{2}\aleph^{2}\gimel^{2},$$
$$\beth \overset{\mathrm{def}}{:=} c\,\aleph\,\sin(\xi), \qquad \hbar \overset{\mathrm{def}}{:=} 1 + \frac{4d^{2}}{3\,\Delta x^{2}}\,\gimel^{2}.$$

By substituting just obtained expressions for Φ_{0} and Φ^{*}_{0} into Eqs. (2.69), (2.70) we obtain two linear equations with respect to amplitudes Λ_{0} and Ψ_{0}:

$$\left[\rho - 1 + \frac{2c^{2}\aleph^{2}(1+\theta)}{\hbar}\,\gimel^{2}\right]\Lambda_{0} + \mathrm{i}\,\aleph\,d\,\sin(\xi)\,\Psi_{0} = 0,$$
$$\mathrm{i}\,\frac{g\,\aleph\,\sin(\xi)}{\hbar}\,\Lambda_{0} + \left[\rho - 1 + \frac{2c^{2}\aleph^{2}(1+\theta)}{\hbar}\,\gimel^{2}\right]\Psi_{0} = 0.$$

The necessary condition to have non-trivial solutions gives us an algebraic equation for the transmission factor ρ:

$$(\rho - 1)^{2} + \frac{\daleth(1+\theta)}{\hbar}\,(\rho - 1) + \frac{\daleth^{2}(1+\theta)^{2}}{4\hbar^{2}} + \frac{\beth^{2}}{\hbar} = 0.$$

This quadratic equation admits two distinct roots:

$$\rho^{\pm} = 1 - \frac{\daleth(1+\theta)}{2\hbar} \pm \mathrm{i}\frac{\beth}{\sqrt{\hbar}}. \tag{2.73}$$

The necessary stability condition $|\rho| \leqslant 1$ is equivalent to the following condition on quadratic equation coefficients:

$$c^2\aleph^2(1+\theta)^2\zeta - \Bigl[1 + \frac{4\varsigma^2}{3}\zeta\Bigr](\zeta+\theta) \leqslant 0, \qquad \varsigma \stackrel{\text{def}}{:=} \frac{d}{\Delta x}, \tag{2.74}$$

which has to be fulfilled for all $\zeta \stackrel{\text{def}}{:=} \gimel^2 \in [0, 1]$. The parameter ς characterizes the grid resolution relative to the mean water depth. This parameter appears in stability condition along with the COURANT ratio $\aleph$. It is one of the differences with non-dispersive equations whose discretization stability depends only on $\aleph$.

Further Thoughts About Stability

When long waves travel towards the shoreline, their shoaling process is often accompanied with the formation of undular bores [153, 263]. Undular bores have dispersive nature and cannot be correctly described by dispersionless models. In [153] it was shown that satisfactory description of dispersive effects in shallow water environments is obtained for $\varsigma \stackrel{\text{def}}{:=} \frac{d}{\Delta x} = 2 \sim 4$. In another study [141] it was shown that for satisfactory modelling of trans-oceanic wave propagation it is sufficient to choose $\varsigma \approx 4$ in deep ocean and $\varsigma \approx 1$ in shallow coastal areas. In other words, it is sufficient to choose the grid size equal to water depth in shallow waters and in deep areas—four times smaller than the water depth. On coarser grids the numerical dispersion may dominate over the physical one [141]. In the present study we shall assume that parameter $\varsigma \geqslant \frac{\sqrt{3}}{2}$.

Substituting into Eq. (2.74) the value $\zeta \equiv 0$ we obtain that for stability reasons necessarily the scheme parameter $\theta \geqslant 0$. Since the predictor layer should be in between time layers $t = t^n$ and $t = t^{n+1}$ we have $\theta \leqslant 1$. Then, for fixed values of parameters ς and θ the stability condition (2.74) takes the following form:

$$c\aleph \leqslant \frac{\sqrt{1 + \frac{4}{3}\varsigma^2\theta}}{1+\theta}.$$

For $\theta \equiv 0$ the last condition simply becomes:

$$c\aleph \leqslant 1,$$

and it does not depend on parameter ς. However, when $\theta > 0$, then the scheme stability depends on the mesh refinement ς relative to the mean water depth. Surprisingly, more we refine the grid, less stringent becomes the stability barrier. In the asymptotic limit $\varsigma \gg 1$ we obtain the following restriction on the time step τ:

$$\tau \leqslant \frac{2\sqrt{\theta}}{\sqrt{3}\,(1+\theta)}\,\tau_0 < \frac{1}{\sqrt{3}}\,\tau_0 \approx 0.58\,\tau_0,$$

where $\tau_0 \stackrel{\text{def}}{:=} \frac{d}{c} \equiv \frac{d}{\sqrt{g\,d}}$ is the characteristic time scale of gravity waves. Above we used the following obvious inequality:

$$1 + \theta \geqslant 2\sqrt{\theta}, \qquad \forall \theta \in \mathbb{R}^{+}.$$

So, in practice for *sufficiently refined* grids the stability condition *de facto* does not involve the grid spacing Δx anymore. This property is very desirable for numerical simulations. For the sake of comparison we give here (without underlying computations) the stability restriction of the same predictor–corrector scheme for NSWE equations:

$$c\,\aleph \leqslant \frac{1}{\sqrt{1+\theta}}.$$

So, another surprising conclusion obtained from this linear stability analysis is that the SGN equations require *in fine* a less stringent condition on the time step than corresponding dispersionless NSWE. Most probably, this conclusion can be explained by the regularization effect of the dispersion. Indeed, the NSWE *bores* are replaced by smooth *undular bores* whose regularity is certainly higher. The smoothness of solutions allows to use a larger time step τ to propagate the numerical solution. This conclusion was checked in (fully nonlinear) numerical experiments (not reported here) where the time step τ was artificially pushed towards the stability limits. In general, the omission of dispersive effects yields a stricter stability condition. The authors of [142] came experimentally to similar conclusions about the time step limit in dispersive and hydrostatic simulations. Our theoretical analysis reported above may serve as a basis of rational explanation of this empirical fact.

This result is to be compared with a numerical scheme proposed in [80] for a weakly nonlinear weakly dispersive water wave model. They used splitting technique and solved an elliptic equation to determine the non-hydrostatic pressure correction. The main drawback of the scheme proposed in [80] is the stability condition:

$$\Delta x \geqslant 1.5\,d.$$

One can easily see that a numerical computation with a sufficiently refined grid is simply impossible with that scheme. Our method is free of such drawbacks.

Discrete Dispersion Relation

The dispersion relation properties are crucial to understand and explain the behaviour of the numerical solution [213]. In this section we perform the dispersion relation analysis of the proposed above predictor–corrector scheme. This analysis is based on the study of elementary plane-wave solutions (2.72). The continuous case was already analysed in Sect. 2.1.7. Dispersive properties of the scheme can be completely characterized by the phase error $\Delta\varphi \stackrel{\text{def}}{:=} \chi - \varphi$ committed during solution transfer from time layer $t = t^n$ to $t = t^{n+1} = t^n + \tau$. Here we denote by *chi* the phase shift due to the SGN equations dynamics and φ is its discrete counterpart. From Eqs. (2.40) and (2.73) we obtain correspondingly:

$$\chi = \arg(\mathrm{e}^{-\mathrm{i}\omega\tau}) \equiv -\omega\tau = \pm\frac{c\aleph\xi}{\sqrt{1 + \dfrac{\varsigma^2\xi^2}{3}}}, \quad \xi \in [0, \pi], \tag{2.75}$$

$$\varphi = \arg\rho = \pm\arccos\Big[\Big(1 - \frac{\daleth(1+\theta)}{2\hbar}\Big)/|\rho|\Big]. \tag{2.76}$$

In other words, the phase change χ is predicted by the 'exact' SGN equations properties, while φ comes from the approximate dynamics as predicted by the predictor–corrector scheme. Since we are interested in long wave modelling, we can consider TAYLOR expansions of the phase shifts in the limit $\xi \to 0$ (assuming that ς and $\aleph$ are kept constant):

$$\chi = \pm\Big[c\aleph\xi - \frac{c\aleph}{6}\varsigma^2\xi^3 + \mathcal{O}(\xi^4)\Big],$$

$$\varphi = \pm\Big[c\aleph\xi + \frac{c\aleph}{6}\big((c\aleph)^2(3\theta+1) - 1 - \varsigma^2\big)\xi^3 + \mathcal{O}(\xi^4)\Big].$$

The asymptotic expression for the phase error is obtained by subtracting above expressions:

$$\Delta\varphi = \mp\frac{c\aleph}{6}\big[(c\aleph)^2(3\theta+1) - 1\big]\xi^3 + \mathcal{O}(\xi^4).$$

From the last relation one can see that the leading part of the phase error has the same asymptotic order as the 'physical' dispersion of the SGN equations. In general, this result is not satisfactory. However, this situation can be improved if for the given scheme parameter $\theta \geqslant 0$, the COURANT ratio $\aleph$ is chosen according to the following formula:

$$c\aleph = \frac{1}{\sqrt{1 + 3\theta}}.$$

In this case the numerical phase error will be one order lower than the physical dispersion of the SGN system.

In Fig. 2.3 we represent graphically phase shifts predicted by various models. The dashed line (1) is the phase shift of the predictor–corrector scheme given by Eq. (2.76) (taken with + sign) for the parameters values $\theta = 0$, $c\aleph = 1$, $\varsigma = 2$. The continuous dispersion relation is shown with the dotted line (3) (the SGN equations, formula (2.75)) and the solid line (4) (full EULER equations):

$$\chi_{\text{Euler}} = \pm c\aleph\,\xi\sqrt{\frac{\tanh(\varsigma\,\xi)}{\varsigma\,\xi}}.$$

It can be seen that our predictor–corrector scheme provides a better approximation to the dispersion relation than the scheme proposed by PEREGRINE [263] (dash-dotted line (2) in Fig. 2.3). The analysis of the discrete dispersion relation of PEREGRINE's scheme is not given here, but we provide only the final result for the phase change:

$$\chi_{\text{Peregrine}} = \pm\arccos\left(1 - \frac{\beth^2}{2\hbar}\right).$$

In Fig. 2.3 one can see that the predictor–corrector scheme (curve (1)) approximates well the dispersion relation of the SGN equations (curve (3)) up to $\xi = k \cdot \Delta x \lesssim \frac{\pi}{4}$. In terms of the wave length λ we obtain that $\lambda \gtrsim 8\,\Delta x$ and for $\varsigma = 2$ we

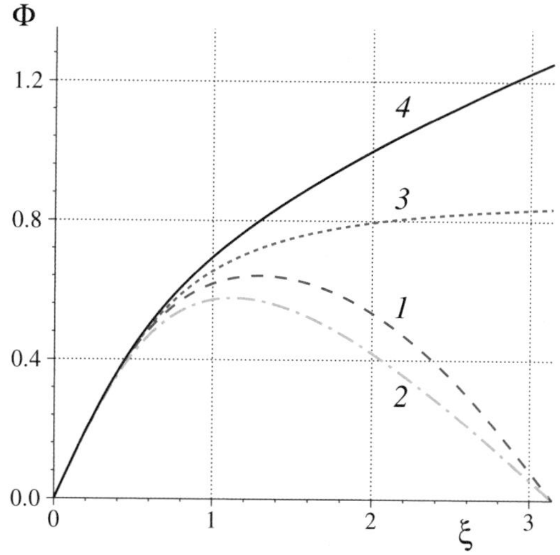

Fig. 2.3 Phase shifts in different models: (1) predictor–corrector scheme; (2) PEREGRINE's numerical scheme [263]; (3) the SGN equations; (4) full EULER equations

obtain the inequality $\lambda \gtrsim 4d$. So, as the main result of the present analysis we conclude that our scheme is able to propagate accurately water waves whose length is four times longer than the mean water depth d.

2.3 Numerical Results

Below we present a certain number of test cases which aim to validate and illustrate the performance of the numerical scheme described above along with our implementation of this method.

2.3.1 *Solitary Wave Propagation Over the Flat Bottom*

As we saw above in Sect. 2.1.5, in a special case of constant water depth $h(x, t) = d$ the SGN equations admit solitary wave solutions (given by explicit simple analytical formulas) which propagate with constant speed without changing their shapes.

Uniform Grid

These analytical solutions can be used to estimate the accuracy of the fully discrete numerical scheme. Consequently, we take a sufficiently large domain $[0, \ell]$ with $\ell = 80$. In this section all lengths are relative to the water depth d, and time is scaled with $\sqrt{g/d}$. For instance, if the solitary wave amplitude $\alpha = 0.7$, then $\alpha_d = 0.7\,d$ in dimensional variables. So, the solitary wave is initially located at $x_0 = 40$. In computations below we take a solitary wave of amplitude $\alpha = 0.4$. In general, the SGN travelling wave solutions approximate fairly well those of the full EULER model up to amplitudes $\alpha \lesssim \frac{1}{2}$ (see [96] for comparisons).

In Fig. 2.4 we show a zoom on free surface profile (a) at $t = 20$ and wave gauge data (b) in a fixed location $x = 60$ for various spatial (and uniform) resolutions. By this time, the solitary wave propagated the horizontal distance of 20 mean water depths. It can be seen that the numerical solution converges to the analytical one.

In order to quantify the accuracy of the numerical solution we measure the relative l_∞ discrete error:

$$\| \varepsilon_h \|_\infty \stackrel{\text{def}}{:=} \alpha^{-1} \| \eta_h - \eta \|_\infty ,$$

where η_h stands for the numerical and η–for the exact free surface profiles. The factor α^{-1} is used to obtain the dimensionless error. Then, the order of convergence $\Bbbk$ can be estimated as

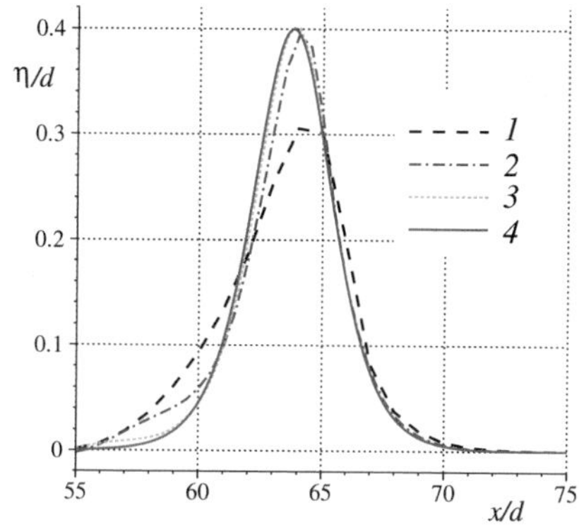

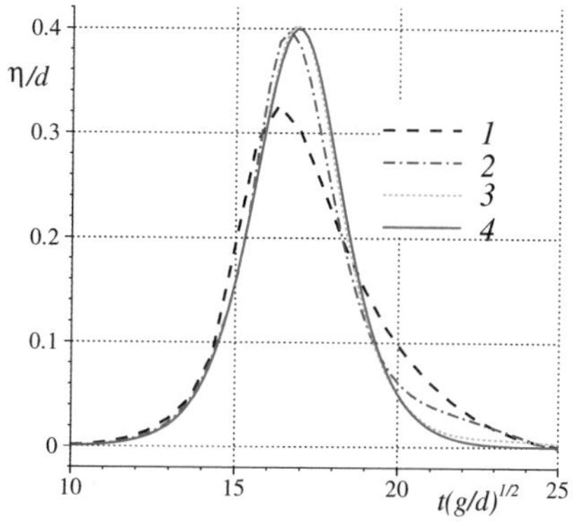

Fig. 2.4 Propagation of a solitary wave over the flat bottom: (**a**) free surface profile at $t = 20$; (**b**) wave gauge data at $x = 60$. Various lines denote: (1) $N = 80$, (2) $N = 160$, (3) $N = 320$, (4) the exact analytical solution given by formula (2.38)

Table 2.1 Numerical estimation of the convergence order for the analytical solitary wave propagation test case

N	ς	$\Vert \varepsilon_h \Vert_\infty$	$\Bbbk$
80	1	0.2442	–
160	2	0.1277	0.94
320	4	0.3344×10^{-1}	1.93
640	8	0.8639×10^{-2}	1.95
1280	16	0.2208×10^{-2}	1.97
2560	32	0.5547×10^{-3}	1.99

The parameter $\varsigma = \frac{d}{\Delta x}$ characterizes the mesh resolution relative to the mean water depth d

$$\Bbbk \simeq \log_2 \left\{ \frac{\Vert \varepsilon_h \Vert_\infty}{\Vert \varepsilon_{h/2} \Vert_\infty} \right\}.$$

The numerical results in Table 2.1 indicate that $\Bbbk \to 2$, when $N \to +\infty$. This validates the proposed scheme and the numerical solver.

Adaptive Grid

In order to show the performance of the adaptive algorithm, we adopt two monitor functions in our computations:

$$\varpi_0[\eta](x, t) = 1 + \vartheta_0 |\eta(x, t)|, \tag{2.77}$$

$$\varpi_1[\eta](x, t) = 1 + \vartheta_0 |\eta(x, t)| + \vartheta_1 |\eta_x(x, t)|, \tag{2.78}$$

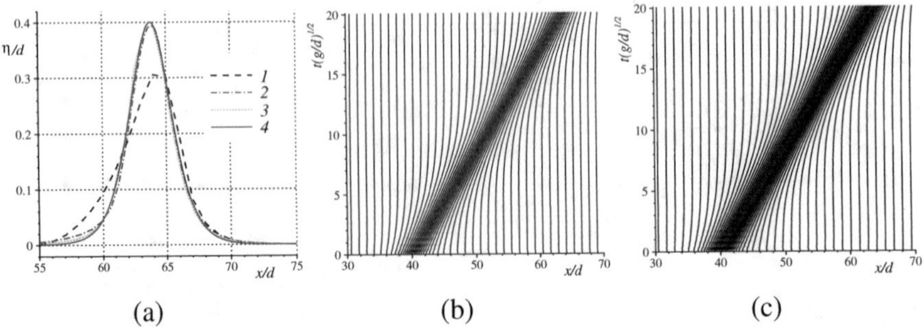

Fig. 2.5 Propagation of a solitary wave over the flat bottom simulated with moving adapted grids: (**a**) free surface profile at $t = 20$; (**b**) trajectory of some grid points predicted with monitor function $\varpi_0(x, t)$; (**c**) the same but with monitor function $\varpi_1(x, t)$. On panel (**a**) the lines are defined as: (1) numerical solution on a uniform fixed grid; (2) numerical solution predicted with monitor function $\varpi_0(x, t)$; (3) the same with $\varpi_1(x, t)$; (4) exact analytical solution

where $\vartheta_{0,1} \geqslant 0$ are some positive constants. In numerical simulations we use $\vartheta_0 = \vartheta_1 = 10$ and only $N = 80$ grid points. Above we showed that numerical results are rather catastrophic when these 80 grid points are distributed uniformly (see Fig. 2.4). Numerical results on adaptive moving grids obtained with monitor functions $\varpi_{0,1}(x, t)$ are shown in Fig. 2.5. The monitor function $\varpi_0(x, t)$ ensures that points concentrate around the wave crest, leaving the areas in front and behind relatively rarefied. The visual comparison of panels 2.5b and c shows that the inclusion of the spatial derivative η_x into the monitor function $\varpi_1(x, t)$ yields the increase of dense zones around the wave crest. With an adaptive grid involving only $N = 80$ points we obtain a numerical solution of quality similar to the uniform grid with $N = 320$ points.

2.3.2 *Solitary Wave/Wall Interaction*

For numerous practical purposes in Coastal Engineering it is important to model correctly wave/structure interaction processes [265]. In this section we apply the above proposed numerical algorithm to the simulation of a simple solitary wave/vertical wall interaction. The reason is twofold:

1. Many coastal structures involve vertical walls as building elements;
2. This problem is well studied by previous investigators and, consequently, there is enough available data/results for comparisons.

We would like to underline that this problem is equivalent to the head-on collision of two equal solitary waves due to simple symmetry considerations. This 'generalized' problem was studied in the past using experimental [236, 279], numerical [53, 134] and analytical techniques [47, 242, 298]. More recently this problem gained again

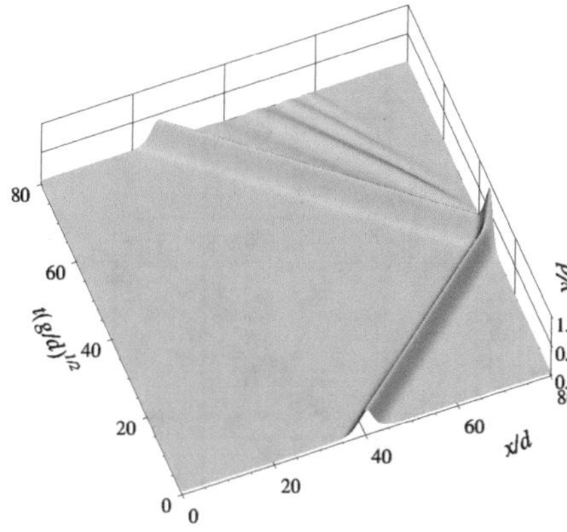

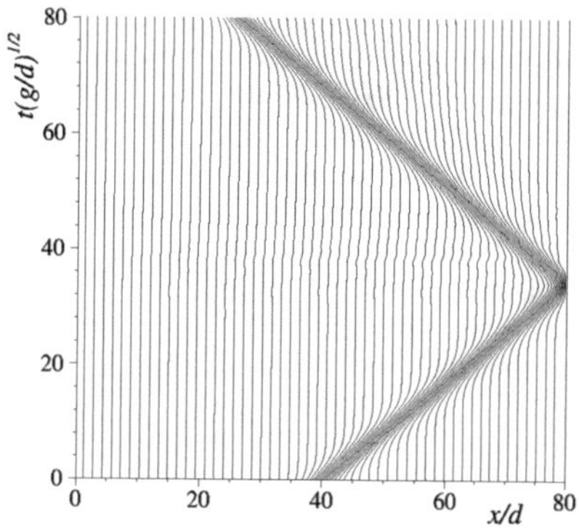

Fig. 2.6 Solitary wave (amplitude $\alpha = 0.4$)/vertical wall interaction in the framework of the SGN equations: (**a**) space-time plot of the free surface elevation; (**b**) nodes trajectories. For the sake of clarity every 5th node is shown only, the total number of nodes $N = 320$

some interest of researchers [48, 52, 75, 112, 116, 117, 232, 312]. Despite the simple form of the obstacle, the interaction process of sufficiently large solitary waves with it takes a highly non-trivial character as it will be highlighted below.

Figure 2.6a shows the free surface dynamics as it is predicted by the SGN equations solved numerically using the moving grid with $N = 320$ nodes. The initial condition consists of an exact analytical solitary wave (2.38) of amplitude $\alpha = 0.4$ moving rightwards to the vertical wall (where the wall boundary condition $u = 0$ is imposed[8] on the velocity, for the pressure see Sect. 2.1.4). The computational domain is chosen to be sufficiently large $[0, \ell] = [0, 80]$, so there is no interaction with the boundaries at $t = 0$. Initially the solitary wave is located at $x_0 = 40$ (right in the middle). The bottom is flat $h(x, t) = d = \text{const}$ in this test case. From Fig. 2.6a it can be clearly seen that the reflection process generates a train of weakly nonlinear waves which propagate with different speeds in agreement with the dispersion relation. The moving grid was constructed using the monitor function $\varpi_1(x, t)$ from the previous section (see the definition in equation (2.78)) with $\vartheta_0 = \vartheta_1 = 10$. The resulting trajectories of mesh nodes are shown in Fig. 2.6b. The grid is clearly refined around the solitary wave and nodes follow it. Moreover, we would like to note also a slight mesh refinement even in the dispersive tail behind the reflected wave (it is not clearly seen in Fig. 2.6b since we show only every 5th node).

One of the main interesting characteristics that we can compute from these numerical experiments is the *maximal* wave run-up $\mathcal{R}$ on the vertical wall:

[8]The same condition is imposed on the left boundary as well, even if during our simulation time there are no visible interactions with the left wall boundary.

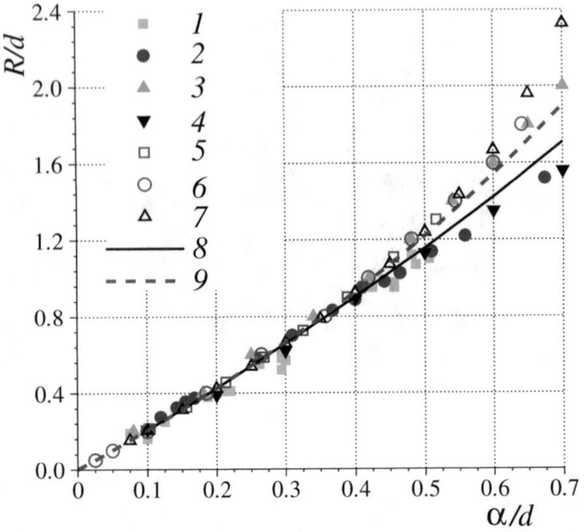

Fig. 2.7 Dependence of the maximal run-up $\mathcal{R}$ on the amplitude α of the incident solitary wave. Experimental data: (1) [335], (2) [236], (3) [81], (4) [233]. Numerical data: (5) [134], (6) [53], (7) [75]. The solid line (8) our numerical results, the dashed line (9) the analytical prediction (2.79)

$$\mathcal{R} \stackrel{\text{def}}{:=} \sup_{0 \leqslant t \leqslant T} \{\eta(\ell, t)\}.$$

The sup is taken in some time window when the wave/wall interaction takes place. For the class of incident solitary wave solutions it is clear that *maximal* run-up $\mathcal{R}$ will depend on the (dimensionless) solitary wave amplitude α. In [298] the following asymptotic formula was derived in the limit $\alpha \to 0$:

$$\mathcal{R}(\alpha) = 2\alpha\left[1 + \tfrac{1}{4}\alpha + \tfrac{3}{8}\alpha^2\right] + \mathcal{O}(\alpha^4). \tag{2.79}$$

The last approximation was already checked against full the EULER simulations [75, 134] and even laboratory experiments [236]. Figure 2.7 shows the dependence of the maximal run-up $\mathcal{R}$ on the incident solitary wave amplitude α as it is predicted by our numerical model, by formula (2.79) and several other experimental [81, 233, 236, 335] and numerical [53, 75, 134] studies. In particular, one can see that almost all models agree fairly well up to the amplitudes $\alpha \lesssim 0.4$. Then, there is an apparent 'separation' of data in two branches. Again, our numerical model gives a very good agreement with experimental data from [233, 236, 335] up to the extreme amplitudes $\alpha \lesssim 0.7$.

Wave Action on the Wall

The nonlinear dispersive SGN model can be used to estimate also the wave force exerted on the vertical wall. Moreover, we shall show below that this model is able to capture the non-monotonic behaviour of the force when the incident wave amplitude is increased. This effect was first observed experimentally [335] and then numerically [343].

For the 2D case with flat bottom the fluid pressure $p(x, y, t)$ can be expressed:

$$\frac{p(x, y, t)}{\rho} = g\big(\mathcal{H} - (y + d)\big) - \left[\frac{\mathcal{H}^2}{2} - \frac{(y + d)^2}{2}\right]\mathscr{R}_1, \quad -d \leqslant y \leqslant \eta(x, t), \tag{2.80}$$

with $\mathscr{R}_1 \stackrel{\text{def}}{:=} u_{xt} + u u_{xx} - u_x^2$. The *horizontal* wave loading exerted on the vertical wall located at $x = \ell$ is given by the following integral:

$$\frac{F_0(t)}{\rho} = \int_{-d}^{\eta(\ell, t)} p(\ell, y, t)\,\mathrm{d}y = \frac{g\,\mathcal{H}^2}{2} - \frac{\mathcal{H}^3}{3}\bar{\mathscr{R}}_1,$$

where due to boundary conditions $\bar{\mathscr{R}}_1 = u_{xt} - u_x^2$. After removing the hydrostatic force, we obtain the dynamic wave loading computed in our simulations:

$$\frac{F(t)}{\rho} = g\left[\frac{\mathcal{H}^2}{2} - \frac{d^2}{2}\right] - \frac{\mathcal{H}^3}{3}\bar{\mathscr{R}}_1.$$

The expression for corresponding tilting moment can be found in [116, Remark 3]. Figure 2.8 shows the wave elevation (*a*) and the dynamic wave loading (*b*) on the vertical wall. From Fig. 2.8b it can be seen that the force has one maximum for small

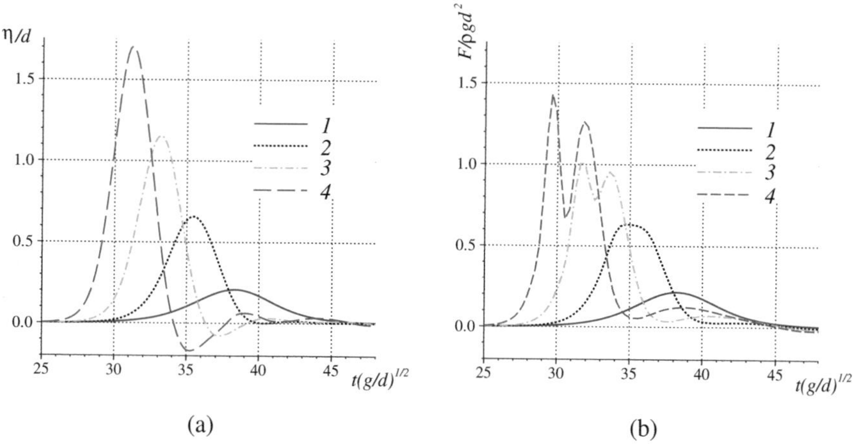

Fig. 2.8 Solitary wave/vertical wall interaction: (**a**) time series of wave run-up on the wall; (**b**) dynamic wave loading on the wall. Different lines correspond to different incident solitary wave amplitudes: (1) $\alpha = 0.1$, (2) $\alpha = 0.3$, (3) $\alpha = 0.5$, (4) $\alpha = 0.7$

amplitude solitary waves. However, when we gradually increase the amplitude (i.e. $\alpha \gtrsim 0.4$), the second (local) maximum appears. For such large solitary waves a slight run-down phenomenon can be noticed in Fig. 2.8a. We reiterate that this behaviour is qualitatively and quantitatively correct comparing to the full EULER equations [52, 75]. However, the complexity of the nonlinear dispersive SGN model and, consequently, the numerical algorithm to solve it, is much lower.

2.3.3 Solitary Wave/Bottom Step Interaction

Water waves undergo continuous changes while propagating over general uneven bottoms. Namely, the wave length and wave amplitude are modified while propagating over bottom irregularities. Such transformations have been studied in the literature [91, 205]. In the present section we focus on the process of a Solitary Wave (SW) transformation over a bottom step. In the early work by Madsen and Mei [229] it was shown using long wave approximation that a solitary wave can be disintegrated into a finite number of SWs with decreasing amplitudes while passing over an underwater step. This conclusion was supported in [229] by laboratory data as well. This test case was used later in many works, see e.g. [58, 102, 207].

We illustrate the behaviour of the adaptive numerical algorithm as well as the SGN model on the solitary wave/bottom interaction problem. The bottom bathymetry is given by the following discontinuous function:

$$y = -h(x) = \begin{cases} -h_0, & 0 \leqslant x \leqslant x_s, \\ -h_s, & x_s < x \leqslant \ell, \end{cases}$$

where ℓ is the numerical wave tank length, h_0 (respectively h_s) are the still water depths on the left (right) of the step located at $x = x_s$. We assume also that $0 < h_s < h_0$. The initial condition is a solitary wave located at $x = x_0$ and propagating rightwards. For the experiment cleanliness we assume that initially the solitary wave does not 'feel' the step. In other words it is located sufficiently far from the abrupt change in bathymetry. In our experiment we choose x_0 so that $\eta(x_s) \lesssim 0.01\,\alpha$, where α is the SW amplitude. The main parameters in this problem are the incident wave amplitude α and the bottom step jump $\Delta b_s = h_0 - h_s$. Various theoretical and experimental studies show that a solitary wave undergoes a splitting into a reflected wave and a finite number of solitary waves after passing over an underwater step. See [205] for a recent review on this topic. Amplitudes and the number of solitary waves over the step were determined in [178] in the framework of the shallow water theory. These expressions were reported later in [278] and this result was improved recently in [262]. However, in the vicinity of the step, one may expect important vertical accelerations of fluid particles, which are simplified (or even neglected) in shallow water type theories. Nevertheless, in [262] a good agreement of this theory with numerical and experimental data was reported.

There is also another difficulty inherent to the bottom step modelling. In various derivations of shallow water models there is an implicit assumption that the bathymetry gradient ∇h is bounded (or even small $|\nabla h| \ll 1$, e.g. in the BOUSSINESQ-type equations [45]). Conversely, numerical tests and comparisons with the full (EULER and even NAVIER–STOKES) equations for finite values of $|\nabla h| \sim \mathcal{O}(1)$ show that resulting approximate models have a larger applicability domain than it was supposed at the outset [45]. In the case of a bottom step, the bathymetry function is even discontinuous which is an extreme case we study in this section.

There are two main approaches to cope with this problem. One consists in running the approximate model directly on discontinuous bathymetry, and the resulting eventual numerical instabilities are damped out by ad-hoc dissipative terms (see e.g. references in [262]). The magnitude of these terms allows to increase the scheme dissipation, and overall computation appears to be stable. The difficulty of this approach consists in the fine-tuning of dissipation, since

- Insufficient dissipation will make the computation unstable,
- Excessive dissipation will yield unphysical damping of the solution.

An alternative approach consists in replacing the discontinuous bathymetry by a smoothed version over certain length $\left[x_s - \frac{\ell_s}{2}, x_s + \frac{\ell_s}{2}\right]$, where ℓ_s is the smoothing length on which the jump from h_0 to h_s is replaced by a smooth variation. For instance, in all numerical computations reported in [278] the smoothing length was chosen to be $\ell_s = 60$ cm independently of the water depths before h_0 and after h_s the step. In another work [132] the smoothing length was arbitrarily set to $\ell_s = 20$ cm independently of other parameters. Unfortunately, in a recent work [341] the smoothing procedure was not described at all. Of course, this method is not perfect since the bathymetry is slightly modified. However, one can expect that sufficiently long waves will not 'notice' this modification. This assumption was confirmed by the numerical simulations reported in [69, 132, 278].

In the present work we also employ the bottom smoothing procedure. However, the smoothing length ℓ_s is chosen in order to have a well-posed problem for the elliptic operator (2.9). For simplicity, we use the sufficient condition (2.18) (obtained under restriction (2.17)), which is not necessarily optimal, but it allows us to invert stably the nonlinear elliptic operator (2.24). Namely, the smoothed step has the following analytical expression:

$$y = -h(x) = \begin{cases} -h_0, & 0 \leqslant x \leqslant x_s - \frac{\ell_s}{2}, \\ -h_0 + \frac{\Delta b_s}{2} \cdot \bigl(1 + \sin \zeta\bigr), & x_s - \frac{\ell_s}{2} \leqslant x \leqslant x_s + \frac{\ell_s}{2}, \\ -h_s, & x_s + \frac{\ell_s}{2} \leqslant x \leqslant \ell, \end{cases} \tag{2.81}$$

where $\zeta \overset{\text{def}}{:=} \dfrac{\pi\,(x - x_s)}{\ell_s}$. For this bottom profile, the inequalities (2.17), (2.18) take the form:

$$\frac{\pi\,\Delta b_s}{2\,\ell_s}\cos\zeta < 1, \quad \forall\zeta \in \Bigl[-\frac{\pi}{2}, \frac{\pi}{2}\Bigr],$$

$$\frac{\pi^2\,\Delta b_s}{2\,\ell_s^2}\sin\zeta > -\frac{2}{\dfrac{h_0 + h_s}{2} - \dfrac{\Delta b_s}{2}\sin\zeta}.$$

These inequalities have corresponding solutions:

$$\ell_s > \frac{\pi\,\Delta b_s}{2}, \qquad \ell_s > \frac{\pi}{2}\sqrt{h_0\,\Delta b_s}.$$

The last inequalities are verified simultaneously if the second inequality is true. If we assume that the bottom step height Δb_s is equal to the half of the water depth before it, then we obtain the following condition:

$$\ell_s > \frac{\pi}{2\sqrt{2}}\,h_0 \approx 1.11\,h_0.$$

We underline that the last condition is only sufficient and stable numerical computations can most probably be performed even for shorter smoothing lengths ℓ_s. For instance, we tested the value $\ell_s = h_0$ and everything went smoothly.

In [278] the results of 80 experiments are reported for various values of α and h_0 (for fixed values of the bottom jump $\Delta b_s = 10$ cm). In our work we repeated all experiments from [278] using the SGN equations solved numerically with the adaptive predictor–corrector algorithm described above. In general, we obtained a very good agreement with experimental data from [278] in terms of the following control parameters:

- number of solitary waves moving over the step;
- amplitudes of solitary waves over the step;
- amplitude of the (main) reflected wave.

We notice that the amplitude of the largest solitary wave over the step corresponds perfectly to the measurements. However, the variation in the amplitude of subsequent solitary waves over the step could reach in certain cases 20%.

Remark 2.6 The conduction of laboratory experiments on the solitary wave/bottom step interaction encounters a certain number of technical difficulties [54, 278] that we would like to mention. First of all, the wave maker generates a solitary wave with some dispersive components. Moreover, one has to take the step sufficiently long so that the transmitted wave has enough time to develop into a finite number of visible well-separated solitary waves. Finally, the reflections of the opposite wave flume's wall are to be avoided as well in order not to pollute the measurements.

Consequently, the successful conduction of experiments and accurate measurement of wave characteristics requires a certain level of technique. We would like to mention the exemplary experimental work [162] on the head-on collision of solitary waves.

Below we focus on one particular case of $\alpha = 3.65$ cm. All other parameters are given in Table 2.2. It corresponds to the experiment N° 24 from [278]. The free surface dynamics is depicted in Fig. 2.9a and the trajectories of every second grid node are shown in Fig. 2.9b. For the mesh adaptation we use the monitor function (2.78) with $\vartheta_1 = \vartheta_2 = 10$. In particular, one can see that three solitary waves are generated over the step. This fact agrees well with the theoretical predictions [178, 262]. Moreover, one can see that the distribution of grid points follows perfectly all generated waves (over the step *and* the reflected wave). Figure 2.10a shows the free surface dynamics in the vicinity of the bottom step. In particular, one can see that

Table 2.2 Values of various numerical parameters used in the solitary wave/bottom step interaction test case

Parameter	Value
Wave tank length, ℓ	35 m
Solitary wave amplitude, α	3.65 cm
Solitary wave initial position, x_0	11 m
Water depth before the step, h_0	20 cm
Water depth after the step, h_s	10 cm
Water depth used in scaling, d	h_0
Bottom step jump, Δb_s	10 cm
Bottom step location, x_s	14 m
Number of grid points, N	350
Simulation time, T	17.6 s

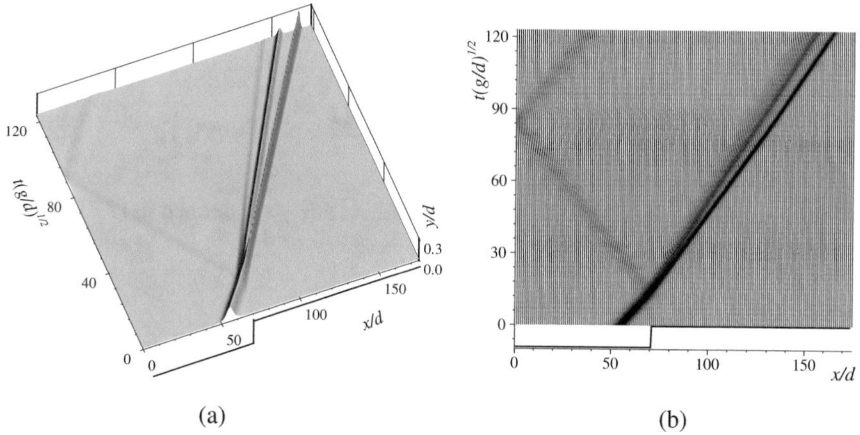

Fig. 2.9 Interaction of a solitary wave with an underwater step: (**a**) space-time plot of the free surface elevation $y = \eta(x, t)$ in the dimensional time interval $[0 \text{ s}, 17.6 \text{ s}]$; (**b**) trajectories of every second grid node. Numerical parameters are provided in Table 2.2

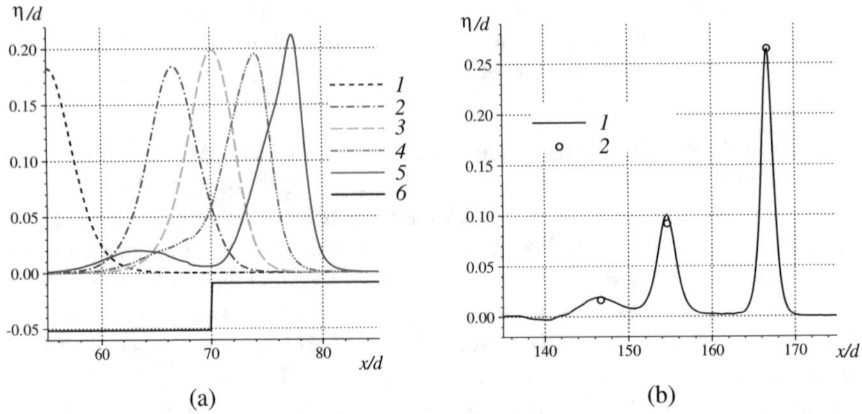

Fig. 2.10 Free surface profiles $y = \eta(x, t)$ during the interaction process of a solitary wave with an underwater step: (**a**) initial condition (1), $t = 1.5$s (2), $t = 2.0$ s (3), $t = 2.5$ s (4), $t = 3.0$ s (5), smoothed bottom profile given by formula (2.81) (6); (**b**) free surface profile $y = \eta(x, T)$ (1) at the final simulation time $t = T$. The experimental points (2) are taken from [278], experiment N° 24. Numerical parameters are provided in Table 2.2

the wave becomes notoriously steep by the time instance $t = 3$ s and during later times it splits into one reflected and three transmitted waves. The free surface profile at the final simulation time $y = \eta(x, T)$ is depicted in Fig. 2.10b. On the same panel the experimental measurements are shown with empty circles $\circ$, which show a very good agreement with our numerical simulations.

In our numerical experiments we go even further since a nonlinear dispersive wave model (such as the SGN equations employed in this study) can provide also information about the internal structure of the flow (i.e. beneath the free surface). For instance, the non-hydrostatic component of the pressure field can be easily reconstructed:[9]

$$\frac{p_d(x, y, t)}{\rho} = -(\eta - y)\Bigl[\frac{\eta + y + 2h}{2} \cdot \mathscr{R}_1 + \mathscr{R}_2\Bigr], \tag{2.82}$$

where $-h(x) \leqslant y \leqslant \eta(x, t)$ and the quantities $\mathscr{R}_{1,2}$ are defined in (2.7) as (see also the complete derivation in Chap. 1):

$$\mathscr{R}_1 = u_{xt} + u u_{xx} - u_x^2,$$
$$\mathscr{R}_2 = u_t h_x + u[u h_x]_x.$$

We do not consider the hydrostatic pressure component since its variation is linear with water depth y:

[9]Please, notice that formula (2.80) is not applicable here, since the bottom is not flat anymore.

$$p_h = \rho g (\eta - y).$$

Even if the non-hydrostatic pressure component p_d might be negligible comparing to the hydrostatic one p_h, its presence is crucial to balance the effects of nonlinearity, which results in the existence of solitary waves, as one of the most widely known effects in dispersive wave propagation [95]. The dynamic pressure field and several other physical quantities under a solitary wave were computed and represented graphically in the framework of the full EULER equations in [100]. A good qualitative agreement with our results can be reported. The balance of dispersive and nonlinear effects results also in the symmetry of the non-hydrostatic pressure distribution with respect to the wave crest. It can be seen in Fig. 2.11a, d before and after the interaction process. However, during the interaction process the symmetry is momentaneously broken (see Fig. 2.11b, c). However, with the time going on, the system relaxes again to a symmetric[10] pressure distribution shown in Fig. 2.11d.

Knowledge of the solution to the SGN equations allows to reconstruct also the velocity field[11] $\bigl(\tilde{u}(x, y, t),\ \tilde{v}(x, y, t)\bigr)$ in the fluid bulk. Under the additional assumption that the flow is potential, one can derive the following asymptotic (neglecting the terms of the order $\mathcal{O}(\mu^4) \equiv \mathcal{O}\left(\frac{d^4}{\lambda^4}\right)$ in the horizontal velocity $\tilde{u}(x, y, t)$ and of the order $\mathcal{O}(\mu^2) \equiv \mathcal{O}\left(\frac{d^2}{\lambda^2}\right)$ for the vertical one $\tilde{v}(x, y, t)$) representation formula [128] (see also the derivation in Chap. 1 for the 3D case with moving bottom):

$$\tilde{u}(x, y, t) = u + \left(\frac{\mathcal{H}}{2} - y - h\right) \cdot \bigl((u h_x)_x + u_x h_x\bigr) + \left(\frac{\mathcal{H}^2}{6} - \frac{(y + h)^2}{2}\right) u_{xx}, \qquad (2.83)$$

$$\tilde{v}(x, y, t) = -u h_x - (y + h) u_x. \qquad (2.84)$$

The formulas above allow to compute the velocity vector field in the fluid domain at any time (when the solution $\bigl(\mathcal{H}(x, t),\ u(x, t)\bigr)$ is available) and in any point (x, y) above the bottom $y = -h(x)$ and beneath the free surface $y = \eta(x, t)$. Figure 2.12 shows a numerical application of this reconstruction technique at two different moments of time $t = 2$ and 3 s during the interaction process with the bathymetry change. In particular, in Fig. 2.12a one can see that important vertical particle velocities emerge during the interaction with the bottom step. In subsequent time moments one can see the division of the flow in two structures (see Fig. 2.12b): the left one corresponds to the reflected wave, while the right structure corresponds to the transmitted wave motion. The reconstructed velocity fields via the SGN model

[10] The symmetry here is understood with respect to the vertical axis passing by the wave crest.

[11] This information can be used later to compute fluid particle trajectories [136], for example.

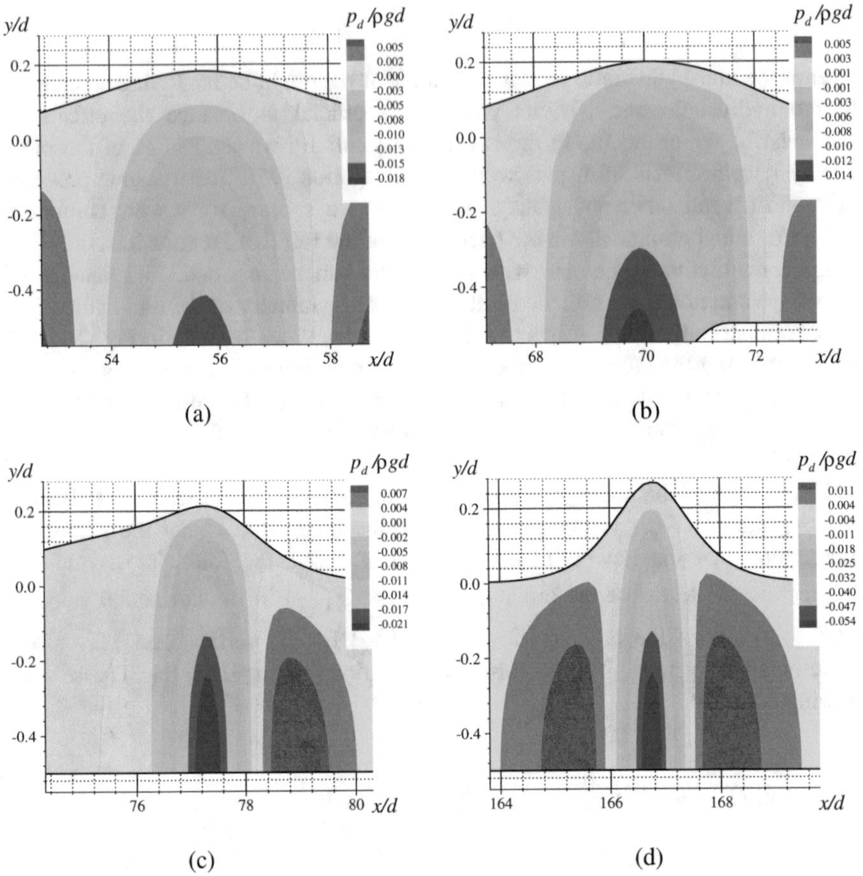

Fig. 2.11 Non-hydrostatic pressure distribution during a solitary wave/underwater step interaction process at different instances of time: (**a**) $t = 0.1$ s, (**b**) $t = 2.0$ s, (**c**) $t = 3.0$ s, (**d**) $t = 17.5$ s. Numerical parameters are provided in Table 2.2

compare fairly well with the 2D NAVIER–STOKES predictions [262]. However, the computational complexity of our approach is significantly lower than the simulation of the full NAVIER–STOKES equations. This is probably the main advantage of the proposed modelling methodology.

2.3.4 *Wave Generation by an Underwater Landslide*

As the last illustration of the proposed above numerical scheme, we model wave generation by the motion of an underwater landslide over uneven bottom. This test-case is very challenging since it involves rapid bottom motion (at least of

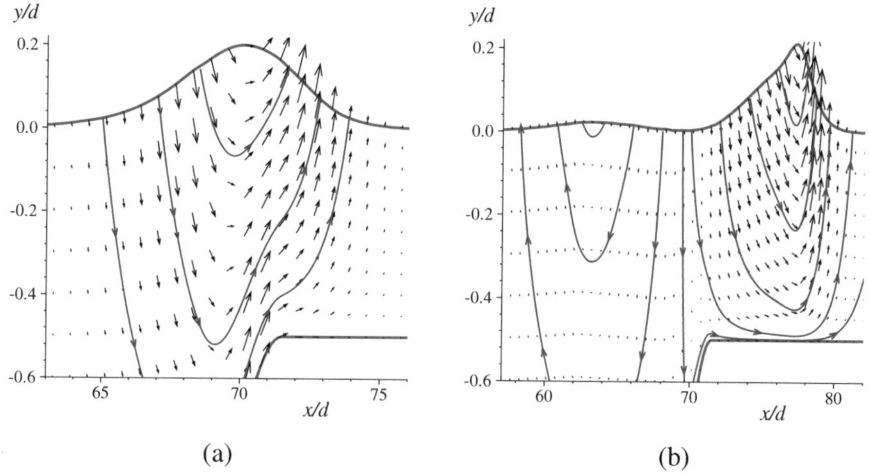

Fig. 2.12 Reconstructed velocity field in the fluid during the solitary wave interaction process with an underwater step: (**a**) $t = 2.0$ s, (**b**) $t = 3.0$ s. Solid blue lines show a few streamlines. Numerical parameters are provided in Table 2.2

its part). We recall that all previous tests were performed on a static bottom (i.e. $h_t \equiv 0$). The numerical simulation of underwater landslides is an important application where the inclusion of non-hydrostatic effects is absolutely crucial [226]. Moreover, the accurate prediction of generated waves allows to assess more accurately the natural hazard induced by unstable sliding masses (rockfalls, debris, ground movements) [321].

Usually, the precise location of unstable underwater masses is unknown and the numerical simulation is a preferred tool to study these processes. The landslide can be modelled as a solid undeformable body moving down the slope [70, 122, 149, 324]. Another possibility consists in representing the landslide as another fluid layer of higher density (possibly also viscosity) located near the bottom [49, 135]. In some works the landslide motion was not simulated (e.g. [176]) and the initial wave field generated by the landslide motion was determined using empirical formulas [150]. Then, this initial condition was propagated using an appropriate water wave model [176]. However, strictly speaking the employed empirical models are valid only for an absolutely rigid landslide sliding down a constant slope. Subsequent numerical simulations showed that the bottom shape influences quite substantially the generated wave field [25]. Consequently, for realistic modelling of real-world cases one needs to take into account the actual bathymetry [220] and even possible deformations of the landslide during its motion [218]. In a recent experimental work [218] the deformability of the landslide was achieved by composing it with four solid parts interconnected by springs. The idea to represent a landslide as a finite number of blocks was used in numerical [267] and theoretical [308] investigations.

In the present study we use the quasi-deformable[12] landslide model [26, 109, 118]. In this model the landslide deforms according to encountered bathymetry changes, however, at every time instance, all components of the velocity vector are the same in all particles which constitute the landslide (as in a solid rigid body). We shall use two long wave models:

- The SGN equations (fully nonlinear non-hydrostatic weakly dispersive model)
- NSWE equations[13] (standard hydrostatic dispersionless model)

The advantage of the SGN equations over other frequently used long wave models [176, 226, 325] are:

- The GALILEAN invariance [96]
- The energy balance equation (consistent with the full EULER [67, 104, 133])

NSWE were employed in [13] to model the real-world 16th October 1979 Nice event. It looks like the consensus on the importance of dispersive effects in landslide modelling is far from being achieved. For example, in [176] the authors affirm that the inclusion of dispersion gives results very similar to NSWE. In other works [142, 223, 303] the authors state that dispersive effects significantly influence the resulting wave field, especially during long time propagation. Consequently, in the present study we use both the SGN and NSWE equations to shed some light on the rôle of dispersive effects.

Consider a 1D fluid domain bounded from below by the solid (static) impermeable bottom given by the following function:

$$h_0(x) = \frac{h_+ + h_-}{2} + \frac{h_+ - h_-}{2} \tanh\bigl[F(x - \xi_F)\bigr], \tag{2.85}$$

where h_+ and h_- are water depths at $\pm\infty$ correspondingly (the domain we take is finite, of course). We assume for definiteness that

$$h_+ < h_- < 0.$$

We have also by definition:

$$F \stackrel{\text{def}}{:=} \frac{2\tan\theta_0}{h_- - h_+} > 0, \qquad \xi_F \stackrel{\text{def}}{:=} \frac{1}{2F}\ln\left[\frac{h_0 - h_+}{h_- - h_+}\right] > 0,$$

where $h_0 \equiv h_0(0)$ is water depth in $x = 0$ and θ_0 is the maximal slope angle, which is reached at the inflection point ξ_F. It can be easily checked that

[12]This model can be visualized if you imagine a landslide composed of infinitely many solid blocks.

[13]The numerical algorithm to solve nonlinear shallow water equations on a moving grid was presented and validated in [192].

$$\frac{h_+ + h_-}{2} < h_0 < h_- .$$

Equation (2.85) gives us the static part of the bottom shape. The following equation prescribes the shape of the bathymetry including the unsteady component:

$$y = -h(x, t) = h_0(x) + \zeta(x, t),$$

where function $\zeta(x, t)$ prescribes the landslide shape. In the present study we assume that the landslide initial shape is given by the following analytical formula:

$$\zeta(x, 0) = \begin{cases} \frac{\hbar}{2}\left[1 + \cos\left[\frac{2\pi\,(x - x_c(0))}{\nu}\right]\right], & |x - x_c(0)| \leqslant \frac{\nu}{2}, \\ 0, & |x - x_c(0)| > \frac{\nu}{2}, \end{cases}$$

where $x_c(0)$, $\hbar$ and ν are initial landslide position, height and width (along the axis Ox) correspondingly. Initially we put the landslide at the unique[14] point where the water depth is equal to $h_0 = 100$ m, i.e.

$$x_c(0) = \xi_F - \frac{1}{2F}\ln\left[\frac{h_0 - h_+}{h_- - h_+}\right] \approx 8\,323.5\ \text{m}.$$

For $t > 0$ the landslide position $x_c(t)$ and its velocity $v(t)$ are determined by solving a second order ordinary differential equation which describes the balance of all the forces acting on the sliding mass [26]. This model is well described in the literature [109, 118] and we do not reproduce the details here.

In Fig. 2.13a we show the dynamics of the moving bottom from the initial condition at $t = 0$ to the final simulation time $t = T$. All parameters are given in Table 2.3. It can be clearly seen that landslide's motion significantly depends on the underlying static bottom shape. In Fig. 2.13b we show landslide's barycentre trajectory $x = x_c(t)$ (line 1), its velocity $v = v(t)$ (line 2) and finally the static bottom profile $y = h_0(x)$ (line 3). From the landslide speed plot in Fig. 2.13b (line 2), one can see that the mass is accelerating during the first 284.2 s and slows down during 613.4 s. The distances travelled by the landslide during these periods have approximatively the same ratio ≈ 2. It is also interesting to notice that the landslide stops abruptly its motion with a negative (i.e. nonzero) acceleration.

In order to simulate water waves generated by the landslide, we take the fluid domain $\mathcal{I} = [0, \ell]$. For simplicity, we prescribe wall boundary conditions[15] at

---

[14]This point is *unique* since the static bathymetry $h_0(x)$ is a monotonically increasing function of its argument x.

[15]It would be better to prescribe transparent boundary conditions here, but this question is totally open for the SGN equations.

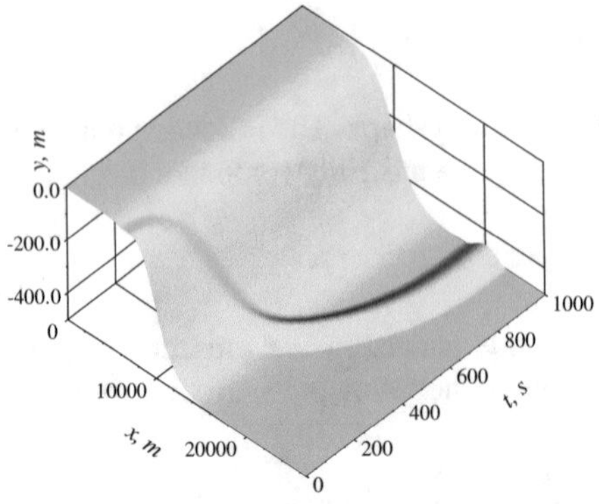

(a)

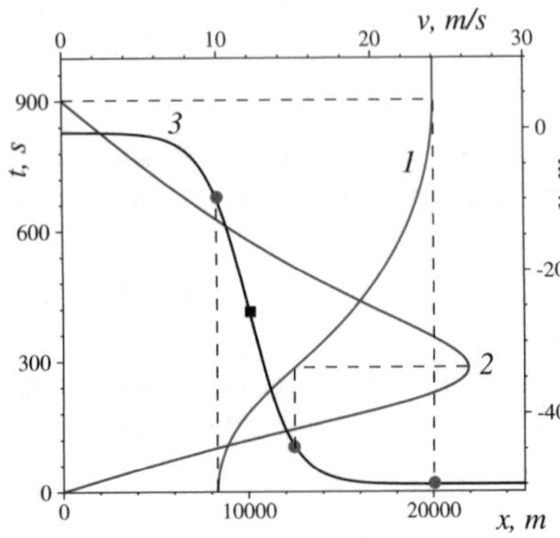

(b)

Fig. 2.13 Generation of surface waves by an underwater landslide motion: (**a**) dynamics of the moving bottom; (**b**) graphics of functions (1) $x = x_c(t)$, (2) $v = v(t)$, (3) $y = h_0(x)$. Two outer red circles denote landslide initial $t = 0$ and terminal $t = 897.6$ s positions. Middle red circle denotes landslide position at the moment of time $t = 284.2$ s where landslide's speed is maximal $v_{\max} \approx 26.3$ m/s. The black square shows the inflection point ξ_F position. The maximal speed is achieved well below the inflection point ξ_F. Numerical parameters are given in Table 2.3

Table 2.3 Numerical and physical parameters used in landslide simulation

Parameter	Value
Fluid domain length, ℓ	80, 000 m
Water depth, $h_0(0)$	– 5.1 m
Rightmost water depth, h_+	– 500 m
Leftmost water depth, h_-	– 5 m
Maximal bottom slope, θ_0	6°
Landslide height, $\hbar$	20 m
Landslide length, ν	5000 m
Initial landslide position, $x_c(0)$	8 323.5 m
Added mass coefficient, C_w	1.0
Hydrodynamic resistance coefficient, C_d	1.0
Landslide density, $\rho_{\mathrm{sl}}/\rho_{\mathrm{w}}$	1.5
Friction angle, θ^*	1°
Final simulation time, T	1000 s
Number of grid points, N	400
Monitor function parameter, ϑ_0	200

$x = 0$ and $x = \ell$. Undisturbed water depth at both ends is h_0 and $\approx h_+$, respectively. The computational domain length ℓ is chosen to be sufficiently large to avoid any kind of reflections from the right boundary. Initially the fluid is at rest with undisturbed free surface, i.e.

$$\eta(x, 0) \equiv 0, \qquad u(x, 0) \equiv 0.$$

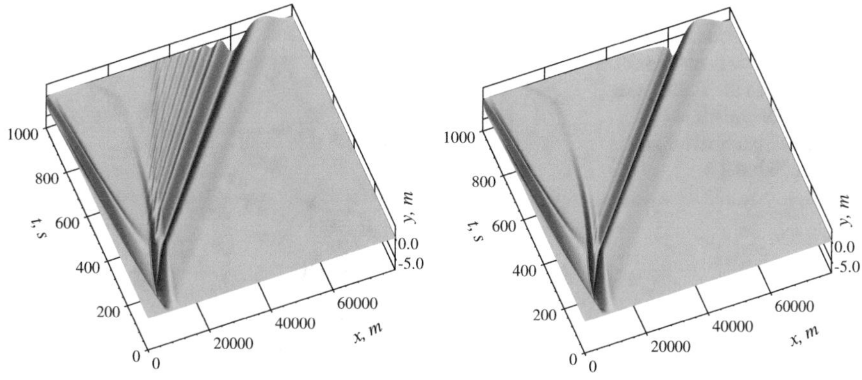

(a) (b)

Fig. 2.14 Generation of surface waves $y = \eta(x, t)$ by an underwater landslide motion: (**a**) the SGN model (dispersive); (**b**) NSWE equations (dispersionless). Numerical parameters are given in Table 2.3

Segment $\mathcal{I}$ is discretized using $N = 400$ points. In order to redistribute optimally mesh nodes, we employ the monitor function defined in Eq. (2.77), which refines the grid where the waves are large (regardless whether they are of elevation or depression type).

In Fig. 2.14 we show the surface $y = \eta(x, t)$ in space-time, which shows the main features of the generated wave field. The left panel (a) is the dispersive SGN prediction, while (b) is the computation with NSWE that we include into this study for the sake of comparison. For instance, one can see that the dispersive wave system is much more complex even if NSWE seem to reproduce the principal wave components. The dispersive components follow the main wave travelling rightwards. There is also at least one depression wave moving towards the shore. The motion of grid points is shown in Fig. 2.15. The initial grid was chosen to be uniform, since the free surface was initially flat. However, during the wave generation process the grid adapts to the solution. The numerical method redistributes the nodes according to the chosen monitor function $\varpi_0[\eta](x, t)$, i.e. where the waves are large (regardless whether they are of elevation or depression type). We would like to underline the fact that in order to achieve a similar accuracy on a uniform grid, one would need about $4N$ points.

In Fig. 2.16 we show two snapshots of the free surface elevation at two moments of time (a) and wave gauge records collected at two different spatial locations (b). In particular, we observe that there is a better agreement between NSWE and the SGN model in shallow regions (i.e. towards $x = 0$), while a certain divergence between two models becomes more apparent in deeper regions (towards the right end $x = \ell$).

In the previous Sect. 2.3.3 we showed the internal flow structure during nonlinear transformations of a solitary wave over a static step. In this section we show that SGN equations can be used to reconstruct and to study the physical fields

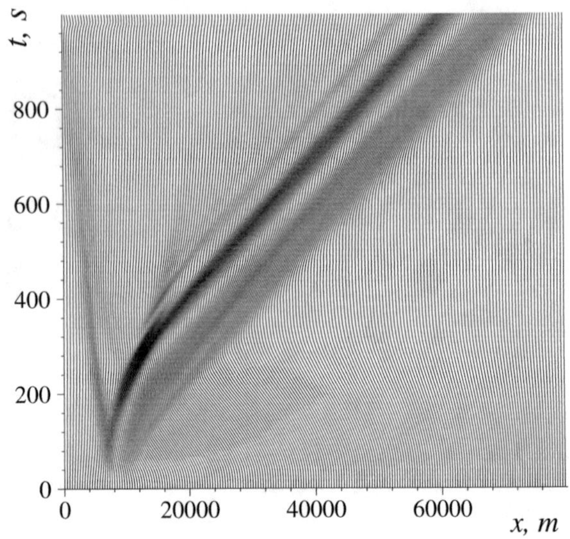

Fig. 2.15 Trajectories of every second grid node during the underwater landslide simulation in the framework of the SGN equations. Numerical parameters are given in Table 2.3

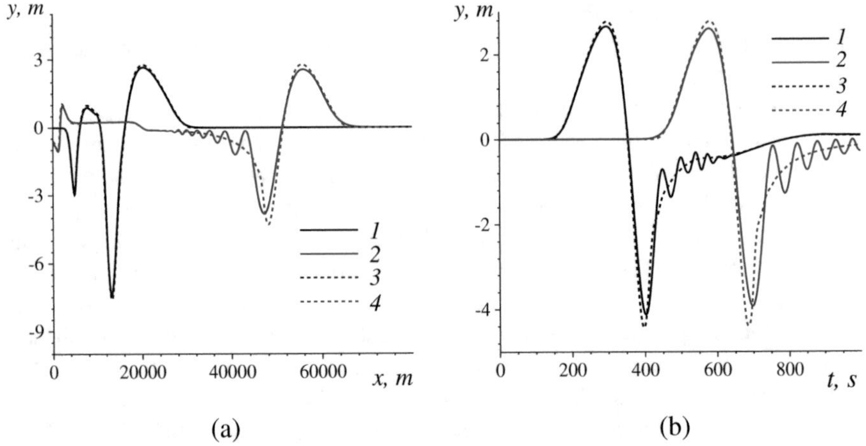

Fig. 2.16 Generation of surface waves by an underwater landslide: (**a**) free surface elevation profiles $y = \eta(x, t_{1,2})$ at $t_1 = 300$ s (1,3) and $t_2 = 800$ s (2,4); (**b**) free surface elevation $y = \eta(x_{1,2}, t)$ as a function of time in two spatial locations $x_1 = 20,000$ m (1,3) and $x_2 = 40,000$ m (2,4). The SGN predictions are represented with solid lines (1,2) and NSWE with dashed lines (3,4). Numerical parameters are given in Table 2.3

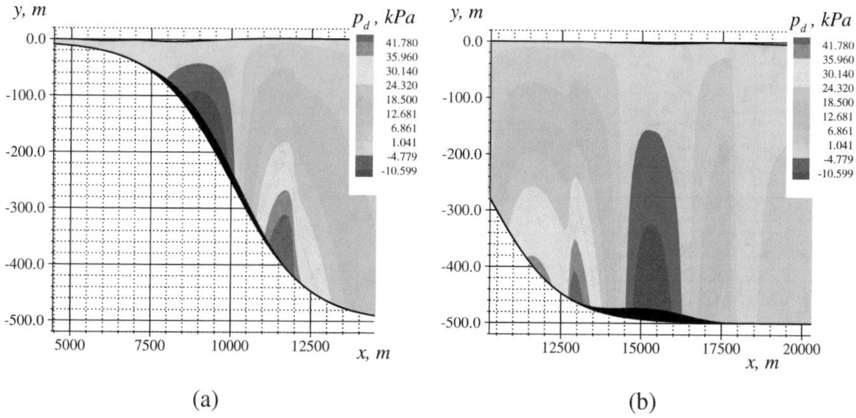

Fig. 2.17 Generation of surface waves by an underwater landslide. Isolines of the non-hydrostatic pressure at two moments of time: $t = 150$ s (**a**); $t = 400$ s (**b**). Numerical parameters are given in Table 2.3

in situations where the bottom moves abruptly. In order to reconstruct the non-hydrostatic field between moving bottom and the free surface, one can use formula (2.82), but the quantity $\mathscr{R}_2$ has some extra terms due to the bottom motion:

$$\mathscr{R}_2 = u_t h_x + u[u h_x]_x + h_{tt} + 2u h_{xt}.$$

In Fig. 2.17 we show the non-hydrostatic pressure field at two different moments of time. More precisely, we show a zoom on the area of interest around the landslide only. In panel (a) $t = t_1 = 150$ s and the landslide barycentre is located at $x_c(t_1) = 9\,456$ m. Landslide moves downhill with the speed $v(t_1) = 15.72$ m/s and it continues to accelerate. In particular, one can see that there is a zone of positive pressure in front of the landslide and a zone of negative pressure just behind. This fact has implications on the fluid particle trajectories around the landslide. In right panel (b) we show the moment of time $t = t_2 = 400$ s. At this moment $x_c(t_2) = 15\,264$ m and $v(t_2) = 21.4$ m/s. The non-hydrostatic pressure distribution qualitatively changed. Zones of positive and negative pressure switched their respective positions. Moreover, in Fig. 2.16 we showed that dispersive effects start to be noticeable at the free surface only after $t \geqslant 400$ s and by $t = 800$ s they are flagrant. In Fig. 2.18 we show the velocity fields in the fluid bulk at corresponding moments of time t_1 and t_2. We notice some similarities between the fluid flow around a landslide with an air flow around an airfoil. To our knowledge the internal hydrodynamics of landslide generated waves on a general non-constant sloping bottom and in the framework of SGN equations has not been shown before.

We remind that in the presence of moving bottom one should use the following reconstruction formulas for the velocity field (which are slightly different from (2.83), (2.84)):

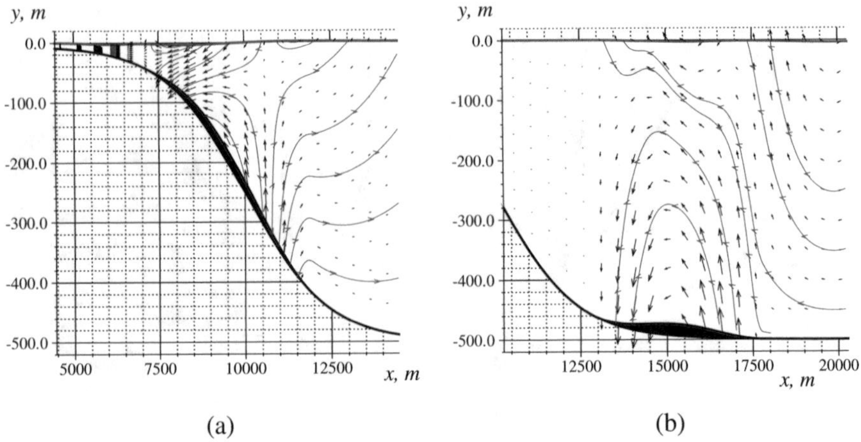

Fig. 2.18 Generation of surface waves by an underwater landslide. The reconstructed velocity field at two moments of time: $t = 150$ s (**a**); $t = 400$ s (**b**). Numerical parameters are given in Table 2.3

$$\tilde{u}(x, y, t) = u + \left(\frac{\mathcal{H}}{2} - y - h\right) \cdot \left([h_t + u h_x]_x + u_x h_x\right) + \left(\frac{\mathcal{H}^2}{6} - \frac{(y+h)^2}{2}\right) u_{xx},$$

$$\tilde{v}(x, y, t) = -h_t - u h_x - (y + h) u_x.$$

These formulas naturally become (2.83), (2.84) if the bottom is static, i.e. $h_t \equiv 0$.

2.3.5 Landslide Modelling

In the previous Sect. 2.3.4 we provided numerical results for the surface waves generated by one-dimensional underwater landslide. In other words, we assumed that the shape of the sliding mass and the bathymetry can be parametrized by one spatial coordinate. So, everything was invariant under translations in the transversal direction. In the present section we shall derive differential equations which describe the motion of the underwater sliding mass along the general two-dimensional sloping bottoms. The one-dimensional reduction of these equations was used in our numerical simulations shown above in Sect. 2.3.4. Moreover, we take into account deformation effects of this sliding body along its trajectory in a specific way. We assume that this continuous body is *quasi-deformable*, which means that the variation of its shape is caused by variations in the bathymetry (as in the usual deformable case), but we assume additionally that horizontal velocity components of this body are the same in all its points (as in the absolutely rigid body). Thus, the concept of quasi-deformability takes an intermediate place between the absolute rigidity and the usual deformability concepts. At the end of this section we present some numerical results obtained with the proposed model for the Paleo mega-landslide which took place in the BLACK SEA in the very distant past. For more numerical results obtained with this landslide model, we refer to [26, 109, 118].

Equations of Motion

As in Sect. 2.3.4, we assume that the solid bottom is described by the equation $y = -h(\mathbf{x}, t) = h_0(\mathbf{x}) + \zeta(\mathbf{x}, t)$, where $h_0(\mathbf{x})$ describes the static part of the bottom on which the mass is sliding:

$$y = h_0(\mathbf{x}). \tag{2.86}$$

The shape of the landslide is described by the non-negative function $y = \zeta(\mathbf{x}, t)$.

We suppose that at the initial moment of time, say $t = 0$, the unstable mass and the fluid have zero initial velocities (this assumption can be relaxed later if necessary). The initial position of the mass is given by the function

$$y = \zeta_0(\mathbf{x}).$$

The function $\mathbf{x} \longmapsto \zeta_0(\mathbf{x}, t)$ has a compact support $\mathcal{D}_0$ to reflect the finite extent property of real landslides, i.e.

$$\zeta(\mathbf{x}, 0) \equiv \zeta_0(\mathbf{x}), \qquad \operatorname{supp}(\zeta_0) = \mathcal{D}_0.$$

The shape and position of the landslide for $t > 0$ will be determined by its law of motion to be specified below. During the derivation of the equations of motion, it will be useful to identify it with a material point $\mathbf{X}_c(t)$, say the barycentre of the sliding body. This point will move along a certain two-dimensional surface under the action of integrated forces applied to all elementary volumes of the sliding body. The shape and position of the body will appear in our model through the domain of the integration to determine the resulting force.

So, we consider that at every moment of time $t > 0$, the location of the landslide is determined by the point $\mathbf{X}_c(t) = \bigl(\mathbf{x}_c(t), y_c(t)\bigr)$, which also moves along the static bottom given in Eq. (2.86) according to the classical second NEWTON law of Mechanics:

$$\mathcal{M}\ddot{x}_{1,c} = \mathcal{F}_1, \quad \mathcal{M}\ddot{x}_{2,c} = \mathcal{F}_2, \quad \mathcal{M}\ddot{y}_c = \mathcal{F}_3. \tag{2.87}$$

We should remember the following holonomic constraint:

$$y_c(t) = h_0\bigl(\mathbf{x}_c(t)\bigr), \tag{2.88}$$

where $\mathbf{x}_c(t) = \bigl(x_{1,c}(t),\, x_{2,c}(t)\bigr)$, $\mathbf{x}_c^0 \stackrel{\text{def}}{:=} \mathbf{x}_c(0) \in \mathcal{D}_0$, $y_c^0 \stackrel{\text{def}}{:=} y_c(0) \equiv h_0(\mathbf{x}_c^0)$. The components $\mathcal{F}_{1,2,3}$ of the force vector $\mathbf{F} = \bigl(\mathcal{F}_1,\, \mathcal{F}_2,\, \mathcal{F}_3\bigr)$ will be determined below. The mass $\mathcal{M}$ incorporates the proper landslide mass along with the added mass effect [21]:

$$\mathcal{M} \stackrel{\text{def}}{:=} (\rho_s + C_w \cdot \rho_w) \cdot V\,,$$

where we introduced the following variables:

- ρ_w : density of the water,
- ρ_s : density of the landslide material ($\rho_s > \rho_w$),
- C_w : added mass coefficient,
- V : volume of the landslide.

Below, the coordinates $\mathbf{x}_c(t)$ will be identified with the landslide barycentre. Moreover, we assume by following [26] that for $\forall t > 0$ the landslide upper boundary is described by the following functional dependence:

$$\zeta(\mathbf{x},\, t) = \zeta_0\bigl(\mathbf{x} + \mathbf{x}_c^0 - \mathbf{x}_c(t)\bigr)\,. \tag{2.89}$$

Henceforth, the support $\mathcal{D}_t$ of the function $\mathbf{x} \longmapsto \zeta(\mathbf{x},\, t)$ is also finite and it can be expressed through the support of the function $\zeta_0(\mathbf{x})$ and the location $\mathbf{x}_c(t)$ as follows:

$$\mathcal{D}_t = \bigl\{\mathbf{x} \in \mathbb{R}^2 \mid \mathbf{x} + \mathbf{x}_c^0 - \mathbf{x}_c(t) \in \mathcal{D}_0\bigr\}\,. \tag{2.90}$$

Let us discuss here some peculiarities of the proposed landslide model. It can be readily seen from Eq. (2.90) that at every instance of time t the landslide is located such that its vertical projection on the plane $y = 0$ coincides with the set $\mathcal{D}_t$, which can be obtained by translating $\mathcal{D}_0$. It means that all landslide points share the same instantaneous horizontal celerity. However, we underline that the vertical projection of the velocity does not have to be the same in all point of the landslide. Moreover, the contact surface between the landslide and solid bottom varies in time. For example, on steep slopes, this surface will be larger than on flatter areas, which reflects the physical reality. Thus, the landslide in our model experiences certain deformation while moving along the slope according to encountered bathymetry variations. In other words, we can say that the motion of a real-world landslide is approximated by the motion of a quasi-deformable body.

Remark 2.7 On a constant slope our approach predicts also the same vertical velocity of landslide points additionally to the shared horizontal velocity components. Thus, on constant slopes our landslide moves as an absolutely rigid body.

Remark 2.8 It is interesting to notice also that despite the deformability of the landslide, its volume remains invariant during the trajectory since

$$V = \int_{\mathcal{D}_t} \zeta(\mathbf{x}, t)\,\mathrm{d}\mathbf{x} \overset{(2.89)}{\equiv} \int_{\mathcal{D}_0} \zeta_0(\mathbf{x})\,\mathrm{d}\mathbf{x}. \tag{2.91}$$

During the derivation procedure, we assume that the two-dimensional surface (2.86) describing the solid bottom has no singularities and admits the following regular parametrization:

$$x_1 = x_1(q^1, q^2), \quad x_2 = x_2(q^1, q^2), \quad y = y(q^1, q^2), \tag{2.92}$$

where $q_{1,2}$ are real parameters. Then, we have

$$\mathbf{x}_c(t) = \mathbf{x}_c\big(q^1(t), q^2(t)\big), \quad y_c(t) = y_c\big(q^1(t), q^2(t)\big)$$

with $\big(q^1(t), q^2(t)\big)$ is the coordinate of the point $\mathbf{X}_c$ at time t in the new parametric form. From parametrization relations (2.92) we can easily obtain the expressions for CARTESIAN components of the velocities $\dot{\mathbf{x}}_c$ and $\dot{y}_c$ in terms of contravariant velocity components $\upsilon^{1,2} \overset{\text{def}}{:=} \dot{q}^{1,2}$:

$$\dot{\mathbf{x}}_c = \mathbf{x}_{q^1}\dot{q}^1 + \mathbf{x}_{q^2}\dot{q}^2,$$
$$\dot{y}_c = y_{q^1}\dot{q}^1 + y_{q^2}\dot{q}^2.$$

In view of this result, Equations of motion (2.87) can be rewritten as follows:

$$\mathcal{M}\frac{\mathrm{d}}{\mathrm{d}t}\big(x_{1,q^1}\dot{q}^1 + x_{1,q^2}\dot{q}^2\big) = \mathcal{F}_1, \tag{2.93}$$

$$\mathcal{M}\frac{\mathrm{d}}{\mathrm{d}t}\big(x_{2,q^1}\dot{q}^1 + x_{2,q^2}\dot{q}^2\big) = \mathcal{F}_2, \tag{2.94}$$

$$\mathcal{M}\frac{\mathrm{d}}{\mathrm{d}t}\big(y_{q^1}\dot{q}^1 + y_{q^2}\dot{q}^2\big) = \mathcal{F}_3. \tag{2.95}$$

By multiplying each equation in (2.93) by the corresponding component x_{1,q^1}, x_{2,q^1}, y_{q^1} of the tangent vector to the surface (2.92), respectively, and summing up the results, we obtain the first equation of motion:

$$\mathcal{M}\frac{\mathrm{d}}{\mathrm{d}t}\big(g_{11}\dot{q}^1 + g_{12}\dot{q}^2\big) =$$
$$\tfrac{1}{2}\,\mathcal{M}\Big(\frac{\partial g_{11}}{\partial q^1}(\dot{q}^1)^2 + 2\,\frac{\partial g_{12}}{\partial q^1}\dot{q}^1\dot{q}^2 + \frac{\partial g_{22}}{\partial q^1}(\dot{q}^2)^2\Big) + \mathcal{F}_{\tau_1}\sqrt{g_{11}}, \tag{2.96}$$

where $g_{\alpha\beta}$ are covariant components of the surface (2.86) metric tensor:

$$g_{11} = \left(\frac{\partial x_1}{\partial q^1}\right)^2 + \left(\frac{\partial x_2}{\partial q^1}\right)^2 + \left(\frac{\partial y}{\partial q^1}\right)^2,$$

$$g_{12} = \frac{\partial x_1}{\partial q^1} \cdot \frac{\partial x_1}{\partial q^2} + \frac{\partial x_2}{\partial q^1} \cdot \frac{\partial x_2}{\partial q^2} + \frac{\partial y}{\partial q^1} \cdot \frac{\partial y}{\partial q^2} = g_{21},$$

$$g_{22} = \left(\frac{\partial x_1}{\partial q^2}\right)^2 + \left(\frac{\partial x_2}{\partial q^2}\right)^2 + \left(\frac{\partial y}{\partial q^2}\right)^2,$$

where $\mathcal{F}_{\boldsymbol{\tau}_1} = \mathbf{F} \cdot \boldsymbol{\tau}_1$ and $\boldsymbol{\tau}_\alpha, \alpha \in \{1, 2\}$ are unitary vectors tangent to the surface (2.92):

$$\boldsymbol{\tau}_\alpha = \frac{1}{\sqrt{g_{\alpha\alpha}}} \left(\frac{\partial x_1}{\partial q^\alpha}, \frac{\partial x_2}{\partial q^\alpha}, \frac{\partial y}{\partial q^\alpha}\right)^\top. \tag{2.97}$$

In a similar way we obtain also the second equation of motion:

$$\mathcal{M} \frac{\mathrm{d}}{\mathrm{d}t}\left(g_{21}\dot{q}^1 + g_{22}\dot{q}^2\right) =$$
$$\frac{1}{2}\mathcal{M}\left(\frac{\partial g_{11}}{\partial q^2}(\dot{q}^1)^2 + 2\frac{\partial g_{12}}{\partial q^2}\dot{q}^1\dot{q}^2 + \frac{\partial g_{22}}{\partial q^2}(\dot{q}^2)^2\right) + \mathcal{F}_{\boldsymbol{\tau}_2}\sqrt{g_{22}}, \tag{2.98}$$

where similarly $\mathcal{F}_{\boldsymbol{\tau}_2} = \mathbf{F} \cdot \boldsymbol{\tau}_2$.

Now we transform Eqs. (2.96) and (2.98) into the form suitable for numerical integration. Let $v_{1,2}$ denote the covariant components of the velocity vector. Then, they can be expressed as

$$v_1 = g_{11}v^1 + g_{12}v^2, \tag{2.99}$$

$$v_2 = g_{21}v^1 + g_{22}v^2. \tag{2.100}$$

The inverse transformation is given by

$$v^1 = g^{11}v_1 + g^{12}v_2, \tag{2.101}$$

$$v^2 = g^{21}v_1 + g^{22}v_2, \tag{2.102}$$

where $g^{\alpha\beta}$, $\alpha, \beta \in \{1, 2\}$ are contravariant components of the metric tensor of the surface (2.86):

$$g^{11} = \frac{g_{22}}{G}, \quad g^{12} = g^{21} = -\frac{g_{12}}{G}, \quad g^{22} = \frac{g_{11}}{G},$$

with G being the following determinant

$$G \overset{\text{def}}{:=} \begin{vmatrix} g_{11} & g_{12} \\ g_{12} & g_{22} \end{vmatrix} .$$

We can ensure that $G \neq 0$ for all points of the surface by assumption that the surface along with its parametrization (2.92) is regular.

By using expressions (2.99), (2.100) and (2.101), (2.102) we can rewrite Eqs. (2.96) and (2.98) as evolution equations for components $v_\alpha, \alpha \in \{1, 2\}$:

$$\mathcal{M} \frac{dv_\alpha}{dt} = \frac{1}{2} \mathcal{M} \Big(\frac{\partial g_{11}}{\partial q^\alpha} (v^1)^2 + 2 \frac{\partial g_{12}}{\partial q^\alpha} v^1 v^2 + \frac{\partial g_{22}}{\partial q^\alpha} (v^2)^2 \Big) + \mathcal{F}_{\tau_\alpha} \sqrt{g_{\alpha\alpha}} . \tag{2.103}$$

Since at the beginning the landslide was at its equilibrium state, we can set zero initial conditions for System (2.103):

$$v_1(0) = v_2(0) = 0 . \tag{2.104}$$

Two equations in (2.103) coming from the dynamics have to be completed by two kinematic differential equations:

$$\frac{dq^\alpha}{dt} = v^\alpha, \qquad \alpha \in \{1, 2\} . \tag{2.105}$$

The right-hand sides in these equations are computed according to Formulas (2.101) and (2.102), which express contravariant velocity components in terms of covariant ones. The last two equations will allow us to determine the coordinates $\big(q^1(t), q^2(t)\big)$ of the point $\mathbf{X}_c(t)$ in the space of parameters. The initial condition $\big(q^1(0), q^2(0)\big)$ is determined as the pre-image of the material point $\mathbf{X}_c(0)$ under the bijective mapping (2.92).

Simplified Parametrization

Some simplification of the expressions given above can be achieved if we choose a particular parametrization of the surface (2.86) instead of the most general one (2.92). We mention also that the following parametric form[16] is enough to solve most of the problems arising in the geophysical applications:

$$x_1 = q^1, \quad x_2 = q^2, \quad y = h_0(\mathbf{x}) . \tag{2.106}$$

[16] Provided that the form of the solid bottom can admit this parametrization.

For this parametrization, the contravariant velocity components v^{α} coincides with CARTESIAN components $u_{\alpha} = \dot{x}_{\alpha,c}$. As a result, Eq. (2.103) take a much simpler form:

$$\mathcal{M}\frac{\mathrm{d}v_{\alpha}}{\mathrm{d}t} = \tfrac{1}{2}\,\mathcal{M}\Big(\frac{\partial g_{11}}{\partial x^{\alpha}}\,u_1^2 + 2\,\frac{\partial g_{12}}{\partial x^{\alpha}}\,u_1 u_2 + \frac{\partial g_{22}}{\partial x^{\alpha}}\,u_2^2\Big) + \mathcal{F}_{\tau_{\alpha}}\sqrt{g_{\alpha\alpha}}, \quad \alpha \in \{1,\,2\}. \qquad (2.107)$$

Moreover, $u_1(0) = 0,\, u_2(0) = 0$ and

$$v_1 = g_{11}u_1 + g_{12}u_2,$$

$$v_2 = g_{12}u_1 + g_{22}u_2.$$

The metric tensor components can be also easily computed:

$$g_{11} = 1 + \Big(\frac{\partial h_0}{\partial x_1}\Big)^2, \quad g_{12} \equiv g_{21} = \frac{\partial h_0}{\partial x_1}\cdot\frac{\partial h_0}{\partial x_2}, \quad g_{22} = 1 + \Big(\frac{\partial h_0}{\partial x_2}\Big)^2.$$

The unitary tangent vectors defined in Eq. (2.97) take the following simple form:

$$\boldsymbol{\tau}_1 = \frac{1}{\sqrt{g_{11}}}\Big(1,\,0,\,\frac{\partial h_0}{\partial x_1}\Big)^{\top},$$

$$\boldsymbol{\tau}_2 = \frac{1}{\sqrt{g_{22}}}\Big(0,\,1,\,\frac{\partial h_0}{\partial x_2}\Big)^{\top}.$$

In case of the simplified parametrization (2.106), the CAUCHY problem for Eq. (2.105) can be simply written as follows:

$$\frac{\mathrm{d}x_{1,c}}{\mathrm{d}t} = u_1, \qquad \frac{\mathrm{d}x_{2,c}}{\mathrm{d}t} = u_2, \qquad (2.108)$$

together with initial conditions $x_{1,c}(0) = x^0_{1,c}$, $x_{2,c}(0) = x^0_{2,c}$, where the numbers $x^0_{1,2,c}$ are known abscissas of the point $\mathbf{X}_c(t)$ at the initial moment of time. The third coordinate is determined from Condition (2.88) stating that point $\mathbf{X}_c(t)$ belongs to the surface (2.86).

Determination of Forces

In this section we clarify how to compute the forces $\mathcal{F}_{\tau_{\alpha}}$ appearing in the dynamic Eq. (2.107) and acting on the landslide with the barycentre located in point $\mathbf{X}_c(t)$.

Gravity Related Forces

Every elementary volume of the landslide with surface element $\mathrm{d}\mathbf{x} = \mathrm{d}x_1\,\mathrm{d}x_2$ experiences two gravity-related forces in the vertical direction: gravity and buoyancy:

$$\mathbf{f}_g(\mathbf{x}) = \bigl(0,\, 0,\, f_g(\mathbf{x})\bigr), \qquad \mathbf{x} \in \mathcal{D}_t, \tag{2.109}$$

where

$$f_g(\mathbf{x}) = -g(\rho_s - \rho_w)\cdot\zeta(\mathbf{x},\, t)\cdot \mathrm{d}\mathbf{x}, \tag{2.110}$$

with g being the acceleration due to gravity. Let us compute the projections of the force vector $\mathbf{f}_g$ on the tangent directions $\boldsymbol{\tau}_{1,2}$ to the bottom surface:

$$f_{g,\boldsymbol{\tau}_1} = \mathbf{f}_g \cdot \boldsymbol{\tau}_1 = \frac{f_g}{\sqrt{g_{11}}}\frac{\partial h_0}{\partial x_1},$$

$$f_{g,\boldsymbol{\tau}_2} = \mathbf{f}_g \cdot \boldsymbol{\tau}_2 = \frac{f_g}{\sqrt{g_{22}}}\frac{\partial h_0}{\partial x_2}.$$

The obtained expressions have to be integrated over the support of the function $\mathbf{x} \longmapsto \zeta(\mathbf{x},\, t)$ to obtain the contribution of gravity and buoyancy forces into components $\mathcal{F}_{\boldsymbol{\tau}_\alpha}$:

$$\mathcal{F}_{g,\boldsymbol{\tau}_\alpha}(t) = \iint\limits_{\mathcal{D}_t} f_{g,\boldsymbol{\tau}_\alpha}\,\mathrm{d}\mathbf{x}, \qquad \alpha \in \{1,\, 2\}. \tag{2.111}$$

Water Resistance

The force vector given in previous Eq. (2.111) contributes to landslide's acceleration. In contrast, the water resistance to the motion and the friction force against the solid bottom will compete to decelerate the landslide. The force of resistance $\mathcal{F}_r$ is directed opposite to the direction of landslide motion. In analogy with [324], we shall assume that the magnitude $f_r \overset{\text{def}}{:=} |\mathcal{F}_r|$ of this force is proportional to the area $\mathfrak{A}$ of the largest intersection of the landslide with a vertical plane, which is defined by its normal vector $(u_1,\, u_2,\, 0)^\top$:

$$f_r = \frac{1}{2}\, C_d\, \rho_w\, \mathfrak{A}\, v_c^2, \tag{2.112}$$

where C_d is the water resistance coefficient to the motion and v_c is the magnitude of the velocity vector $\mathbf{v}_c(t)$ of the point $\mathbf{X}_c(t)$:

$$v_c \stackrel{\text{def}}{:=} |\mathbf{v}_c| = \sqrt{g_{11} u_1^2 + 2 g_{12} u_1 u_2 + g_{22} u_2^2}.$$

Let us stress out that the resistance force $\mathfrak{F}_r$ vanishes for $v_c = 0$. If the motion takes place, then its expression in vector form can be easily obtained:

$$\mathfrak{F}_r = -\frac{\mathbf{v}_c}{v_c} f_r .$$

The projections of this force onto tangent vectors $\boldsymbol{\tau}_{1,2}$ read:

$$\mathfrak{F}_{r,\tau_\alpha} = \mathfrak{F}_r \cdot \boldsymbol{\tau}_\alpha = -\frac{1}{2} \frac{C_d \rho_w \mathfrak{A}}{\sqrt{g_{\alpha\alpha}}} v_c (g_{\alpha 1} u_1 + g_{\alpha 2} u_2) = -\frac{1}{2} \frac{C_d \rho_w \mathfrak{A}}{\sqrt{g_{\alpha\alpha}}} v_c v_\alpha . \tag{2.113}$$

Friction Forces

In order to compute the friction forces, we consider again an elementary volume of the landslide with section area $\mathrm{d}\mathbf{x}$ and mass $m = \rho_s \zeta(\mathbf{x}, t)\,\mathrm{d}\mathbf{x}$. The friction force acting on this elementary volume can be determined through the normal reaction force N, which acts on this element from the solid bottom. We shall compute the quantity N by assuming that this elementary volume with mass m moves according to governing equations of the type (2.93) written for the parametrization (2.106):

$$m \frac{\mathrm{d}u_1}{\mathrm{d}t} = f_1 ,$$
$$m \frac{\mathrm{d}u_2}{\mathrm{d}t} = f_2 ,$$
$$m \frac{\mathrm{d}}{\mathrm{d}t}\Big(u_1 \frac{\partial h_0}{\partial x_1} + u_2 \frac{\partial h_0}{\partial x_2} \Big) = f_3 .$$

We multiply the first two equations by $-\frac{\partial h_0}{\partial x_1}$ and $-\frac{\partial h_0}{\partial x_2}$, respectively, and sum up the result with the third equation to obtain:

$$-m \frac{\partial h_0}{\partial x_1} \frac{\mathrm{d}u_1}{\mathrm{d}t} - m \frac{\partial h_0}{\partial x_2} \frac{\mathrm{d}u_2}{\mathrm{d}t} + m \frac{\mathrm{d}}{\mathrm{d}t}\Big(u_1 \frac{\partial h_0}{\partial x_1} + u_2 \frac{\partial h_0}{\partial x_2} \Big) = f_n \sqrt{G} , \tag{2.114}$$

where $f_n \stackrel{\text{def}}{:=} \mathbf{f} \cdot \mathbf{n}$ with $\mathbf{f} \stackrel{\text{def}}{:=} (f_1, f_2, f_3)^\top$ and $\mathbf{n} \stackrel{\text{def}}{:=} (n_1, n_2, n_3)^\top$ is the unitary normal vector to Surface (2.86) whose components are given by

$$n_1 = -\frac{1}{\sqrt{G}} \frac{\partial h_0}{\partial x_1}, \qquad n_2 = -\frac{1}{\sqrt{G}} \frac{\partial h_0}{\partial x_2}, \qquad n_3 = \frac{1}{\sqrt{G}} . \tag{2.115}$$

After some simplifications, Eq. (2.114) take the following form:

$$\frac{m}{\sqrt{G}}\left(u_1^2\,\frac{\partial^2 h_0}{\partial x_1^2} + 2u_1u_2\,\frac{\partial^2 h_0}{\partial x_1\partial x_2} + u_2^2\,\frac{\partial^2 h_0}{\partial x_2^2}\right) = f_n\,. \tag{2.116}$$

For an elementary volume of the landslide, the quantity f_n is composed of the normal component $\mathbf{f}_g\cdot\mathbf{n}$ of gravity-related forces (2.109) and of the normal reaction force N, which acts on this elementary volume. Then, the friction force magnitude f_{fr} can be obtained as follows:

$$f_{\mathrm{fr}} = C_{\mathrm{fr}}\cdot N = C_{\mathrm{fr}}\cdot\left(f_n - \mathbf{f}_g\cdot\mathbf{n}\right),$$

where $C_{\mathrm{fr}} = \tan\theta_*$ is the sliding friction coefficient and $\theta_* > 0$ is the friction angle. Then, from Eqs. (2.116), (2.109), (2.110) and (2.115) we obtain the following expression for the friction component:

$$f_{\mathrm{fr}} = \frac{C_{\mathrm{fr}}}{\sqrt{G}}\Bigg[g\,(\rho_s - \rho_w) + \rho_s\left(u_1^2\,\frac{\partial^2 h_0}{\partial x_1^2} + 2u_1u_2\,\frac{\partial^2 h_0}{\partial x_1\partial x_2} + u_2^2\,\frac{\partial^2 h_0}{\partial x_2^2}\right)\Bigg]\,\zeta(\mathbf{x},\,t)\,\mathrm{d}\mathbf{x}\,,$$

where $u_{1,2}$ are first two components of the vector $\mathbf{v}_c$. Vector of the friction force acting on the landslide $\mathbf{F}_{\mathrm{fr}}(t)$, acting against the motion, can be obtained by integrating all elementary forces over the support $\mathcal{D}_t$:

$$\mathbf{F}_{\mathrm{fr}}(t) = -\frac{\mathbf{v}_c}{v_c}\iint_{\mathcal{D}_t} f_{\mathrm{fr}}(\mathbf{x},\,t)\,\mathrm{d}\mathbf{x}\,.$$

Knowing the force vector $\mathbf{F}_{\mathrm{fr}}$, it is not difficult to determine its projections on tangent vectors $\boldsymbol{\tau}_{1,2}$:

$$\mathcal{F}_{\mathrm{fr},\boldsymbol{\tau}_\alpha} = \mathbf{F}_{\mathrm{fr}}\cdot\boldsymbol{\tau}_\alpha = -\frac{v_\alpha}{\sqrt{g_{\alpha\alpha}}\,v_c}\iint_{\mathcal{D}_t} f_{\mathrm{fr}}(\mathbf{x},\,t)\,\mathrm{d}\mathbf{x}\,, \qquad \alpha\in\{1,\,2\}\,. \tag{2.117}$$

Remark 2.9 We would like to mention that the computation of the friction force for $v_c = 0$ (e.g. at the initial and final instances of our simulation) is done according to a different formula, which will be specified below.

Governing Equations

The summation of components (2.111), (2.113) and (2.117) gives us the required total force component $\mathcal{F}_{\tau_\alpha}$ acting on the landslide. Henceforth, the final form of the equations of motion takes the following form:

$$\frac{dv_\alpha}{dt} = \frac{R_\alpha}{2} + \Big[g\,(\gamma - 1)\Big(\mathcal{I}_{1,\alpha} - \sigma_\alpha C_{\mathrm{fr}} \mathcal{I}_2\Big) - \sigma_\alpha \Big(\gamma\, C_{\mathrm{fr}} \big(u_1^2 \mathcal{I}_{3,11} + 2u_1 u_2 \mathcal{I}_{3,12} + u_2^2 \mathcal{I}_{3,22}\big) + \frac{C_d}{2}\,\mathfrak{A}\, v_c^2\Big)\Big] \frac{\sqrt{g_{\alpha\alpha}}}{(\gamma + C_w)\,V}, \quad \alpha \in \{1, 2\}, \tag{2.118}$$

where $\gamma \stackrel{\mathrm{def}}{:=} \dfrac{\rho_s}{\rho_w} > 1$ is the ratio of densities and

$$\sigma_\alpha \stackrel{\mathrm{def}}{:=} \frac{v_\alpha}{\sqrt{g_{\alpha\alpha}}\, v_c},$$

$$R_\alpha \stackrel{\mathrm{def}}{:=} \frac{\partial g_{11}}{\partial x^\alpha} u_1^2 + 2\,\frac{\partial g_{12}}{\partial x^\alpha} u_1 u_2 + \frac{\partial g_{22}}{\partial x^\alpha} u_2^2,$$

and the integrals $\mathcal{I}_{1,2,3}$ are defined as follows:

$$\mathcal{I}_{1,\alpha}(t) \stackrel{\mathrm{def}}{:=} -\iint_{\mathcal{D}_t} \frac{\zeta(\mathbf{x}, t)}{\sqrt{g_{\alpha\alpha}(\mathbf{x})}} \frac{\partial h_0}{\partial x^\alpha}(\mathbf{x})\,\mathrm{d}\mathbf{x},$$

$$\mathcal{I}_2(t) \stackrel{\mathrm{def}}{:=} \iint_{\mathcal{D}_t} \frac{\zeta(\mathbf{x}, t)}{\sqrt{G(\mathbf{x})}}\,\mathrm{d}\mathbf{x} > 0,$$

$$\mathcal{I}_{3,\alpha\beta}(t) \stackrel{\mathrm{def}}{:=} \iint_{\mathcal{D}_t} \frac{\zeta(\mathbf{x}, t)}{\sqrt{G(\mathbf{x})}} \frac{\partial^2 h_0}{\partial x^\alpha \partial x^\beta}(\mathbf{x})\,\mathrm{d}\mathbf{x}, \quad \alpha, \beta \in \{1, 2\}.$$

From the first sight, Eq. (2.118) may appear to be a second order ordinary differential equations. However, it is not exactly the case because of the presence of integrals in the equation, which depend on the solution. Consequently, strictly speaking, we deal with integro-differential equations in this section.

Intermediate Conclusions

So, in order to determine the instantaneous shape of the bottom (i.e. the lower boundary of the fluid layer), we have to accomplish the following three steps:

- We have to solve first the nonlinear System (2.118) of ODEs with the initial condition (2.104) to determine the components $t \longmapsto v_{\alpha}(t)$.
- Thanks to Formulas (2.101), (2.102) we determine the CARTESIAN components[17] of the velocity $u_{1,2} \equiv v^{1,2}$.
- Finally, we solve the CAUCHY problem (2.108) to compute the coordinates $x_{1,2,c}(t)$ of the moving barycentre $\mathbf{X}_c(t)$. Thanks to Formula (2.89), we are finally able to compute the instantaneous shape of the bottom.

All these computations are performed until the moment when the landslide stops its motion, i.e. the instance of time when v_c vanishes. We would like to stress out that the fact that $v_c = 0$ does not necessarily mean that the landslide finished completely its motion. It may happen that the landslide takes speed during the sliding down phase and, by inertia, it is even able to go up to the opposite slope of the bottom up to a certain extent. At the highest point on the opposite slope, it will momentaneously stop its motion before moving back to its equilibrium position in the deepest parts of the reservoir. We have to mention that the fact of the continuation of motion depends on the ratio between the friction coefficient to the curvature of the slope where the slide stopped its motion. Consequently, in order to compute the entire landslide trajectory, it is absolutely necessary to check the possibility of a new departure after each event when $v_c = 0$. Now we explain how we perform it in our code.

Continuation of Motion Criterium

At the initial time instance, but also at every stopping time, the landslide experiences the action of two opposite forces whose vectors lie in the tangent space generated by two vectors $\boldsymbol{\tau}_1$ and $\boldsymbol{\tau}_2$. Here, we have in mind the friction force $\mathbf{F}_{\mathrm{fr}}$ and the gravity-related force $\mathbf{F}_{g,\tau}$ entirely determined by its projection components given in Eq. (2.111). The landslide at rest will start its motion only under the following condition:

$$|\mathbf{F}_{g,\tau}| > |\mathbf{F}_{\mathrm{fr}}|, \tag{2.119}$$

where

$$\mathbf{F}_{g,\tau} \stackrel{\mathrm{def}}{:=} \frac{\mathcal{F}_{g,\boldsymbol{\tau}_1} - \mathcal{F}_{g,\boldsymbol{\tau}_2}(\boldsymbol{\tau}_1 \cdot \boldsymbol{\tau}_2)}{1 - (\boldsymbol{\tau}_1 \cdot \boldsymbol{\tau}_2)^2}\,\boldsymbol{\tau}_1 + \frac{\mathcal{F}_{g,\boldsymbol{\tau}_2} - \mathcal{F}_{g,\boldsymbol{\tau}_1}(\boldsymbol{\tau}_1 \cdot \boldsymbol{\tau}_2)}{1 - (\boldsymbol{\tau}_1 \cdot \boldsymbol{\tau}_2)^2}\,\boldsymbol{\tau}_2,$$

and

[17] We remind that for the simplified parametrization (2.106), the CARTESIAN velocity components $u_{1,2}$ coincide with the contravariant velocity components $v^{1,2}$.

$$\mathbf{F}_{\text{fr}} \overset{\text{def}}{:=} -g\,(\rho_s - \rho_w) \cdot C_{\text{fr}} \cdot \mathfrak{I}_2\, \frac{\mathbf{F}_{g,\tau}}{|\mathbf{F}_{g,\tau}|}\,.$$

Taking into account Eq. (2.111) and the following identity

$$\boldsymbol{\tau}_1 \cdot \boldsymbol{\tau}_2 = \frac{g_{12}}{\sqrt{g_{11} g_{12}}}\,,$$

we obtain a simplified expression for the Criterium (2.119) to decide the possibility to continue the motion from the motionless state:

$$\sqrt{\frac{g_{11} g_{22}}{G}\left(\mathfrak{I}_{1,1}^2 - 2\,\frac{g_{12}}{\sqrt{g_{11} g_{22}}}\,\mathfrak{I}_{1,1}\mathfrak{I}_{1,2} + \mathfrak{I}_{1,2}^2\right)} > C_{\text{fr}} \cdot \mathfrak{I}_2\,, \qquad (2.120)$$

which translates the fact the tangent component of the sum of the gravity and buoyancy forces is greater than the friction force. We underline the fact that Condition (2.120) has to be verified also at the initial moment of our simulation. If it is not initially verified, the landslide motion cannot be initiated.

Landslide Motion in a Model 2D Reservoir

Let us consider a model 2D reservoir whose static solid bottom Surface (2.86) is given by the following function:

$$\mathbf{x} \longmapsto h_0(\mathbf{x}) = h_\xi + (h_b - h_\xi) \cdot \begin{cases} \left(\dfrac{x_1 - \xi(x_2)}{x_{\text{sh}}(x_2) - \xi(x_2)}\right)^2, & x_1 \in \left[x_{\text{sh}}(x_2),\, \xi(x_2)\right], \\[2ex] \left(\dfrac{x_1 - \xi(x_2)}{L_1 - \xi(x_2)}\right)^2, & x_1 \in \left[\xi(x_2),\, L_1\right], \end{cases} \qquad (2.121)$$

where $x_2 = \left[0,\, L_2\right]$ and $x_1 \in \left[x_{\text{sh}}(x_2),\, L_1\right]$. The left reservoir shore (see Fig. 2.19) is a curvilinear cylindrical surface with the directrix parallel to the axis $O\,y$ and the generatrix given by the following equation:

$$x_1 = x_{\text{sh}}(x_2)\,.$$

The right shore is given by the flat vertical wall $x_1 = L_1$. At the state of the rest, the water depth at these vertical walls is equal to $h_b < 0$ or, in other words:

$$h_b \equiv h_0\big(x_{\text{sh}}(x_2),\, x_2\big) = h_0(L_1,\, x_2) = \text{const} < 0\,, \qquad \forall x_2 \in \left[0,\, L_2\right].$$

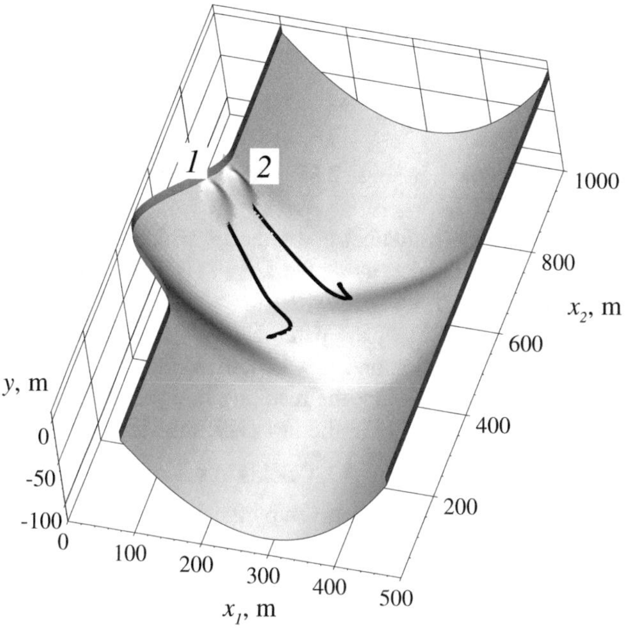

Fig. 2.19 Bathymetry of a model two-dimensional reservoir together with the trajectories of two landslides differing by their initial location: (1) $\left(x_{1,c}^0, x_{2,c}^0\right) = (91.5, 620)$ m and (2) $\left(x_{1,c}^0, x_{2,c}^0\right) = (115.2, 620)$ m

The curve $x_1 = \xi(x_2)$ is an isobath curve corresponding to the maximal reservoir depth $h_\xi < h_b$. Moreover, $\forall x_2 \in [0, L_2]$ we have $x_{\mathrm{sh}}(x_2) < \xi(x_2) < L_1$. Every section of Surface (2.121) by the plane $x_2 = \mathrm{const}$ is composed of two parabolas, which are smoothly glued together at points with abscissa $x_1 = \xi(x_2)$.

By choosing various (continuous) functions $x \longmapsto \xi(x)$ and $x \longmapsto x_{\mathrm{sh}}(x)$ one can obtain various curvilinear bottoms of our 2D reservoir. In the present section we made the following choice to illustrate the capabilities of our model:

$$\xi(x_2) = \mathcal{C}\left(x_2;\, \xi_1,\, \xi_2,\, y_\xi,\, \delta_\xi\right), \tag{2.122}$$

$$x_{\mathrm{sh}}(x_2) = \mathcal{C}\left(x_2;\, \varsigma_1,\, \varsigma_2,\, y_{\mathrm{sh}},\, \delta_{\mathrm{sh}}\right), \tag{2.123}$$

where we assume that $\delta_{\xi,\mathrm{sh}} > 0$ and $y_{\xi,\mathrm{sh}} \in (0, L_2)$. Finally, we used the following function parametrized by four real numbers:

$$x \longmapsto \mathcal{C}\left(x;\, p_1,\, p_2,\, x_0,\, \delta\right) \stackrel{\mathrm{def}}{:=}$$

$$\begin{cases} p_1 + \dfrac{p_2 - p_1}{2}\left[1 + \cos\left(\dfrac{2\pi(x - x_0)}{\delta}\right)\right], & |x - x_0| \leqslant \dfrac{\delta}{2}, \\ p_1, & |x - x_0| > \dfrac{\delta}{2}. \end{cases}$$

The bottom shape depicted in Fig. 2.19 was constructed using the values of parameters given in Table 2.4.

This model reservoir has a distinct particularity, which explains our choice for this bathymetry: in the middle part along the $O\,x_2$ axis there is a symmetric canyon with smooth sloping sides. If we specify the function $x \longmapsto x_{\text{sh}}(x)$ according to Eq. (2.123) with zero values of the parameters $\varsigma_{1,2}$ we shall obtain a simpler bathymetry with rectilinear left shore. If we choose *additionally* $\xi_1 \equiv \xi_2 = \xi$, then from Eq. (2.122) it follows that the resulting bathymetry will be invariant by translations along the variable x_2. In the last case, the simplest bathymetry shape can be obtained by choosing $\xi = \dfrac{L_1}{2}$. It yields to the bottom having the form of a cylindrical surface with the generatrix given by the following parabola:

Table 2.4 Physical parameters used to construct the reservoir bathymetry shown in Fig. 2.19

Parameter	Value
Water depth, h_b	-10 m
Maximal depth, h_ξ	-100 m
Reservoir width, L_1	500 m
Reservoir length, L_2	1 000 m
Bathymetry parameter, ξ_1	300 m
Bathymetry parameter, ξ_2	150 m
Bathymetry parameter, y_ξ	$\frac{L_2}{2}$
Bathymetry parameter, δ_ξ	$0.3 \cdot L_2$
Canyon parameter, ς_1	100 m
Canyon parameter, ς_2	0 m
Canyon parameter, y_{sh}	$\frac{L_2}{2}$
Canyon parameter, δ_{sh}	$0.4 \cdot L_2$
Landslide height, H	10 m
Landslide width, W_1	50 m
Landslide width, W_2	50 m
Initial landslide position (1), $x_{1,c}^0$	91.5 m
Initial landslide position (2), $x_{1,c}^0$	115.2 m
Initial landslide position (1), $x_{2,c}^0$	$y_{\text{sh}} + 0.3 \cdot \delta_{\text{sh}}$
Initial landslide position (2), $x_{2,c}^0$	$y_{\text{sh}} + 0.4 \cdot \delta_{\text{sh}}$
Initial landslide depth, y_c^0	-30 m
Densities ratio, γ	2
Water resistance coefficient, C_w	1
Friction angle, θ_*	5°
Drag coefficient, C_d	1

$$y = h_0(x_1) = (h_b - h_\xi) \cdot \left(\frac{x_1}{\xi} - 1\right)^2 + h_\xi, \qquad (2.124)$$

and having the directrix parallel to the axis $O\,x_2$.

In Fig. 2.19 we show also two simulated landslide trajectories differing by their starting point. The initial shape of the landslide was given by the following formula:

$$\zeta_0(\mathbf{x}) = H \cdot \mathcal{C}(x_1;\, 0,\, 1,\, x_{1,c}^0,\, W_1) \cdot \mathcal{C}(x_2;\, 0,\, 1,\, x_{2,c}^0,\, W_2), \qquad (2.125)$$

where $\left(x_{1,c}^0,\, x_{2,c}^0\right)$ are the coordinates of landslide's barycentre at $t = 0$, H is the landslide height, $W_{1,2}$ are landslide widths along $O\,x_{1,2}$ axes correspondingly. The volume of this landslide is also determined by Eq. (2.91) and an explicit expression for the volume V can be easily obtained:

$$V = \frac{1}{4}\, H \cdot W_1 \cdot W_2\,.$$

According to the Definition (2.90), the support of function $\zeta\,(\mathbf{x},\, t)$ is the following rectangular area:

$$\mathcal{D}_t = \left[x_{1,c}(t) - \frac{W_1}{2},\, x_{1,c}(t) + \frac{W_1}{2}\right] \times \left[x_{2,c}(t) - \frac{W_2}{2},\, x_{2,c}(t) + \frac{W_2}{2}\right].$$

The last result allows us to compute also the area $\mathfrak{A}$ which is needed to determine the water resistance force according to Eq. (2.112). This area depends on the values of the first two components $u_{1,2}$ of the velocity vector $\mathbf{v}_c(t)$. If both components $u_{1,2} \equiv 0$ vanish, there is no resistance to the motion because there is no motion. If only one of the components vanishes, we obtain a particularly simple expression for $\mathfrak{A}$:

$$\mathfrak{A} = \frac{H}{2} \cdot \begin{cases} W_1, & u_1 = 0,\ u_2 \neq 0, \\ W_2, & u_1 \neq 0,\ u_2 = 0. \end{cases}$$

Finally, if both components $u_{1,2}$ are different from zero, then the area is determined by the following formula:

$$\mathfrak{A} = \frac{H}{4} \begin{cases} \sqrt{W_1^2 + (k\,W_2)^2} \cdot \left(1 + \dfrac{\sin(\pi\,k)}{\pi\,k\,(1 - k^2)}\right), & |k| < 1, \\ \sqrt{(k'\,W_1)^2 + W_2^2} \cdot \left(1 + \dfrac{\sin(\pi\,k')}{\pi\,k'\left(1 - (k')^2\right)}\right), & |k| > 1, \\ \dfrac{3}{2}\sqrt{W_1^2 + W_2^2}, & |k| = 1, \end{cases}$$

where we assumed that $u_1^2 + u_2^2 \neq 0$ and we defined also the following quantities:

$$k \overset{\text{def}}{:=} \frac{u_1 W_1}{u_2 W_2}, \qquad k' \overset{\text{def}}{:=} \frac{1}{k}.$$

Now it is clear that to define k and k', one needs the condition $u_1^2 + u_2^2 \neq 0$. In our computations shown in Fig. 2.19, we used the values of parameters H, $W_{1,2}$ reported in Table 2.4.

Regarding the initial landslide position $\mathbf{X}_c(0)$, we fixed the coordinate $x_{2,c}^0$ together with the landslide depth y_c^0 (also reported in Table 2.4), while abscissa $x_{1,c}^0$ was found from the condition that the point $(x_{1,c}^0, x_{2,c}^0, y_c^0)$ belongs to Surface (2.121). The trajectories (1) and (2) depicted in Fig. 2.19 were obtained from the following choices:

- $(1): x_{2,c}^0 = y_{\text{sh}} + 0.3 \cdot \delta_{\text{sh}}$;
- $(2): x_{2,c}^0 = y_{\text{sh}} + 0.4 \cdot \delta_{\text{sh}}$.

As a result, in the first case (1) we obtain the value

$$x_{1,c}^0 = \xi(x_{2,c}^0) + \bigl(x_{\text{sh}}(x_{2,c}^0) - \xi(x_{2,c}^0)\bigr)\sqrt{\frac{y_c^0 - h_\xi}{h_b - h_\xi}} \approx 91.5\ \text{m}.$$

In the second scenario (2) one obtains the value for the coordinate $x_{1,c}^0 \approx 115.2$ m in a similar way.

From Fig. 2.19 it can be seen that the landslide sliding down the slope climbs up the ridge located on the opposite side due to its inertia. At the highest point of its trajectory, the landslide stops momentaneously to depart again. However, the direction of this new departure is almost opposite in two cases (1) and (2) (see Fig. 2.19). This 'bifurcating' behaviour explains our choice for these two initial positions aiming at illustrating the variability of landslide trajectories. By tuning the landslide parameters (see Table 2.4), one can easily obtain other interesting trajectories, especially for small values of the friction angle θ_*. In this way, the proposed model reservoir is sufficiently complex to show quite different underwater landslide behaviours (depending on the physical parameters and initial conditions). Thus, it can be used in practice to study the generated wave field on the free surface and its variability under various scenarii of the subaqueous motion.

One of the main features of the proposed landslide model is that the landslide trajectory is not arbitrarily prescribed by the investigator such as it was done in [273], but the trajectory is computed by solving exact equations of the Classical Mechanics. Some authors take an intermediate position, where they simplify the governing equations until they can be solved analytically to provide a closed form expression for the landslide trajectory in terms of a superposition of elementary functions [323, 326]. It is possible to do for constant slopes essentially.

Reduction to the One-Dimensional Space

In the previous section we could witness the variety of possible behaviours of landslides in 2D reservoirs depending on the initial conditions. The changes in parameters[18] result also in potentially new behaviours, without speaking of various bathymetric features also influencing the landslide trajectory. As a result, we estimate that a simplified one-dimensional landslide model is needed to gain some understanding in a simpler geometrical setting. Moreover, the computational complexity of a 1D problem is not comparable to the 2D situation described above, especially if we include into the simulation a wave propagation module. In some problems, the ability to obtain very quickly the first estimations can be appreciated. As a final motivational remark, we can add that the one-dimensional configuration presented below describes a two-dimensional problem enjoying the symmetry with respect to the translations along the axis $O\,x_2$.

In this section we shall simply denote x_1 by x and we assume that the bathymetry (2.86) along with the slide shape (2.89) do not depend on the coordinate x_2. It means that the bathymetry profile can be simply described by the following equation:

$$y = h_0(x), \tag{2.126}$$

while the slide shape is given by

$$y = \zeta(x, t) = \zeta^0\bigl(x + x_c^0 - x_c(t)\bigr). \tag{2.127}$$

By analogy with x_1, we denote u_1 by u, $x_{1,c}^0$ by x_c^0 and $x_{1,c}(t)$ simply by $x_c(t)$. Under these assumptions, the landslide trajectory will belong to the plane parallel to $x\,O\,y$, i.e. $u_2 \equiv 0$. It is not difficult to see that in our situation we have the following identities:

$$\frac{\partial h_0}{\partial x_2} \equiv 0,$$

$$g_{12} = g_{21} \equiv 0, \quad g_{22} \equiv 1, \quad G = g_{11},$$

$$v_1 = g_{11}u, \quad v_c = \sqrt{g_{11}}\,|u|.$$

The landslide biggest section area $\mathfrak{A} = H \cdot W_2$, where H is the maximal sliding body height in the vertical direction.

The landslide motion is described by the first (in the sense of $\alpha = 1$) ordinary differential Eq. (2.118), which takes the following simplified form in our 1D case:

[18]Especially in parameters which control the friction and other energy loss processes.

$$\frac{\mathrm{d}g_{11}u}{\mathrm{d}t} = \frac{R}{2} + \Big[(\gamma - 1)\,g\,(\mathcal{I}_1 - \sigma\,C_{\mathrm{fr}}\,\mathcal{I}_2) - \sigma\Big(\gamma\,C_{\mathrm{fr}}\,\mathcal{I}_3\,u^2 + \frac{C_d}{2}\,H\,W_2\,v_c^2\Big)\Big]\frac{\sqrt{g_{11}}}{(\gamma + C_w)\cdot V}, \tag{2.128}$$

where $\sigma \overset{\mathrm{def}}{:=} \operatorname{sign} u$ and

$$R = \frac{\mathrm{d}g_{11}}{\mathrm{d}x}\,u^2, \qquad g_{11} = 1 + \Big(\frac{\mathrm{d}h_0}{\mathrm{d}x}\Big)^2.$$

To shorten the notation, the integrals $\mathcal{I}_{1,1}$ and $\mathcal{I}_{3,11}$ were denoted simply by $\mathcal{I}_1$ and $\mathcal{I}_3$, respectively.

Independence of the function $x \longmapsto \zeta(x, t)$ on the coordinate x_2 allows to compute easily its supports for $t = 0$ and for $t > 0$:

$$\mathcal{D}_0 = \big[x_l(0), x_r(0)\big] \times \big[x_{2,l}, x_{2,r}\big],$$
$$\mathcal{D}_t = \big[x_l(t), x_r(t)\big] \times \big[x_{2,l}, x_{2,r}\big].$$

Moreover, $\forall t \geqslant 0$ we have the following inclusions:

$$x_c(t) \in \big(x_l(t), x_r(t)\big), \qquad x_{2,c}(t) \equiv x_{2,c}^0 \in \big(x_{2,l}, x_{2,r}\big)$$

where $x_{2,r} \equiv x_{2,l} + W_2$. The following equalities do also hold $\forall t \geqslant 0$:

$$x_l(t) = x_l(0) + x_c(t) - x_c^0, \qquad x_r(t) = x_l(t) + W_1.$$

Henceforth, in the case of the 1D landslide, its volume (2.91) is determined as $V = W_2 \cdot S_0$, where S_0 is independent of x_2 and it is defined as the surface of the landslide section by a plane orthogonal to the axis $O\,x_2$, which can be computed as follows:

$$S_0 = \int_{x_l(t)}^{x_r(t)} \zeta(x, t)\,\mathrm{d}x = \int_{x_l(t)}^{x_r(t)} \zeta_0\big(x + x_c^0 - x_c(t)\big)\,\mathrm{d}x = \int_{x_l(0)}^{x_r(0)} \zeta_0(x)\,\mathrm{d}x = \text{const.}$$

The integrals appearing in Eq. (2.128) can be computed as follows:

$$\mathcal{I}_1(t) = W_2 \int_{x_l(t)}^{x_r(t)} \zeta(x, t)\,\sin\theta(x)\,\mathrm{d}x,$$

$$\mathfrak{I}_2(t) = W_2 \int_{x_l(t)}^{x_r(t)} \zeta(x, t) \cos\theta(x)\,\mathrm{d}x,$$

$$\mathfrak{I}_3(t) = W_2 \int_{x_l(t)}^{x_r(t)} \zeta(x, t) \frac{\frac{\mathrm{d}^2 h_0}{\mathrm{d}x^2}}{\sqrt{g_{11}}}\,\mathrm{d}x,$$

where $\theta(x) \stackrel{\text{def}}{:=} -\arctan\left(\frac{\mathrm{d}h_0}{\mathrm{d}x}\right)$ is the local slope angle of the bottom defined in Eq. (2.126). By using elementary formulas of the trigonometry, we obtain three relations:

$$\tan\theta(x) = -\frac{\mathrm{d}h_0}{\mathrm{d}x},$$

$$\sin\theta(x) = -\frac{\frac{\mathrm{d}h_0}{\mathrm{d}x}}{\sqrt{g_{11}}},$$

$$\cos\theta(x) = \frac{1}{\sqrt{g_{11}}}.$$

All these formulas show that the law of landslide motion does not depend on W_2 since the parameter simplifies in all terms, thanks to the division by $V = W_2 \cdot S_0$.

Nonlinear (integro-)differential Eq. (2.128) has to be completed by an extra equation $\dot{x}_c = u$ to determine the abscissa $x_c(t)$ of the moving point $\mathbf{X}_c(t)$. The appropriate initial conditions are also needed:

$$x_c(0) = x_c^0, \qquad u(0) = 0. \tag{2.129}$$

Having this solution allows us, thanks to Formula (2.127), to reconstruct the complete moving bottom shape. The knowledge of this underlying surface is needed to solve the hydrodynamic problem in the fluid layer to determine the generated wave field [103, 106].

Constant Slope Case: Analytics

It is not difficult to see that in the case of the constant slope bathymetry given by the formula

$$h = h_0(x) = h_b - x\tan\theta, \qquad x \geqslant 0, \qquad \theta = \text{const},$$

our governing Eq. (2.128) reduces to the governing equation from [149]. Moreover, if one neglects the friction force, we may obtain the reduction to the governing equation employed earlier in [324]. With a little bit of calculus, it is not difficult to obtain the exact analytical solution to the initial value problem (2.128), (2.129):

$$u(t) = \cos\theta \sqrt{\frac{2\,S_0\,Q}{C_d\,H}}\,\tanh\left(\frac{t}{\gamma + C_w}\sqrt{\frac{C_d\,H\,Q}{2\,S_0}}\right), \tag{2.130}$$

$$x_c(t) = x_c^0 + \cos\theta\,\frac{2\,S_0\,(\gamma + C_w)}{C_d\,H}\,\ln\left[\cosh\left(\frac{t}{\gamma + C_w}\sqrt{\frac{C_d\,H\,Q}{2\,S_0}}\right)\right], \tag{2.131}$$

where $Q \stackrel{\text{def}}{:=} g\,(\gamma - 1)\,(\sin\theta - C_{\text{fr}}\cos\theta)$ and $h_b < 0$ is the still water depth at the point $x = 0$. So that the motion can take place, we have to require that the bottom slope angle be greater than the friction angle, i.e. $\theta > \theta_*$.

From Eq. (2.130) we may conclude that for small times t we have an approximate asymptotic solution:

$$u(t) \approx \frac{Q\cos\theta}{\gamma + C_w}\,t\,.$$

In other words, for small times the landslide motion is close to be rectilinear with uniform acceleration. Also, from the same Eq. (2.130), we can obtain the large time asymptotics:

$$u(t) \approx \cos\theta\sqrt{\frac{2\,S_0\,Q}{C_d\,H}}\,.$$

The last formula indicates that for sufficiently large times, the landslide will move down the slope with an approximatively constant velocity given by the expression above.

Curvilinear Slope: Numerics

For a general curvilinear bottom profile, finding a solution to CAUCHY problem (2.128), (2.129) requires an extensive application of numerical methods. In Fig. 2.20 we show a few landslide barycentre trajectories in our model 1D reservoir. The initial shape of the landslide is given by the following formula:

$$\zeta_0(x) = H \cdot \mathcal{C}\left(x;\,0,\,1,\,x_c^0,\,W_1\right). \tag{2.132}$$

The landslide is moving on a curvilinear bottom whose profile is given by a paraboloid (2.124). On lateral points $x = 0$ and $x = 500$ m, the reservoir is bounded by vertical impermeable walls. All physical landslide and bathymetry parameters are provided in Table 2.5. At the initial instant of time $t = 0$, the landslide depth was set to $y_c^0 = -30$ m. Consequently, we may determine its abscissa also:

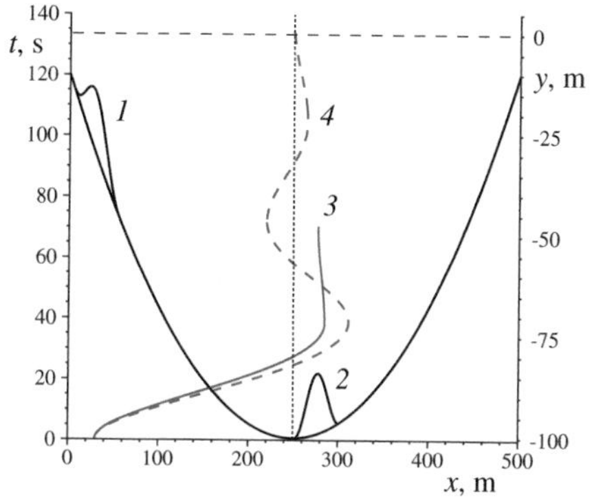

(a)

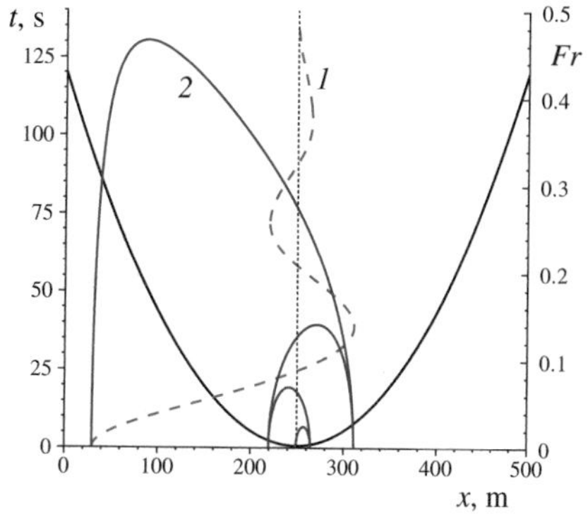

(b)

Fig. 2.20 Landslide motion along the curvilinear bottom of a model 1D reservoir. Left panel (**a**): slide shape at the initial moment of time $t = 0$ (1) and at the final stopping time for the friction angle $\theta_* = 5°$ (2). A few graphs of the function $t \longmapsto x_c(t)$ for the friction angle $\theta_* = 5°$ (3) and $\theta_* = 1°$ (4). Right panel (**b**): Graph of the function $t \longmapsto x_c(t)$ for $\theta_* = 1°$ (1) and the (multi-valued) dependence of the FROUDE number Fr on the coordinate $x = x_c(t)$ (2)

Table 2.5 Physical and geometrical parameters used for the numerical simulation of a 1D landslide in Sect. 2.3.5

Parameter	Value
Water depth, h_ξ	-100 m
Water depth, h_b	-10 m
Trench location, ξ	250 m
Initial slide depth, y_c^0	-30 m
Initial slide location, x_c^0	29.5 m
Computational domain, $[0, L_1]$	$[0, 500]$ m
Slide height, H	10 m
Slide width, W_1	50 m
Densities ratio, γ	2
Friction angle, θ_*	5°
Water resistance coefficient, C_w	1
Drag coefficient, C_d	1

$$x_c^0 = \xi \cdot \left(1 - \sqrt{\frac{y_c^0 - h_\xi}{h_b - h_\xi}}\right) \approx 29.5 \text{ m}.$$

The parameters employed in Eq. (2.128) are also provided in Table 2.5. We have to mention also that for the landslide (2.127), (2.132), it can be shown that

$$S_0 = \frac{H B_1}{2}$$

and the quantity H enters as a linear multiplier in all integrals $\mathfrak{I}_{1,2,3}$. Thus, this parameter will be cancelled in Eq. (2.128). Thus, the law of landslide motion will not depend on this parameter H.

A visual inspection of the function $t \longmapsto x_c(t)$ (see e.g. the graph (3) in Fig. 2.20a) shows that with the values of parameters employed in our simulations, the landslide crosses the deepest point $x = \xi$ and climbs up the opposite slope to certain height. Then, it stops at its highest point (on the opposite slope), slides back and stops somewhere around the deepest point $x = \xi$ (see curve (3) in Fig. 2.20a). When we decrease the friction angle θ_*, more complex trajectories may emerge. Namely, multiple changes in the motion direction may happen. See, for example, the curve (4) in Fig. 2.20a which was obtained with $\theta_* = 1°$. Nevertheless, even for very small values of the friction angle θ_*, the motion along the bathymetry (2.124) happens in the sub-critical regime. We can see it on curve (2) in Fig. 2.20b, which shows the dependence of the FROUDE number $\mathrm{Fr} = \frac{v_c(x_c)}{\sqrt{g h_0(x_c)}}$ on the abscissa $x_c(t)$ of the barycentre $\mathbf{X}_c(t)$. We remind also that at every stopping point, we have to check the criterium (2.120) of the continuation of motion, which takes the following form in the 1D case:

$$|\mathfrak{I}_1| > C_{\mathrm{fr}} \cdot \mathfrak{I}_2. \tag{2.133}$$

The last condition (2.133) has to be also checked at the initial instance of time $t = 0$. To give an example, we can mention that the unstable mass with initial shape (2.132) posed on the bathymetry (2.124) together with parameters given in Table 2.5 will not initiate the motion provided that $\theta_* \geqslant 32.5°$.

Tsunami Generation by a Landslide in the Black Sea

The tsunami wave hazard in the BLACK SEA is related to the landslide generation mechanism. We have very limited data regarding the nature and properties of the sedimentary deposits at the BLACK SEA deposits. Meanwhile, the sediments are well-known for triggering self-sustained underwater avalanches. This process has been recently investigated in [216, 217]. Moreover, it has been shown that sediments are even able to amplify the tsunami generation process by earthquakes due to their specific elastic properties [107]. Very often, sediment deposits contain also the very precious information regarding the paleo-tsunami events [24, 55, 56].

In this section we shall apply extensively the methods of numerical simulation to study the wave field generated by hypothetical landslides in the BLACK SEA. We shall present the numerical results obtained with the celebrated SGN model along with the classical dispersionless NSWE. The comparison of these two predictions will allow us to draw some conclusions regarding the importance of dispersive effects in tsunami generation by an underwater landslide but also during its propagation to the shore [26, 90, 109, 118]. These results are of particular interest because they are obtained in a real sea bathymetry configuration. In the computations presented in this section, we did not take into account the EARTH sphericity and rotation effect. The landslide trajectory was computed according to Eq. (2.118). The numerical algorithm employed to solve SGN and NSWE is a straightforward extension to two spatial dimensions of the method presented in this chapter (in the linear case this method was described in [193]).

The main idea behind initiating this study [186] was to investigate the properties of the generated waves in the near- and far-field depending on the physical parameters and initial locations of hypothetical landslides in the BLACK SEA. Three landslide scenarii (denoted below by $\mathfrak{L}_{1,2,3}$) were considered. We assumed that all landslides have the same initial height H_0. The initial location of three landslides $\mathfrak{L}_{1,2,3}$ is shown in Fig. 2.21.

Three landslides $\mathfrak{L}_{1,2,3}$ were initially located near the continental slope in the region of the giant paleo-landslide [180] whose approximate volume was estimated to be $V_0 \approx 40$ km^3, height to be $H_0 \approx 200$ m and the base surface area — $S_0 \approx 200$ km^2. The landslides $\mathfrak{L}_{2,3}$ had approximatively the same initial barycentre location (37.15 °E, 44.4 °N) and the depth $y_c^0 = -1\,769$ m as the giant

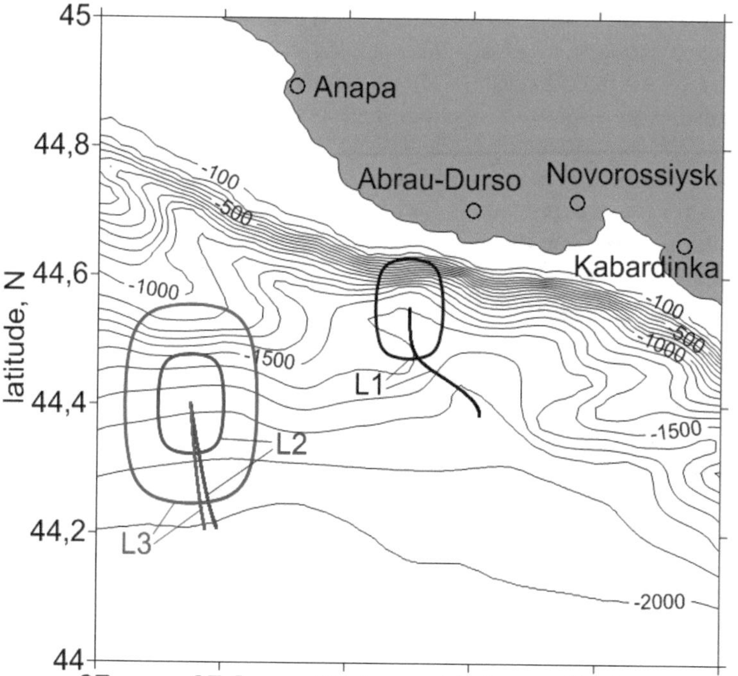

Fig. 2.21 A fragment of the computational domain with superposed isobath lines, landslide $\mathfrak{L}_{1,2,3}$ shape contours and the trajectories of their motion

paleo-landslide [180]. The difference consisted[19] in the slide surface area S_0 and volume V_0. According to Equation (2.125), for any time $t \geqslant 0$ we have $V = \frac{HS}{4}$. Our hypothetical landslide $\mathfrak{L}_2$ had the initial surface area S_0 equal to the one of the giant paleo-landslide. Thus, the volume of $\mathfrak{L}_2$ was only one quarter of paleo-landslide. For the landslide $\mathfrak{L}_3$, we chose the initial surface area $S = 4\,S_0$. Thus, the equality of volumes was respected (but, obviously, not of the areas). Finally, the landslide $\mathfrak{L}_1$ had the same linear dimensions as $\mathfrak{L}_2$, but its barycentre was initially located at (37.5 °E, 44.55 °N) farther upslope with initial depth $y_c^0 = -1\,598$ m.

In Fig. 2.21 we also show the trajectories of these three hypothetical landslides computed with the following parameters:

$$C_w = 1, \quad C_d = 1, \quad \gamma = 2.$$

The friction angle θ_* was specifically chosen to match the trajectory length of the paleo-landslide, which was approximatively equal to 22 km [180]. Since the landslides $\mathfrak{L}_{2,3}$ were initially located in the constant sloping bathymetry region of the continental slope, their respective trajectories were very close to simple rectilinear lines. We can mention also that the weak velocities and the smooth ending of motion generate surface waves of only small and moderate amplitudes.

Let us describe, first of all, the common characteristics of the wave fields generated by all three landslides $\mathfrak{L}_{1,2,3}$ before addressing the particularities. In the beginning of a landslide motion, it generates above, on the free surface, a crescent wave of positive amplitude with a slight wave of depression behind. This coherent structure moves in the direction of landslide trajectory. Later, above the landslide a wave of depression is formed. This wave then gradually disappears as the landslide moves into the deeper parts of the sea. More precisely, this wave of negative amplitude travels in the direction opposite to landslide motion. Eventually, the wave reaches the opposite shore and a portion of its energy is reflected as a wave of positive amplitude.

More exciting waves are generated in course of the landslide $\mathfrak{L}_1$ motion. Initially, this landslide was located on a more inclined part of the continental slope. As a result, this landslide departure was much more accelerated, thanks to the slope, but also to the estimated[20] friction angle $\theta_* = 0.8°$. The slide $\mathfrak{L}_1$ could even achieve the maximal speed of 18 $\frac{\text{m}}{\text{s}}$ during its trajectory. Once in the canyon, the landslide $\mathfrak{L}_1$ changed its direction of sliding twice before abruptly stopping the motion. The abrupt termination of motion generated an additional significant wave of positive polarity. The last wave largely contributed to the measured wave height along the shoreline (see Fig. 2.22b). The curvilinear character of the landslide $\mathfrak{L}_1$

[19] We would like to mention also that computations with dispersive models often require the knowledge of higher derivatives of the bathymetry function. Thus, we have to impose an additional smoothness requirement on the sliding mass upper surface.

[20] We would like to remind that the friction angle was chosen so that the hypothetical landslide trajectory length matches the historical paleo-landslide event [180].

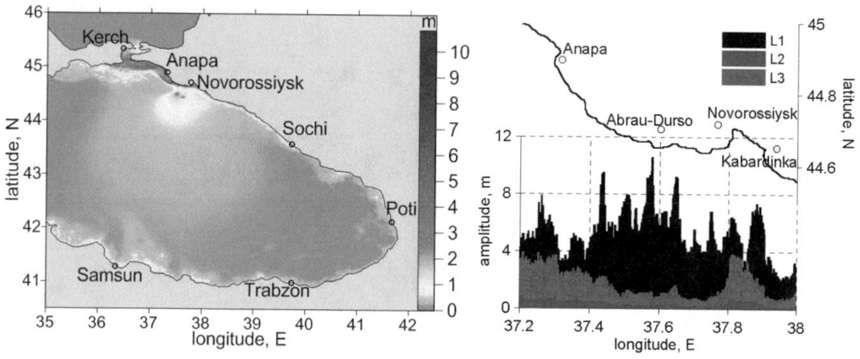

(a) (b)

Fig. 2.22 (**a**) The distribution of maximal tsunami wave amplitudes (left panel) generated by hypothetical landslide $\mathfrak{L}_1$ whose propagation was computed during 3 h of the physical time. On the right panel (**b**) we represent the distribution of maximal wave heights along the shore according to the SGN model prediction

trajectory makes the analysis of the wave generation process more complicated because of the multi-directional (and eventually nonlinear) interaction of generated waves on the free surface. The important initial acceleration of the landslide $\mathfrak{L}_1$ was responsible for high amplitude waves near the coasts, which greatly contributed to the intensification of the problem nonlinearity. We mention also that the maximal amplitudes were not observed in the generation region in the proximity of the landslide, but near the coasts where the generated crescent wave arrived. The distribution of maximal wave heights along the coastline turns out to be highly irregular (see Fig. 2.22b). We believe that this distribution greatly depends on the continental shelf extent, local bathymetric features and the shoreline geometry. In the case of our hypothetical landslide $\mathfrak{L}_1$, the most affected areas were located along the coastline closest to the generation region as it can be seen in Fig. 2.22a, where we represent the radiation diagram[21] in the Eastern part of the Black Sea. In particular, one can notice in Fig. 2.22a the absence of radial symmetry in the propagation of the wave energy. Indeed, most of the generated tsunami wave energy was projected by this landslide towards the North.

The results presented in this section were obtained on a uniform grid, covering the areas of Black and Azov Seas. More precisely, the following rectangular area served for us as the computational domain:

- 27 °E to 42.5 °E in longitude;
- 40.5 °N to 47.5 °N in latitude.

[21] We recall here that the radiation diagram shows the maximal amplitude distribution over the computational domain with the maximum taken over all simulation times.

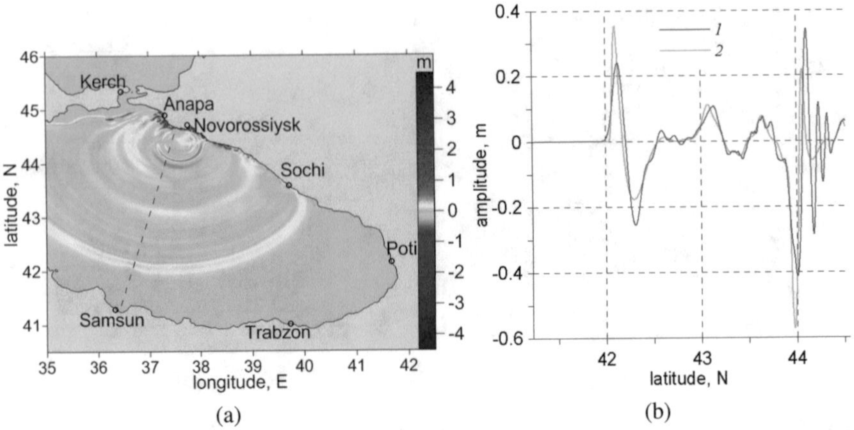

Fig. 2.23 Tsunami wave generated by the landslide $\mathfrak{L}_1$. Left panel (**a**): a snapshot of the free surface elevation computed according to the SGN model. With a dashed line we show the cut along which we extract the one-dimensional profiles. Right panel (**b**): one-dimensional profile of the free surface elevation predicted according to the SGN model (1) and NSWE (2). The snapshots were taken at time $t = 2000$ s from the beginning of the landslide $\mathfrak{L}_1$ motion

The preliminary computations have been performed on a coarse grid based on the digital freely available bathymetry data "The GEBCO One Minute Grid-2008". Then, these simulations were repeated on two successively refined grids: on each level of the grid refinement, the discretization step was halved. The comparison of obtained results on different grids at several spatial locations of the computational domain allowed us to constate the convergence of the numerical solution. However, the detailed reproduction of the wave interaction with the coast[22] requires the computations on even more refined grids since finer grid unveils new bathymetric features which might have effect on the computed wave field (at least locally and mostly short (dispersive) waves get affected by successive refinements). We have to admit that the methods of nested or locally adapted grids seem to be very pertinent in this context [139, 158].

A general idea about the landslide generated wave field dynamics can be obtained by taking several snapshots of the free surface elevation at well-chosen time instances. One of such snapshots is shown in Fig. 2.23a. The colour is used to encode the free surface deflection from the still water level. As it can be clearly seen from this Fig. 2.23a, 33 min after the beginning of the event, the wave field is essentially composed of three elements: the main wave near the coastline and two waves already reflected from the (North) coast and propagating in the Southward direction. When the wave propagates further from the generation area, its amplitude is generally decreasing (mainly due to the geometrical spreading effect). However,

[22]We remind that in this section the coastline was replaced for simplicity by a vertical wall located at the water depth $h_b = 20$ m.

when the wave enters into the continental shelf area, its amplitude starts to grow again to reach sometimes the value of 2 m near the shoreline (see Fig. 2.22a). We would like to mention also the wave front deformation effect which occurs during the propagation due to the differences in the wave travelling celerities. The latter is caused by the heterogeneities in the Sea bathymetry distribution.

In order to estimate the frequency dispersion effect on the predicted wave field, we considered the one-dimensional cuts of the free surface along the main propagation axes and taken at different moments of time. The comparison of the (non-dispersive) NSWE prediction with weakly dispersive SGN model showed that in this situation the dispersion had an overall *weak effect* on the wave generation and on the wave interaction with the coasts which are located in the vicinity of the generation area. Indeed, the travel time towards the closest portions of the coast was too short for the dispersion to produce its effect. Moreover, the landslides $\mathfrak{L}_{2,3}$ generate such long waves that the dispersive effects are not expected to be observed in the BLACK SEA. For the landslide $\mathfrak{L}_1$ this effect was visible, without being dominant.

In Fig. 2.23a we show with a dashed line the location of the one-dimensional free surface cut, which passes through the initial location of the landslide $\mathfrak{L}_1$ barycentre. At the moment of time, when we take the snapshot depicted in Fig. 2.23a, the main wave already travelled most of the distance towards the South coast leaving behind a dispersive tail of small amplitude waves. The dispersive effects become clearly visible around $t = 1\,700$ s provoked by the abrupt termination of the landslide $\mathfrak{L}_1$ trajectory. The steep wave, which forms by the termination, breaks very quickly and develops into a wave train of short waves clearly visible in rightmost part of Fig. 2.23b. It is clear also that the non-dispersive NSWE model cannot reproduce this effect. We should stress out also the rôle played by the elongated shallow continental shelf, which is present in some areas of the Southern parts of the BLACK SEA. This shelf filters out some of short waves generated in the deeper parts of the BLACK SEA. As a result of this filtering process, some of the (synthetic) wave gauges located along the Southern coasts of the BLACK SEA showed a good agreement of numerical results between NSWE and SGN models. More differences can be observed on the wave gauges located in more distant[23] locations of the BLACK SEA. However, even those apparent divergences do not enable us to speak about the dominant effect of the dispersion during the considered hypothetical tsunami event. We underline the fact that the landslide $\mathfrak{L}_1$ was the most dispersive among three landslide scenarii $\mathfrak{L}_{1,2,3}$ considered in this section.

Above we discussed the effect along the coasts. More visible dispersive effects can be seen in deeper parts of the BLACK SEA. In particular, the main wave amplitude predicted with the SGN model is significantly smaller than the amplitude obtained with the NSWE model. Moreover, we can see that its crest celerity is slightly smaller also. We would like to add also that during the wave propagation

[23]The distance is measured from the generation region.

over large distances, sometimes the maximum of the amplitude may even move into the dispersive tail of the group [142, 223, 224].

2.4 Discussion

Above we presented a detailed description of the numerical algorithm and a number of numerical tests which illustrate its performance. The main conclusions and perspectives of this chapter are outlined below.

2.4.1 Conclusions

In this chapter we focused on the development of numerical algorithms for shallow water propagation over globally flat spaces (i.e. we allow some variations of the bathymetry in the limits discussed in Sect. 2.1.2). The main distinction of our work is that the proposed algorithm allows for local mesh adaptivity by moving the grid points where they are needed. The performance of our method was illustrated on several test cases ranging from purely academic ones (e.g. propagation of a solitary waves, which allowed us to estimate the overall accuracy of the scheme) to more realistic applications with landslide-generated waves [26]. The mathematical model chosen in this study allows us to have a look into the distribution of various physical fields in the fluid bulk. In particular, in some interesting cases we reconstructed the velocity field and the non-hydrostatic pressure distribution beneath the free surface.

We studied the linear stability of the proposed finite volume discretization. It was shown that the resulting scheme possesses an interesting and possible counter-intuitive property: the smaller we take the spatial discretization step Δx, the less restrictive is becoming the stability CFL-type condition on the time step τ. This result was obtained using the classical VON NEUMANN analysis [57]. However, we show (and we compute it) that there exists the upper limit of allowed time steps. Numerical schemes with such properties seem to be new.

We considered also in great detail the question of wall boundary conditions[24] for the SGN system. It seems that this issue was not properly addressed before. The wall boundary condition for the elliptic part of equations follows naturally from the form of the momentum equation we chose in this study.

Finally, in numerical experiments we showed how depth-integrated SGN equations can be used to study nonlinear transformations of water waves over some bathymetric features (such as an underwater step or a sliding mass). Moreover, we

[24]The wall boundary condition for the velocity component $u(x, t)$ is straightforward, i.e. $u(x, t)|_{x=0}^{x=\ell} = 0$. However, there was an open question of how to prescribe the boundary conditions for the elliptic part of the equations.

illustrated clearly that SGN equations (and several other approximate dispersive wave models) can be successfully used to reconstruct the flow field under the wave. The accuracy of this reconstruction will be studied in future works by direct comparisons with the full EULER equations where these quantities are resolved. Moreover, we described in great detail the landslide model including the derivation of its physical laws of motion. This model was applied to study hypothetical landslide events in the BLACK SEA. This study was inspired by the giant paleo-landslide which took place in this sea [180].

2.4.2 Perspectives

The main focus of our study was set on the adaptive spatial discretization. The first natural continuation of our study is the generalization to 3D physical problems (i.e. involving two horizontal dimensions). The main difficulty is to generalize the mesh motion algorithm to this case, even if some ideas have been proposed in the literature [14].

In the present computations the time step was chosen to ensure the linear CFL condition. In other words, it was chosen in order to satisfy the numerical solution stability. In future works we would like to incorporate an adaptive time stepping procedure along the lines of e.g. [294] aimed to meet the prescribed error tolerance. Of course, the extension of the numerical method presented in this study to three-dimensional flows (i.e. two horizontal dimensions) represents the main important extension of our work. Further improvement of the numerical algorithm can be expected if we include also some bathymetric features (such as ∇h) into the monitor function $\varpi\,[\eta,\, h]\,(x,\, t)$. Physically this improvement is fully justified since water waves undergo constant transformations over bottom irregularities (as illustrated in Sects. 2.3.3 and 2.3.4). A priori, everything is ready to perform these further numerical experiments.

Ideally, we would like to generalize the algorithm presented in this study for the SERRE–GREEN–NAGHDI (SGN) equations to the base model in its most general form (2.3), (2.4). In this way we would be able to incorporate several fully nonlinear shallow water models (discussed in the Chap. 3) in the same numerical framework. It would allow the great flexibility in applications to choose and to assess the performance of various approximate models. With minimal modifications, the numerical strategy described in this study can be also applied to the numerical simulation of long waves in plasma using the models described in [85, 86, 140]. The main point is that shallow magneto-hydrodynamic models have the same mathematical structure as our base model (2.3), (2.4).

Moreover, in the present chapter we raised the question of boundary conditions for SGN equations. However, non-reflecting (or transparent) boundary conditions would allow to take much smaller domains in many applications. Unfortunately, this question is totally open to our knowledge for the SGN equations (however, it is well understood for NSWE). In future works we plan to fill in this gap as well.

Finally, the SGN equations possess a number of variational structures. The HAMILTONIAN formulation can be found e.g. in [177]. Various LAGRANGIANS can be found in [72, 127, 196, 239]. Recently, a multi-symplectic formulation for SGN equations has been proposed [64]. All these available variational structures raise an important question: after the discretization can we preserve them at the discrete level as well? It opens beautiful perspectives for the development of structure-preserving numerical methods as it was done for the classical KORTEWEG–DE VRIES [115] and nonlinear SCHRÖDINGER (NLS) [61] equations.

In the following chapters we shall discuss the derivation of the SGN equations on a sphere and their numerical simulation using the finite volume method.

Chapter 3
Model Derivation on a Globally Spherical Geometry

In this chapter we investigate the derivation of some long wave models on a deformed sphere. We propose first a suitable for our purposes formulation of the full EULER equations on a sphere. Then, by applying the depth-averaging procedure we derive first a new fully nonlinear weakly dispersive base model. After this step we show how to obtain some weakly nonlinear models on the sphere in the so-called BOUSSINESQ regime. We have to say that the proposed base model contains an additional velocity variable which has to be specified by a closure relation. Physically, it represents a dispersive correction to the velocity vector. So, the main outcome of this chapter should be rather considered as a whole family of long wave models.

Recent mega-tsunami events in 2004 [5, 211, 299] and in TOHOKU, JAPAN 2011 [152, 248] required the simulation of tsunami wave propagation on the global trans-oceanic scale. Moreover, similar catastrophic events in the future are to be expected in these regions [237]. The potential tsunami hazard caused by various seismic scenarii can be estimated by extensive numerical simulations. During recent years the modelling challenges of tsunami waves have been extensively discussed [79, 250]. On such scales the effects of EARTH's rotation and geometry might become important. Several authors arrived to this conclusion, see e.g. [80, 151]. There is an intermediate stage where the model is written on a tangent plane to the sphere in a well-chosen point. In the present study we consider the globally spherical geometry without such local simplifications.

The direct application of full hydrodynamic models such as EULER or NAVIER–STOKES equations does not seem realistic nowadays. Consequently, approximate mathematical models for free surface hydrodynamics on rotating spherical geometries have to be proposed. This is the main goal of the present chapter. The existing (dispersive and non-dispersive) shallow water wave models on a sphere will be reviewed below. Nowadays, hydrostatic models are mostly used on a sphere [311, 340]. The importance of frequency dispersion effects was underlined in e.g. [303]. Their importance has been realized for tsunami waves generated by

G. Khakimzyanov et al., *Dispersive Shallow Water Waves*, Lecture Notes in Geosystems Mathematics and Computing, https://doi.org/10.1007/978-3-030-46267-3_3

sliding/falling masses [26, 109, 118, 321]. However, we believe that on global trans-oceanic scales frequency dispersion effects might have enough time to accumulate and, hence, to play a certain rôle. Finally, the topic of numerical simulation of these equations on a sphere is another important practical issue. It will be addressed in some detail in the following chapter.

Shallow water equations describing long wave dynamics on a (rotating) sphere have been routinely used in the fields of Meteorology and Climatology [340]. Indeed, there exist many similarities in the construction of approximate models of atmosphere and ocean dynamics [234]. The derivation of these equations by depth-averaging can be found in the classical monograph [161]. The main numerical difficulties here consist mainly in (structured) mesh generation on a sphere and treating the degeneration of governing equations at poles (the so-called poles problem). So far, the finite differences [210, 219] and spectral methods [44] were the most successful in the numerical solution of these equations. Our approach to these problems will be described in [190].

It is difficult to say who was the first to apply Nonlinear Shallow Water Equations (NSWE) on a sphere to the problems of Hydrodynamics. Contrary to the Meteorology, where the scales are planetary from the outset and the spherical coordinates are introduced even on local scales [225], in surface wave dynamics people historically tended to use local CARTESIAN coordinates. However, the need to simulate trans-oceanic tsunami wave propagation obliges us to consider spherical and EARTH's rotation effects. We would like to mention that in numerical modelling of water waves on the planetary scale the problem of poles does not arise since these regions are covered with the ice. Thus, the flow cannot take place there.

In [332] one can find various forms of shallow water equations on a sphere along with standard test cases to validate numerical algorithms. The standard form of NSWE in the spherical coordinates $O\,\lambda\,\phi\,r$, where λ is the longitude, ϕ is the latitude, r is the radial distance, is

$$\begin{aligned}
\mathcal{H}_t \;+\; \nabla\cdot[\,\mathcal{H}\mathbf{u}\,] \;&=\; 0\,,\\
(\mathcal{H}u)_t \;+\; \nabla\cdot[\,\mathcal{H}u\,\mathbf{u}\,] \;&=\; \Bigl(\mathcal{F} \;+\; \frac{u}{R}\,\tan\phi\Bigr)\,\mathcal{H}v \;-\; \frac{g\,\mathcal{H}}{R\cos\phi}\,\frac{\partial\eta}{\partial\lambda}\,,\\
(\mathcal{H}v)_t \;+\; \nabla\cdot[\,\mathcal{H}v\,\mathbf{u}\,] \;&=\; -\Bigl(\mathcal{F} \;+\; \frac{u}{R}\,\tan\phi\Bigr)\,\mathcal{H}u \;-\; \frac{g\,\mathcal{H}}{R}\,\frac{\partial\eta}{\partial\phi}\,.
\end{aligned}$$

Here $\mathcal{H}(\lambda,\,\phi,\,t) \stackrel{\text{def}}{:=} \bigl(\eta \,+\, h\bigr)(\lambda,\,\phi,\,t)$ is the total water depth and $\mathbf{u}(\lambda,\,\phi,\,t)$ is the linear velocity vector with components

$$\mathbf{u}(\lambda,\,\phi,\,t) \stackrel{\text{def}}{:=} \Bigl(R\,\cos\phi\cdot\dot{\lambda},\;R\,\dot{\phi}\Bigr)\,,$$

where the over dot denotes the usual derivative with respect to time, i.e. $\dot{(\cdot)} \stackrel{\text{def}}{:=} \frac{\mathrm{d}(\cdot)}{\mathrm{d}t}$. Function h specifies the bottom bathymetry shape and $\mathcal{F} \stackrel{\text{def}}{:=} 2\,\Omega\,\sin\phi$ is CORIO-

LIS's parameter, Ω being the EARTH constant angular velocity. The constant g is the absolute value of usual gravity acceleration. The divergence operator in spherical coordinates is computed as

$$\nabla \cdot \bigl((\cdot)_1, (\cdot)_2\bigr) \stackrel{\text{def}}{:=} \frac{1}{R \cos\phi} \left[\frac{\partial(\cdot)_1}{\partial\lambda} + \frac{\partial\bigl(\cos\phi\,(\cdot)_2\bigr)}{\partial\phi} \right].$$

The right-hand sides of the last two NSWE contain the CORIOLIS effect, additional terms due to rotating coordinate system and hydrostatic pressure gradient. Recently a new set of NSWE on a sphere was derived [62, 63] including also the centrifugal force due to the EARTH rotation. The applicability range of this model was discussed [62] and some stationary solutions are provided [63]. NSWE on a sphere are reported in [250] in a non-conservative form and including the bottom friction effects. The derivation of these equations can be found in [199, 202]. In earlier attempts such as [249] NSWE did not include terms

$$\Bigl(\frac{u}{R}\tan\phi\Bigr)\,\mathcal{H}\,v \qquad \text{and} \qquad \Bigl(\frac{u}{R}\tan\phi\Bigr)\,\mathcal{H}\,u\,.$$

We remark however that the contribution of these terms might be negligible for tsunami propagation problems. This system of NSWE is implemented, for example, in the code MOST[1] [309]. The need to include dispersive effects was mentioned in several works. In [80] linear dispersive terms were added to NSWE and this model was integrated in TUNAMI-N2 code. This numerical model allowed the authors to model the celebrated SUMATRA 2004 event [299]. In another work published the same year Grilli et al. [151] outlined the importance to work in spherical coordinates even if in [151] they used the CARTESIAN version of the code FUNWAVE. This goal was achieved 6 years later and published in [197]. Kirby et al. used scaling arguments to introduce two small parameters (ς and μ in notation of our study). A weakly nonlinear and weakly dispersive BOUSSINESQ-type model was given and effectively used in [223, 224]. However, the authors did not publish the derivation of these equations. Moreover, they included a free parameter which can be used to improve the dispersion relation properties, even if this modification may appear to be rather ad-hoc without a proper derivation to justify it.

The systematic derivation of fully nonlinear models on a sphere was initiated in our previous works [129–131, 290]. In this work we would like to combine and generalize the existing knowledge on the derivation of dispersive long wave models in the spherical geometry including rotation effects. We cover the fully and weakly nonlinear cases. The relation of our developments to existing models is outlined whenever it is possible. The derivation in the present generality has not been reported in the literature before.

[1] Method of Splitting Tsunami (MOST).

The present chapter is organized as follows. In Sect. 3.2 we present the full EULER equations on an arbitrary moving coordinate system. The modified scaled EULER equations are given in Sect. 3.2.3. The base nonlinear dispersive wave model is then derived in Sect. 3.3 from the modified EULER equations. The base model has to be provided with a closure relation. Two particular and popular choices are given in Sects. 3.4 and 3.6. Finally, the main conclusions and perspectives of the present study are outlined in Sect. 3.7. As a reminder, in the following Sect. 3.1 we explain the notations and provide all necessary information from tensor analysis used in our study.

3.1 Basic Tensor Analysis

In this section (as well as in our study above) we adopt EINSTEIN's summation convention, i.e. the summation is performed over repeating lower and upper indices. Moreover, indices denoted with Latin letters i, j, k, etc. vary from 0 to 3, while indices written with Greek letters α, β, γ, etc. vary from 1 to 3. Some indices will be supplied with a prime, e.g. α, α'. In this case α and α' should be considered as independent indices. Along with the diffeomorphism

$$x^0 = q^0 = t, \qquad x^\alpha = x^\alpha\left(q^0, q^1, q^2, q^3\right), \qquad \alpha = 1, 2, 3. \tag{3.1}$$

We shall consider also the inverse transformation of coordinates:

$$q^0 = x^0 = t, \qquad q^\alpha = q^\alpha\left(x^0, x^1, x^2, x^3\right), \qquad \alpha = 1, 2, 3. \tag{3.2}$$

For the sake of convenience we introduce also short-hand notations for partial derivatives of the direct (3.1) and inverse transformations (3.2):

$$\mathscr{D}_i^{i'} \stackrel{\text{def}}{:=} \frac{\partial x^{i'}}{\partial q^i}, \qquad \mathscr{D}^i_{i'} \stackrel{\text{def}}{:=} \frac{\partial q^i}{\partial x^{i'}},$$

which are equivalent to the series of definitions

$$\mathscr{D}_0^{0'} = 1, \quad \mathscr{D}_\alpha^{0'} = 0, \quad \mathscr{D}_0^{\alpha'} = \frac{\partial x^{\alpha'}}{\partial t} \quad \mathscr{D}_\alpha^{\alpha'} = \frac{\partial x^{\alpha'}}{\partial q^\alpha}, \tag{3.3}$$

$$\mathscr{D}^0_{0'} = 1, \quad \mathscr{D}^0_{\alpha'} = 0, \quad \mathscr{D}^\alpha_{0'} = \frac{\partial q^\alpha}{\partial t} \quad \mathscr{D}^\alpha_{\alpha'} = \frac{\partial q^\alpha}{\partial x^{\alpha'}}. \tag{3.4}$$

We also have the following obvious identities:

$$\mathscr{D}^i_{i'} \cdot \mathscr{D}^{i'}_j = \delta^i_j, \qquad \mathscr{D}^{i'}_i \cdot \mathscr{D}^i_{j'} = \delta^{i'}_{j'}, \tag{3.5}$$

where δ_i^j is the KRONECKER symbol. We remind again that in the first formula there is an implicit summation over index i' and over i in the second one.

In the sequel we complete the set of CARTESIAN basis vectors $\{\mathbf{i}_\alpha\}_{\alpha=1}^3$ with an additional vector $\mathbf{i}_0 = (1, 0, 0, 0)$. In other words, we use the standard orthonormal basis $\{\mathbf{i}_i\}_{i=0}^3$ in the EUCLIDEAN space $\mathbb{R}^4 = \{(t, x^1, x^2, x^3)\}$. Sometimes, in order to introduce the summation, we shall equivalently employ basis vectors $\{\mathbf{i}^i\}_{i=0}^3$ with the upper index notation.

The transformation of coordinates (3.1) along with the inverse transformation (3.2) induces two new bases in $\mathbb{R}^4$:

$\{\mathbf{e}_i\}_{i=0}^3$: moving *covariant* basis
$\{\mathbf{e}^i\}_{i=0}^3$: moving *contravariant* basis

By definition, the components of covariant and contravariant basis vectors can be expressed as follows:

$$\mathbf{e}_i \equiv \mathscr{D}_i^{i'} \cdot \mathbf{i}_{i'}, \qquad \mathbf{e}^i \equiv \mathscr{D}^i_{i'} \cdot \mathbf{i}^{i'}, \qquad i = 0, \ldots, 3.$$

Right from this definition and relations (3.5) we can write down the connection among all the bases introduced so far:

$$\mathbf{i}_{i'} = \mathscr{D}^i_{i'} \mathbf{e}_i = \mathbf{i}^{i'} = \mathscr{D}_i^{i'} \mathbf{e}^i,$$

$$\mathbf{e}_i = \delta_{i'j'} \cdot \mathscr{D}_i^{i'} \mathscr{D}_j^{j'} \mathbf{e}^j, \qquad \mathbf{e}^i = \delta^{i'j'} \cdot \mathscr{D}^i_{i'} \mathscr{D}^j_{j'} \mathbf{e}_j, \qquad \mathbf{e}_i \cdot \mathbf{e}^j = \delta_i^j.$$

Using vectors of these new bases we can define covariant $\{g_{ij}\}_{i,j=0}^3$ and contravariant $\{g^{ij}\}_{i,j=0}^3$ components of the metric tensor as scalar products of corresponding basis vectors:

$$g_{ij} \stackrel{\text{def}}{:=} \mathbf{e}_i \cdot \mathbf{e}_j \equiv \mathscr{D}_i^{i'} \mathscr{D}_j^{j'} \cdot \delta_{i'j'} \equiv g_{ji},$$

$$g^{ij} \stackrel{\text{def}}{:=} \mathbf{e}^i \cdot \mathbf{e}^j \equiv \mathscr{D}^i_{i'} \mathscr{D}^j_{j'} \cdot \delta^{i'j'} \equiv g^{ji}.$$

Moreover, thanks to (3.5), one can easily check that

$$g_{ij} \cdot g^{jk} = \delta_i^k. \tag{3.6}$$

Using Formulas (3.3) and (3.4) we obtain that

$$g_{00} = 1 + \mathscr{D}_0^{\alpha'} \mathscr{D}_0^{\alpha'} \delta_{\alpha'\alpha'}, \qquad g_{0\alpha} = \mathscr{D}_0^{\alpha'} \mathscr{D}_\alpha^{\alpha'} \delta_{\alpha'\alpha'},$$

$$g_{\alpha\beta} = \mathscr{D}_\alpha^{\alpha'} \mathscr{D}_\beta^{\alpha'} \delta_{\alpha'\alpha'}, \tag{3.7}$$

$$g^{00} = 1, \quad g^{0\alpha} = \mathscr{D}^{\alpha}_{0'}, \quad g^{\alpha\beta} = \mathscr{D}^{\alpha}_{0'}\mathscr{D}^{\beta}_{0'} + \mathscr{D}^{\alpha}_{\alpha'}\mathscr{D}^{\beta}_{\alpha'} \cdot \delta^{\alpha'\alpha'}. \tag{3.8}$$

For a CARTESIAN coordinate system the metric tensor components coincide with the KRONECKER symbol. Indeed,

$$g_{i'j'} \stackrel{\text{def}}{:=} \mathbf{i}_{i'} \cdot \mathbf{i}_{j'} \equiv \delta_{i'j'}, \qquad g^{i'j'} \stackrel{\text{def}}{:=} \mathbf{i}^{i'} \cdot \mathbf{i}^{j'} \equiv \delta^{i'j'}.$$

Consequently, when we change the coordinates from CARTESIAN to curvilinear, the metric tensor components are transformed according to the following rule:

$$g_{ij} = \mathscr{D}^{i'}_{i}\mathscr{D}^{j'}_{j} \cdot g_{i'j'}, \qquad g^{ij} = \mathscr{D}^{i}_{i'}\mathscr{D}^{j}_{j'} \cdot g^{i'j'}.$$

Let us take an arbitrary vector $\mathbf{v} \in \mathbb{R}^4$. As an element of a vector space it does not depend on the chosen coordinate basis. This object remains the same in any basis.[2] However, for simplicity, it is easier to work with vector $\mathbf{v}$ coordinates, which are changing from one basis to another. Let us develop vector $\mathbf{v}$ in three coordinate systems:

$$\mathbf{v} = v_{i'}\mathbf{i}^{i'} = \upsilon_i \mathbf{e}^i = \upsilon^i \mathbf{e}_i .$$

Consequently, $\{v_{i'}\}_{i'=0}^{3}$ are CARTESIAN, $\{\upsilon_i\}_{i=0}^{3}$ are covariant and $\{\upsilon^i\}_{i=0}^{3}$ are contravariant components of the same vector $\mathbf{v} \in \mathbb{R}^4$. As we said, a vector $\mathbf{v}$ as an element of a vector space is invariant [316]. However, sometimes we shall say that a vector is covariant (contravariant) by meaning that this vector is given in terms of its covariant (contravariant) components. Using aforementioned relations between various bases vectors we can obtain relations among the covariant and contravariant components solely [316]:

$$\upsilon_i = g_{ij}\upsilon^j, \qquad \upsilon^i = g^{ij}\upsilon_j, \tag{3.9}$$

or express covariant (contravariant) components through respective CARTESIAN coordinates CARTESIAN COORDINATES:

$$\upsilon_i = \mathscr{D}^{i'}_{i} v_{i'}, \qquad \upsilon^i = \mathscr{D}^{i}_{i'} v^{i'},$$

and inversely we can express the CARTESIAN coordinates through the covariant (contravariant) components of the vector $\mathbf{v}$:

$$v_{i'} = \mathscr{D}^{i}_{i'}\upsilon_i \equiv v^{i'} = \mathscr{D}^{i'}_{i}\upsilon^i .$$

[2]This statement applies to any other tensor of the first rank.

Two last sets of relations express actually the transformation rules of a general 1−tensor from a coordinate system to another one. Notice also that

$$v_{0'} = v^{0'} = \frac{\mathrm{d}x^0}{\mathrm{d}t} = 1, \qquad \text{thus,} \qquad \upsilon^0 \equiv 1.$$

Let us denote by $\mathcal{J}$ the JACOBIAN of the transformation (3.1):

$$\mathcal{J} \stackrel{\text{def}}{:=} \det\left\{\mathscr{D}_i^{i'}\right\}. \tag{3.10}$$

It is not difficult to show that we have also

$$\mathcal{J} = \det\left\{\mathscr{D}_\alpha^{\alpha'}\right\}, \qquad |\mathcal{J}| = \sqrt{\det\left\{g_{ij}\right\}} = \frac{1}{\sqrt{\det\left\{g^{ij}\right\}}}.$$

Also from CRAMER's rule we have

$$\mathscr{D}^{\alpha}_{\;\alpha'} = (-1)^{\alpha-\alpha'}\,\frac{\det\left\{\mathscr{D}_\gamma^{\gamma'}\right\}}{\mathcal{J}}, \qquad \gamma \neq \alpha, \quad \gamma' \neq \alpha'.$$

For the sequel we shall need to introduce the so-called CHRISTOFFEL symbols. First we introduce vectors

$$\mathbf{e}_{ik} \equiv \mathbf{e}_{ki} \stackrel{\text{def}}{:=} \frac{\partial\,\mathbf{e}_i}{\partial q^k} \equiv \frac{\partial\,\mathbf{e}_k}{\partial q^i}.$$

The expansion coefficients of these vectors in contravariant and covariant bases

$$\mathbf{e}_{ij} = \Upsilon_{ij,k}\,\mathbf{e}^k = \Upsilon_{ij}^k\,\mathbf{e}_k$$

are called CHRISTOFFEL symbols of the first and second kind correspondingly:

$$\Upsilon_{ij,k} \stackrel{\text{def}}{:=} \mathbf{e}_{ij}\cdot\mathbf{e}_k, \qquad \Upsilon_{ij}^k \stackrel{\text{def}}{:=} \mathbf{e}_{ij}\cdot\mathbf{e}^k.$$

It is not difficult to see that for a CARTESIAN coordinate system all CHRISTOFFEL symbols are identically zero. In the general case CHRISTOFFEL symbols satisfy the following relations

$$\Upsilon_{ij,k} \equiv \Upsilon_{ji,k}, \qquad \Upsilon_{ij}^k \equiv \Upsilon_{ji}^k, \qquad \Upsilon_{jk}^0 \equiv 0,$$

$$\Upsilon_{kj,i} + \Upsilon_{ij,k} = \frac{\partial g_{ik}}{\partial q^j}, \qquad \Upsilon_{jk}^i = \mathscr{D}^i_{\;j'}\cdot\frac{\partial \mathscr{D}_j^{j'}}{\partial q^k},$$

$$\Upsilon_{ik,\,j} = \frac{1}{2}\left[\frac{\partial g_{ij}}{\partial q^k} + \frac{\partial g_{kj}}{\partial q^i} - \frac{\partial g_{ik}}{\partial q^j}\right], \quad \Upsilon_{ik,\,j} = g_{\ell j}\,\Upsilon^{\ell}_{ik}, \quad \Upsilon^{\ell}_{ik} = g^{\ell j}\,\Upsilon_{ik,\,j}\,.$$

Using the definition of the JACOBIAN $\mathcal{J}$, it is straightforward to show the following important identity:

$$\Upsilon^{i}_{ik} = \frac{1}{\mathcal{J}}\,\frac{\partial \mathcal{J}}{\partial q^k}\,. \tag{3.11}$$

In curvilinear coordinates the analogue of a partial derivative with respect to a coordinate (i.e. an independent variable) is a covariant derivative over curvilinear coordinates which is defined using CHRISTOFFEL symbols. For example, the covariant derivative ∇_k of a covariant component υ_i of vector $\mathbf{v}$ is defined as

$$\nabla_k\,\upsilon_i = \frac{\partial\,\upsilon_i}{\partial q^k} - \Upsilon^{j}_{ik}\,\upsilon_j\,.$$

Similarly one can define the covariant derivative of a contravariant component υ^i of vector $\mathbf{v}$:

$$\nabla_k\,\upsilon^i = \frac{\partial\,\upsilon^i}{\partial q^k} + \Upsilon^{i}_{kj}\,\upsilon^j\,.$$

We remind that in CARTESIAN coordinates the covariant derivatives coincide with usual partial derivatives since CHRISTOFFEL's symbols vanish. For instance, the divergence of a vector field $\mathbf{v}(q^0,\,q^1,\,q^2,\,q^3) \in \mathbb{R}^4$ can be readily obtained:

$$\nabla\cdot\mathbf{v} = \nabla_i\,\upsilon^i = \frac{\partial\,\upsilon^i}{\partial q^i} + \Upsilon^{i}_{ik}\,\upsilon^k\,.$$

The last equation can be equivalently rewritten using formula (3.11) as

$$\nabla\cdot\mathbf{v} \overset{\text{def}}{:=} \frac{1}{\mathcal{J}}\,\frac{\partial(\mathcal{J}\,\upsilon^i)}{\partial q^i}\,. \tag{3.12}$$

In order to recast EULER equations in arbitrary moving frames of reference, we shall have to work with 2−tensors as well. Similarly to vectors (or 1−tensors) these objects are independent from the frame of reference. However, when we change the coordinates, tensor components change accordingly. Suppose that we have a 2−tensor $\mathbb{T} = \left\{T_{i'\,j'}\right\}_{i'\,j'\,=\,0}^{3}$ and we know its components $T_{i'\,j'} \equiv T^{\,i'\,j'}$ in a CARTESIAN frame of reference. Then, in any other curvilinear coordinates these components can be computed as

$$T_{ij} = \mathscr{D}_i^{\,i'}\,\mathscr{D}_j^{\,j'}\,T_{i'\,j'}, \qquad T^{\,ij} = \mathscr{D}^{i}_{\;i'}\,\mathscr{D}^{j}_{\;j'}\,T^{\,i'\,j'}\,. \tag{3.13}$$

Similarly, covariant derivatives of a 2−tensor components are defined as

$$\nabla_k T_{ij} = \frac{\partial T_{ij}}{\partial q^k} - \Upsilon^{\ell}_{ik} T_{\ell j} - \Upsilon^{\ell}_{jk} T_{i\ell},$$

$$\nabla_k T^{ij} = \frac{\partial T^{ij}}{\partial q^k} + \Upsilon^{i}_{k\ell} T^{\ell j} + \Upsilon^{j}_{k\ell} T^{i\ell}.$$

Now one can show that covariant derivatives of the metric tensor vanish, i.e.

$$\nabla_k g_{ij} = 0, \qquad \nabla_k g^{ij} = 0. \tag{3.14}$$

Finally, we can compute the divergence of a 2−tensor, which is by definition a 1−tensor defined as

$$\{\nabla \cdot \mathbb{T}\}^i = \nabla_j T^{ij}.$$

Similarly to the divergence of a vector field, the divergence of a 2−tensor can be also expressed through the JACOBIAN thanks to the aforementioned definition and formula (3.11) as

$$\{\nabla \cdot \mathbb{T}\}^i = \frac{1}{\mathfrak{J}} \frac{\partial (\mathfrak{J} T^{ij})}{\partial q^j} + \Upsilon^{i}_{jk} T^{jk}. \tag{3.15}$$

3.1.1 Flow Vorticity

Using the LEVI-CIVITA tensor $\varepsilon^{\alpha\beta\gamma}$, which is defined as [280]:

$$\varepsilon^{\alpha\beta\gamma} = \begin{cases} \dfrac{1}{\mathfrak{J}}, & \operatorname{sign}\begin{pmatrix} 1 & 2 & 3 \\ \alpha & \beta & \gamma \end{pmatrix} = 1, \\ -\dfrac{1}{\mathfrak{J}}, & \operatorname{sign}\begin{pmatrix} 1 & 2 & 3 \\ \alpha & \beta & \gamma \end{pmatrix} = -1, \\ 0, & (\alpha = \beta) \vee (\alpha = \gamma) \vee (\beta = \gamma), \end{cases}$$

where sign(·) is the signature of a permutation. In other words the sign of LEVI-CIVITA tensor components changes the sign depending on if the permutation $(\alpha\ \beta\ \gamma)$ is odd or even. Using this tensor, we can define the *rotor* of a vector $\mathbf{v}$ as:

$$\boldsymbol{\omega} = \operatorname{rot} \mathbf{v} = \omega^{\gamma} \mathbf{e}_{\gamma}, \qquad \omega^{\gamma} \stackrel{\text{def}}{:=} \varepsilon^{\alpha\beta\gamma} \cdot \nabla_{\alpha} \upsilon_{\beta}.$$

We can write explicitly the contravariant components of vector $\boldsymbol{\omega}$:

$$\omega^1 = \frac{1}{\mathfrak{J}}\left(\frac{\partial \upsilon_3}{\partial q^2} - \frac{\partial \upsilon_2}{\partial q^3}\right), \quad \omega^2 = \frac{1}{\mathfrak{J}}\left(\frac{\partial \upsilon_1}{\partial q^3} - \frac{\partial \upsilon_3}{\partial q^1}\right),$$
$$\omega^3 = \frac{1}{\mathfrak{J}}\left(\frac{\partial \upsilon_2}{\partial q^1} - \frac{\partial \upsilon_1}{\partial q^2}\right). \qquad (3.16)$$

These technicalities are used in our paper in order to reformulate the full EULER equations in an arbitrary moving curvilinear frame of reference. As a practical application we employ these techniques to the globally spherical geometry due to obvious applications in Geophysical Fluid Dynamics (GFD) on the planetary scale.

3.2 Euler Equations

The full EULER equations in spherical coordinates can be found in many works (see e.g. the classical book [198]). However, for our purposes we prefer to have a more compact form of these equations. It will be derived in the present section departing from EULER equations written in a standard CARTESIAN coordinate system $O\,x^1\,x^2\,x^3$. We assume that the axis $O\,x^3$ coincides with the rotation axis and points vertically upwards to the North pole. In this setting the coordinate plane $O\,x^1\,x^2$ coincides with the celestial equator. The definition of the employed CARTESIAN and spherical (curvilinear) coordinate systems is illustrated in Fig. 3.1.

Moreover, we introduce a virtual sphere of radius R whose center coincides with the center of EARTH, R being the mean EARTH's radius. This sphere rotates with the

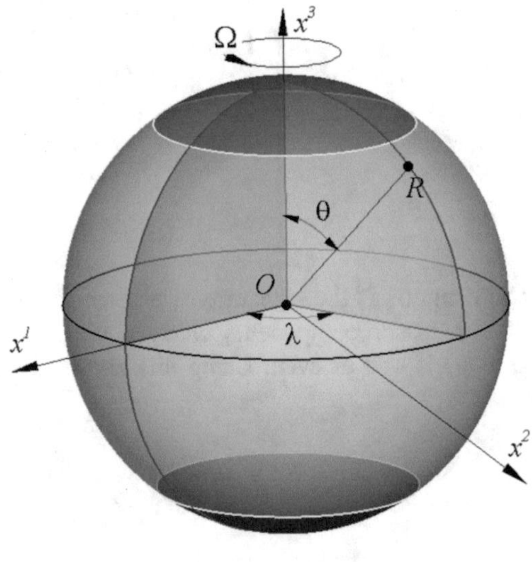

Fig. 3.1 Cartesian and spherical coordinates used in this study

angular velocity Ω. We shall need this object below for the derivation of the base shallow water model. The real planet shape does not have to be spherical. We only assume that its geometry is globally spherical and can be obtained as a continuous deformation of the virtual sphere (shown in blue in Fig. 3.1).

Among all volumetric forces we consider only the NEWTONIAN gravity $\mathbf{g}$ directed towards the center of the rotating sphere. In other words, the force acting on a fluid particle located at the point $\mathbf{x} = (x^1, x^2, x^3)$ has the following expression:

$$\mathbf{g} = -g \frac{\mathbf{x}}{|\mathbf{x}|} = -\frac{g}{|\mathbf{x}|}\left(x^1 \mathbf{i}_1 + x^2 \mathbf{i}_2 + x^3 \mathbf{i}_3\right),$$

where $\{\mathbf{i}_\alpha\}_{\alpha=1}^3$ are unitary vectors of the CARTESIAN coordinate system. In the derivation of the base model we shall assume that the liquid layer depth is much smaller than EARTH's (mean) radius R. However, for the EULER equations this assumption is not really needed. Moreover, we assume that the liquid is homogeneous, thus liquid density $\rho \equiv$ const. Moreover, for the sake of simplicity we assume that the gravity acceleration $g = |\mathbf{g}|$ is also constant throughout the fluid bulk.[3] Under these conditions the equations which describe the motion of an ideal incompressible fluid are well known:

$$\frac{\partial(\rho U_\alpha)}{\partial x^\alpha} = 0, \tag{3.17}$$

$$\frac{\partial(\rho U_\beta)}{\partial t} + \frac{\partial(\rho U_\beta U_\alpha)}{\partial x^\alpha} + \frac{\partial p}{\partial x^\beta} = -\rho g \frac{x^\beta}{|\mathbf{x}|}, \qquad \beta = 1, 2, 3. \tag{3.18}$$

In equations above and throughout this study we adopt the summation convention over repeating lower and upper indices. Functions $\{U_\alpha\}_{\alpha=1}^3$ are CARTESIAN components of the fluid particles velocity vector and p is the fluid pressure.

3.2.1 *Euler Equations in Arbitrary Moving Frames of Reference*

A curvilinear coordinate system (q^0, q^1, q^2, q^3) is given by a regular bijective homomorphism (or even a diffeomorphism) onto a certain domain with CARTESIAN coordinates (x^0, x^1, x^2, x^3). In the present study for the sake of convenience we give a different treatment to time and space coordinates, i.e.

[3] The authors are not aware of any study in the field of Hydrodynamics where this assumption was not adopted.

$$x^0 = q^0 = t\,, \qquad x^\alpha = x^\alpha\left(q^0, q^1, q^2, q^3\right), \qquad \alpha = 1, 2, 3\,. \tag{3.19}$$

More precisely, we assume that the mapping above satisfies the following conditions [240]:

- The map is bijective.
- The map *and* its inverse are at least continuous (or even smooth).
- The JACOBIAN of this map is non-vanishing.

More information on curvilinear coordinate systems is given in Sect. 3.1. As one can see, the time variable t is chosen to be the same in both coordinate systems. It does not have to change (at least in the Classical Mechanics). Consequently, a point P having CARTESIAN (spatial) coordinates (x^1, x^2, x^3) will have curvilinear coordinates (q^1, q^2, q^3).

Remark 3.1 Before reading the sequel of this chapter we strongly recommend to read first Sect. 3.1 where we provide all necessary information from tensor analysis and we explain the system of notations used below.

The system of Eqs. (3.17) and (3.18) can be recast in a compact tensorial form as follows [201]:

$$\nabla \cdot \mathbb{T} = \rho\,\boldsymbol{F}\,,$$

where $\mathbb{T} = \{T^{i'j'}\}_{i',\,j'=0}^{3}$ is the 2−tensor and $\boldsymbol{F} = \{F^{i'}\}_{i'=0}^{3}$ is the 1−tensor. The last form has the advantage of being coordinate frame invariant (i.e. independent of coordinates provided that the components of tensors $\mathbb{T}$ and $\boldsymbol{F}$ are transformed according to some well-established rules). However, in the perspective of numerical discretization [190], one needs to introduce explicitly the coordinate system into notation to work with. It can be done starting from the components of tensors $\mathbb{T}$ and $\boldsymbol{F}$ in a CARTESIAN frame of reference:

$$T^{0'j'} = \rho\,U^{0'}U^{j'} \equiv \rho\,U^{j'}\,, \qquad T^{\alpha' j'} = \rho\,U^{\alpha'}U^{\upsilon j'} + p\,\delta^{\alpha' j'}\,,$$

$$F^{0'} = 0\,, \qquad F^{\alpha'} = -g\,\frac{x^{\alpha'}}{|\mathbf{x}|}\,.$$

We employ indices with primes in order to denote CARTESIAN components. For instance, $U^{j'}$ is the jth component of the velocity vector in a CARTESIAN frame of reference and it can be computed as

$$U^{j'} = \dot{x}^{j'} \stackrel{\text{def}}{:=} \frac{\mathrm{d}x^{j'}}{\mathrm{d}t}\,,$$

$U^{0'}$ being equal to 1 thanks to the choice (3.19). In any other moving curvilinear frame of reference the components of tensor $\mathbb{T}$ can be computed according to formulas (3.13):

$$T^{0j} = \mathscr{D}^0_{i'}\mathscr{D}^j_{j'}T^{i'j'} = \mathscr{D}^0_{0'}\mathscr{D}^j_{j'}T^{0'j'} = \mathscr{D}^j_{j'}T^{0'j'} = \mathscr{D}^j_{j'}(\rho U^{j'}) = \rho \mathcal{V}^j,$$

$$T^{\alpha 0} = \mathscr{D}^\alpha_{i'}\mathscr{D}^0_{j'}T^{i'j'} = \mathscr{D}^\alpha_{i'}\mathscr{D}^0_{0'}T^{i'0'} = \mathscr{D}^\alpha_{i'}(\rho U^{i'}) = \rho \mathcal{V}^\alpha,$$

$$\begin{aligned}
T^{\alpha\beta} = \mathscr{D}^\alpha_{i'}\,\mathscr{D}^\beta_{j'}T^{i'j'} &= \mathscr{D}^\alpha_{0'}\,\mathscr{D}^\beta_{j'}T^{0'j'} + \mathscr{D}^\alpha_{\alpha'}\,\mathscr{D}^\beta_{j'}T^{\alpha'j'} \\
&= \mathscr{D}^\alpha_{0'}(\rho\,\mathcal{V}^\beta) + \mathscr{D}^\alpha_{\alpha'}(\rho U^{\alpha'}\,\mathcal{V}^\beta) + p\cdot\mathscr{D}^\alpha_{\alpha'}\,\mathscr{D}^\beta_{j'}\delta^{\alpha'j'} \\
&= \rho\,\mathcal{V}^\alpha\,\mathcal{V}^\beta + p\cdot(\mathscr{D}^\alpha_{i'}\,\mathscr{D}^\beta_{j'}\delta^{i'j'} - \mathscr{D}^\alpha_{0'}\,\mathscr{D}^\beta_{j'}\delta^{0'j'}) \\
&\equiv \rho\,\mathcal{V}^\alpha\,\mathcal{V}^\beta + p\cdot(g^{\alpha\beta} - g^{\alpha 0}g^{\beta 0}),
\end{aligned}$$

where $\{g^{ij}\}^3_{i,j=0}$ are components of the contravariant metric tensor (defined in Sect. 3.1) and $\mathcal{V}^j$ is the jth contravariant component of velocity in a curvilinear frame of reference:

$$\mathcal{V}^j = \dot{q}^j \overset{\text{def}}{:=} \frac{\mathrm{d}q^j}{\mathrm{d}t} = \mathscr{D}^j_{j'}U^{j'}. \tag{3.20}$$

The components of tensor $\boldsymbol{F}$ are transformed as

$$F^0 = \mathscr{D}^0_{i'}F^{i'} \equiv 0, \qquad F^\alpha = \mathscr{D}^\alpha_{i'}F^{i'} = -g\,\frac{x^{\alpha'}}{|\mathbf{x}|}\,\mathscr{D}^\alpha_{\alpha'}.$$

We have obviously that $\mathcal{V}^0 = 1$, $T^{00} = \rho$ and in Sect. 3.1 we show that

$$g^{0\alpha} \equiv q^\alpha_t, \qquad g^{0\beta} \equiv q^\beta_t.$$

The expressions of 2−tensor $\mathbb{T}$ elements along with the 1−tensor $\boldsymbol{F}$ are used to write the full Euler equations in an arbitrary curvilinear coordinate system. The following compact notation is already familiar to us:

$$(\boldsymbol{\nabla}\cdot\mathbb{T})^i = \rho F^i, \qquad i = 0, \ldots, 3,$$

or using formula (3.15) for the divergence operator we have

$$\frac{\partial\,(\mathcal{J}\,T^{ij})}{\partial q^j} + \mathcal{J}\,\Upsilon^i_{jk}\,T^{jk} = \rho\,\mathcal{J}\,F^i, \qquad i = 0, \ldots, 3. \tag{3.21}$$

For instance, for $i = 0$ we obtain the mass conservation equation in an arbitrary frame of reference:

$$\frac{\partial\,(\rho\,\mathcal{J})}{\partial t} + \frac{\partial\,(\rho\,\mathcal{J}\,\mathcal{V}^{\alpha})}{\partial q^{\alpha}} = 0\,. \tag{3.22}$$

For $i = 1, 2, 3$ from Eq. (3.21) one obtains the momentum conservation equations, which can be expanded by inserting expressions of 2−tensor components T^{ij}:

$$\begin{aligned}&\frac{\partial\,(\rho\,\mathcal{J}\,\mathcal{V}^{\beta})}{\partial t}+\frac{\partial\,(\rho\,\mathcal{J}\,\mathcal{V}^{\alpha}\,\mathcal{V}^{\beta})}{\partial q^{\alpha}}+\mathcal{J}\,\bigl(g^{\alpha\beta}-g^{\alpha 0}\,g^{\beta 0}\bigr)\,\frac{\partial p}{\partial q^{\alpha}}+\rho\,\mathcal{J}\,\Upsilon^{\beta}_{jk}\,\mathcal{V}^{j}\,\mathcal{V}^{k}\\&+p\cdot\underbrace{\Bigl[\mathcal{J}\,\frac{\partial\,(g^{\alpha\beta}-g^{\alpha 0}\,g^{\beta 0})}{\partial q^{\alpha}}+\bigl(g^{\alpha\beta}-g^{\alpha 0}\,g^{\beta 0}\bigr)\,\frac{\partial\,\mathcal{J}}{\partial q^{\alpha}}+\mathcal{J}\,\Upsilon^{\beta}_{\alpha\gamma}\,\bigl(g^{\alpha\gamma}-g^{\alpha 0}\,g^{\gamma 0}\bigr)\Bigr]}_{(\bigstar)}\\&=\rho\,\mathcal{J}\,F^{\beta}\,,\end{aligned} \tag{3.23}$$

where $\beta \in \{1, 2, 3\}$. By using Formula (3.14) to differentiate the components of the contravariant metric components $g^{\alpha\beta}$ along with formula (3.11) one can show that expression $(\bigstar) \equiv 0$. Consequently, the momentum conservation equations simplify substantially. However, this set of equations still represents an important drawback: each equation contains the derivative of the pressure with respect to all three coordinate directions q^{α}, $\alpha = 1, 2, 3$. So, we continue to modify the governing equations. We take index $\nu \in \{1, 2, 3\}$ and we multiply the continuity equation (3.22) by the covariant metric tensor component $g_{0\nu}$. Then, momentum conservation Eq. (3.23) is multiplied by $g_{\beta\nu}$ and we sum up obtained expressions. As a result, we obtain the following equation: Then, momentum conservation

$$\begin{aligned}&\frac{\partial\,(\rho\,g_{j\nu}\,\mathcal{J}\,\mathcal{V}^{j})}{\partial t}+\frac{\partial\,(\rho\,g_{j\nu}\,\mathcal{J}\,\mathcal{V}^{\alpha}\,\mathcal{V}^{j})}{\partial q^{\alpha}}+\mathcal{J}\,g_{\beta\nu}\cdot\bigl(g^{\alpha\beta}-g^{\alpha 0}\,g^{\beta 0}\bigr)\,\frac{\partial p}{\partial q^{\alpha}}\\&-\rho\,\mathcal{J}\,\mathcal{V}^{j}\,\frac{\partial g_{j\nu}}{\partial t}-\rho\,\mathcal{J}\,\mathcal{V}^{\alpha}\,\mathcal{V}^{j}\,\frac{\partial g_{j\nu}}{\partial q^{\alpha}}+\rho\,g_{\beta\nu}\,\mathcal{J}\,\Upsilon^{\beta}_{jk}\,\mathcal{V}^{j}\,\mathcal{V}^{k}=\rho\,g_{j\nu}\,\mathcal{J}\,F^{j}\,.\end{aligned}$$

Using relation (3.2) we can show that

$$g_{\beta\nu}\cdot\bigl(g^{\alpha\beta}-g^{\alpha 0}\,g^{\beta 0}\bigr)\equiv\delta^{\alpha}_{\nu}\,.$$

Using the relations (3.9) between covariant and contravariant components of a vector one obtains also

$$g_{j\nu}\,F^{j} = F_{\nu}\,, \qquad g_{j\nu}\,\mathcal{V}^{j} = \mathcal{V}_{\nu}\,. \tag{3.24}$$

Above, F_{ν} and $\mathcal{V}_{\nu}$ are covariant components of the force and velocity vectors (or 1−tensors), respectively. Finally, we obtain the *conservative* form the full EULER momentum equations in an arbitrary frame of reference:

$$\frac{\partial\,(\rho\,\mathfrak{J}\,\mathcal{V}_\nu)}{\partial t} + \frac{\partial\,(\rho\,\mathfrak{J}\,\mathcal{V}^\alpha\,\mathcal{V}_\nu)}{\partial q^\alpha} + \mathfrak{J}\,\frac{\partial p}{\partial q^\nu} + \rho\,\mathfrak{J}\,\mathcal{V}^j\,\mathcal{V}^k\left[g_{\beta\nu}\,\Upsilon^\beta_{jk} - \frac{\partial\,g_{j\nu}}{\partial q^k}\right]$$
$$= \rho\,\mathfrak{J}\,F_\nu\,,\qquad \nu = 1,\,2,\,3\,.$$

By using the continuity equation (3.22), one can derive similarly the non-conservative form of the momentum equation:

$$\frac{\partial\mathcal{V}_\nu}{\partial t} + \mathcal{V}^\alpha\,\frac{\partial\mathcal{V}_\nu}{\partial q^\alpha} + \frac{1}{\rho}\,\frac{\partial p}{\partial q^\nu} + \mathcal{V}^j\,\mathcal{V}^k\left[\Upsilon_{jk,\,\nu} - \frac{\partial g_{j\nu}}{\partial q^k}\right] = F_\nu\,,$$
$$\nu = 1,\,2,\,3\,, \qquad (3.25)$$

where we used CHRISTOFFEL symbols of the first kind for the sake of simplicity.

3.2.2 Euler Equations in Spherical Coordinates

From now on we choose to work in spherical coordinates since the main applications of our work aim the Geophysical Fluid Dynamics on planetary scales [260]. As we know the planets are not exactly spheres. Nevertheless, the introduction of spherical coordinates still simplifies a lot the analytical work.

Consider a spherical coordinate system $O\,\lambda\,\theta\,r$ with the origin placed in the center of a virtual sphere of radius R rotating with constant angular speed Ω. By λ we denote the longitude whose zero value coincides with a chosen meridian. Angle θ is the colatitude defined as $\theta \overset{\text{def}}{:=} \frac{\pi}{2} - \phi$, where ϕ is the geographical latitude. Finally, $r > 0$ is the radial coordinate. Since latitude $-\frac{\pi}{2} < \phi < \frac{\pi}{2}$, we have that $0 < \theta < \pi$. However, we assume additionally that

$$\theta_0 \leqslant \theta \leqslant \pi - \theta_0\,,$$

where $1 \gg \theta_0 = \text{const} > 0$ is a small angle. In other words, we exclude the poles with their small neighbourhood.[4] Spherical coordinates $q^0 = t$, $q^1 = \lambda$, $q^2 = \theta$, $q^3 = r$ and CARTESIAN coordinates x^0, x^1, x^2, x^3 are related by the following formulas:

$$x^0 = t\,,$$
$$x^1 = r\,\cos\,(\lambda + \Omega\,t)\,\sin\,\theta\,,$$

[4]In Atmospheric sciences this assumption is not realistic, of course. However, in Hydrodynamics it is justified by natural ice covers around pole regions—Arctic and Antarctic. So, water wave phenomena do not take place near Earth's poles.

$$x^2 = r \sin(\lambda + \Omega t) \sin\theta,$$

$$x^3 = r \cos\theta.$$

Using Formula (3.10) it is not difficult to show that JACOBIAN of the transformation above is

$$\mathfrak{J} = -r^2 \sin\theta. \tag{3.26}$$

Similarly, using Formulas (3.7) and (3.8), one can compute covariant components of the metric tensor:

$$g_{00} = 1 + \Omega^2 r^2 \sin^2\theta, \quad g_{10} \equiv g_{01} = \Omega r^2 \sin^2\theta,$$

$$g_{20} = g_{02} = g_{30} = g_{03} \equiv 0,$$

$$g_{11} = r^2 \sin^2\theta, \quad g_{22} = r^2, \quad g_{33} = 1, \quad g_{\alpha\beta} = g_{\beta\alpha} \equiv 0,$$

with $\alpha, \beta = 1, 2, 3$ and $\alpha \neq \beta$. From formulas (3.20) we compute contravariant components of the velocity vector:

$$\mathcal{V}^0 = 1, \quad \mathcal{V}^1 = \dot{\lambda}, \quad \mathcal{V}^2 = \dot{\theta}, \quad \mathcal{V}^3 = \dot{r}.$$

The covariant components of the velocity vector $\{\mathcal{V}_\alpha\}_{\alpha=1}^3$ and the exterior volume force $\{F_\alpha\}_{\alpha=1}^3$ are computed thanks to relations (3.24):

$$\mathcal{V}_1 = g_{10} + g_{11}\mathcal{V}^1 = \Omega r^2 \sin^2\theta + r^2 \sin^2\theta\, \dot{\lambda},$$

$$\mathcal{V}_2 = g_{22}\mathcal{V}^2 = r^2\dot{\theta},$$

$$\mathcal{V}_3 = g_{33}\mathcal{V}^3 = \dot{r},$$

and the force components are

$$F_1 = F_2 \equiv 0, \qquad F_3 = -g.$$

Finally, by using the definition of CHRISTOFFEL symbols of the first kind, we obtain the sequence of the following relations for the term

$$\mathcal{V}^j\mathcal{V}^k\left[\Upsilon_{jk,\nu} - \frac{\partial g_{j\nu}}{\partial q^k}\right]$$

$$\equiv \sum_{k=1}^{3}\sum_{j=0}^{k-1}\left[2\Upsilon_{jk,\nu} - \frac{\partial g_{j\nu}}{\partial q^k} - \frac{\partial g_{k\nu}}{\partial q^j}\right]\mathcal{V}^j\mathcal{V}^k + \sum_{j=0}^{3}\left[\Upsilon_{jj,\nu} - \frac{\partial g_{j\nu}}{\partial q^j}\right](\mathcal{V}^j)^2$$

$$= -\sum_{k=1}^{3}\sum_{j=0}^{k-1} \frac{\partial g_{jk}}{\partial q^{\nu}} \mathcal{V}^j \mathcal{V}^k - \frac{1}{2}\sum_{j=0}^{3} \frac{\partial g_{jj}}{\partial q^{\nu}} (\mathcal{V}^j)^2 .$$

Using the fact that the JACOBIAN $\mathcal{J}$ is time-independent (see Formula (3.26)), we obtain the full EULER equations governing the flow of a homogeneous incompressible fluid in spherical coordinates:

$$\frac{\partial(\rho \mathcal{J} \mathcal{V}^{\alpha})}{\partial q^{\alpha}} = 0, \tag{3.27}$$

$$\frac{\partial \mathcal{V}_{\beta}}{\partial t} + \mathcal{V}^{\alpha} \frac{\partial \mathcal{V}_{\beta}}{\partial q^{\alpha}} + \frac{1}{\rho} \cdot \frac{\partial p}{\partial q^{\beta}} = \mathcal{S}_{\beta}, \qquad \beta = 1, 2, 3, \tag{3.28}$$

where

$$\mathcal{S}_{\beta} = \begin{cases} 0, & \beta = 1, \\ \dfrac{(\Omega + \mathcal{V}^1)^2}{2} \cdot \dfrac{\partial g_{11}}{\partial \theta}, & \beta = 2, \\ -g + \dfrac{(\Omega + \mathcal{V}^1)^2}{2} \cdot \dfrac{\partial g_{11}}{\partial r} + \dfrac{(\mathcal{V}^2)^2}{2} \cdot \dfrac{\partial g_{22}}{\partial r}, & \beta = 3. \end{cases}$$

The derivatives of covariant components of the metric tensor can be explicitly computed to give:

$$\frac{\partial g_{11}}{\partial \theta} = 2r^2 \sin\theta \cos\theta,$$

$$\frac{\partial g_{11}}{\partial r} = 2r \sin^2\theta,$$

$$\frac{\partial g_{22}}{\partial r} = 2r.$$

We notice that components $\mathcal{S}_{2,3}$ contain correspondingly the terms $r^2\Omega^2 \sin\theta \cos\theta$ and $r\Omega^2 \sin^2\theta$. They are due to the centrifugal force coming from the EARTH rotation. The presence of this force causes, for instance, the deviation of the pressure gradient from the radial direction even in the quiescent fluid layer. This effect will be examined in the following section.

Equilibrium Free Surface Shape

When we worked on a globally flat space [189], the free surface elevation $y = \eta(\mathbf{x}, t)$ was measured as the excursion of fluid particles from the coordinate plane $y = 0$. This plane is chosen to coincide with the free surface profile of a quiescent fluid at rest. On a rotating sphere, the situation is more complex since the equilibrium free surface shape does not coincide, in general, with any virtual sphere of a radius R. It is the centrifugal force which causes the divergence from the perfectly symmetric spherical profile. So, when we work on a sphere, the free surface elevation will be also measured as the deviation from the equilibrium shape. In this section we shall determine the equilibrium free surface profile $r = R + \eta_{00}(\lambda, \theta)$ by using two natural conditions:

- The equilibrium profile along with bottom is steady.
- The pressure p on the free surface is constant. Since the flow is incompressible, this constant can be set to zero without any loss of generality.

In the case of a quiescent fluid (i.e. all $\mathcal{V}^{\alpha} \equiv 0$), the full EULER equations of motion simply become

$$\frac{\partial p}{\partial \lambda} = 0\,,$$
$$\frac{\partial p}{\partial \theta} = \rho\, \Omega^2 r^2 \sin\theta \cos\theta\,,$$
$$\frac{\partial p}{\partial r} = -\rho g + \rho\, \Omega^2 r \sin^2\theta\,.$$

The solution of these equations can be trivially obtained by successive integrations in each of spherical independent variables:

$$p = -\rho g r + \frac{\rho r^2 \Omega^2}{2} \sin^2\theta + C\,,$$

where $C = \text{const} \in \mathbb{R}$ is an arbitrary integration constant which is to be specified later. Now we can enforce the dynamic boundary condition on the free surface, which states that the pressure $p = 0$ vanishes at the free surface $r = R + \eta_{00}(\lambda, \theta)$. This gives us an algebraic equation to determine the required profile $\eta_{00}(\lambda, \theta)$:

$$-\rho g (R + \eta_{00}) + \frac{\rho (R + \eta_{00})^2 \Omega^2}{2} \sin^2\theta + C = 0\,.$$

The constant C is determined from the condition that on the (North) pole $\theta = 0$ the free surface elevation is fixed. For simplicity we choose $\eta_{00}(\lambda, 0) = 0$. Then, the constant C can be readily computed by evaluating the equation above at the North pole:

$$C = \rho g R\,.$$

With this value of C in hands, the algebraic equation to determine the function $\eta_{00}(\lambda, \theta)$ simply becomes:

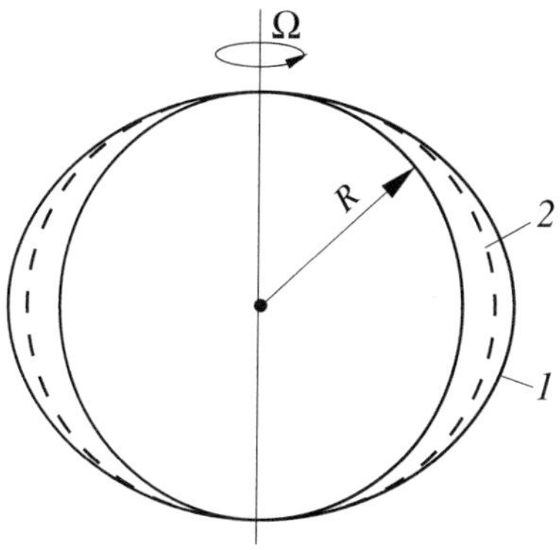

Fig. 3.2 Stationary free surface profiles of a liquid layer over a virtual sphere of radius R and rotating with constant angular speed Ω: (1) solution of EULER equations; (2) solution of modified EULER equations

$$-g\,\eta_{00} + \frac{(R + \eta_{00})^2\,\Omega^2}{2}\,\sin^2\theta = 0.$$

The physical sense has the following solution to the last equation:

$$\eta_{00}(\theta) = \frac{2}{g}\,\Omega^2 R^2 \sin^2\theta \cdot \left[1 + \sqrt{1 - \frac{2}{g}\,\Omega^2 R \sin^2\theta}\right]^{-2}. \qquad (3.29)$$

This solution is represented in Fig. 3.2(1). This solution will be used below as zero level $y = 0$ in free surface flows on globally flat geometries (see, for example, Chap. 1). For example, the solid impermeable bottom of constant depth d is given by the following equation

$$r = R + \eta_{00}(\theta) - d.$$

Boundary Conditions

When we model surface water waves, it is standard to use the full EULER equations as the governing equations. However, in the presence of impermeable bottom and free surface, the corresponding boundary conditions have to be specified [296]. From now on, we assume that the solid uneven moving bottom is given by the following equation:

$$r = R + \eta_{00}(\theta) - h(t, \lambda, \theta) \stackrel{\text{def}}{=:} \check{r}(t, \lambda, \theta),$$

and the free surface is given by

$$r = R + \eta_{00}(\theta) + \eta(t, \lambda, \theta) \stackrel{\text{def}}{=:} \tilde{r}(t, \lambda, \theta).$$

By taking the full (material) derivative of the last two equations with respect to time, we obtain two kinematic boundary conditions on the free surface and bottom, respectively:

$$\eta_t + \mathcal{V}^1 \eta_\lambda + \mathcal{V}^2 (\eta + \eta_{00})_\theta - \mathcal{V}^3 = 0, \qquad r = \tilde{r}, \tag{3.30}$$

$$h_t + \mathcal{V}^1 h_\lambda + \mathcal{V}^2 (h - \eta_{00})_\theta + \mathcal{V}^3 = 0, \qquad r = \check{r}. \tag{3.31}$$

Finally, on the free surface we also have the following dynamic condition:

$$p = 0, \qquad r = \tilde{r}.$$

The last condition expresses the fact that the free surface is an isobar and the constant pressure is chosen to be zero without any loss of generality. Lateral boundary conditions are dependent on the application in hands and they have to be discussed separately.

3.2.3 *Modified Euler Equations*

In order to derive shallow water equations in a moving curvilinear frame of reference, we have to estimate the relative importance of various terms already at the level of the full EULER equations (3.22) and (3.25). Consequently, we have to pass to dimensionless variables. Let ℓ and d be characteristic flow scales in horizontal and vertical directions correspondingly. Let α be the typical wave amplitude. Then, we can form three important dimensionless numbers:

$\varepsilon \stackrel{\text{def}}{:=} \dfrac{\alpha}{d}$: Measure of the nonlinearity

$\mu \stackrel{\text{def}}{:=} \dfrac{d}{\ell}$: Measure of the frequency dispersion

$\varsigma \stackrel{\text{def}}{:=} \dfrac{d}{R}$: Measure of flow thickness.

Parameters ε and μ are well known in long wave modelling (see, for example, [189]), while parameter ς is specific to globally spherical geometries. The values of all three parameters $(\varepsilon, \mu, \varsigma)$ characterize the aspect ratios of the flow. Various assumptions on the (relative) magnitude of these parameters allow to simplify more or less significantly the governing equations of Hydrodynamics.

Dimensionless Variables

We shall use also some 'derived' dimensionless quantities. Another dimensionless horizontal scale can be introduced on a sphere as

$$\gamma \stackrel{\text{def}}{:=} \frac{\varsigma}{\mu} \equiv \frac{\ell}{R}.$$

The characteristic time scale τ is introduced as follows:

$$\tau \stackrel{\text{def}}{:=} \frac{\ell}{c},$$

where $c \stackrel{\text{def}}{:=} \sqrt{g\,d}$ is the usual (linear) gravity wave speed. Then, we can easily introduce the characteristic angular velocity of wave propagation:

$$\omega \stackrel{\text{def}}{:=} \frac{\gamma}{\tau} \equiv \frac{c}{R}.$$

Finally, from now on we introduce also a new (independent) signed radial variable:

$$\mathring{r} \stackrel{\text{def}}{:=} r - R.$$

Using the characteristic scales introduced above, we can scale all dependent and independent variables in our mathematical formulation:

$$\{\lambda^\star, \theta^\star\} \ltimes \frac{\{\lambda, \theta\}}{\gamma}, \quad \{h^\star, \mathring{r}^\star, \mathcal{H}^\star\} \ltimes \frac{\{h, \mathring{r}, \mathcal{H}\}}{d}, \quad \eta^\star \ltimes \frac{\eta}{\alpha},$$

$$t^\star \ltimes \frac{t}{\tau}, \qquad \Omega^\star \ltimes \frac{\Omega}{\omega}, \qquad p^\star \ltimes \frac{p}{\rho\, c^2}.$$

Contravariant components of the velocity vector are scaled as follows:

$$\mathcal{V}^{\beta,\,\star} \ltimes \frac{\mathcal{V}^\beta}{\omega}, \qquad \mathcal{V}^{3,\,\star} \ltimes \frac{\mathcal{V}^3}{\mu\, c},$$

and covariant coordinates scale as

$$\mathcal{V}^\star_\beta \ltimes \frac{\mathcal{V}_\beta}{R\, c}, \qquad \mathcal{V}^\star_3 \ltimes \frac{\mathcal{V}_3}{\mu\, c},$$

where $\beta = 1, 2$. Finally, we adimensionalize also the transformation JACOBIAN along with non-zero components of the covariant metric tensor:

$$\bigl\{\mathcal{J}^\star, g^\star_{11}, g^\star_{22}\bigr\} \ltimes \frac{\bigl\{\mathcal{J}, g_{11}, g_{22}\bigr\}}{R^2}, \qquad g^\star_{01} \ltimes \frac{g_{01}}{R\, c}.$$

This dimensionless formulation will be used below to perform an important simplification of full spherical EULER equations.

Modification of Euler Equations

The parameter ς represents the relative thickness of the liquid layer and usually in geophysical applications it rarely exceeds $\varsigma \lesssim 10^{-3}$. Consequently, the presence of this factor in front of $\mathcal{O}(1)$ terms shows their negligible importance. The modification of EULER equations consists in omitting such terms in Eqs. (3.27) and (3.28). Later, from these modified equations we shall derive the base wave model in Sect. 3.3. The main difference between the original and modified EULER equations is that the JACOBIAN along with metric tensor components lose their dependence on the radial coordinate r (or equivalently $\mathring{r}$) after the modification.

Consider, for example, the case of the JACOBIAN $\mathcal{J}$, which is computed using formula (3.26) in the original EULER equations. In dimensionless variables the JACOBIAN becomes

$$\mathcal{J}^\star = \frac{\mathcal{J}}{R^2} = -\left(\frac{r}{R}\right)^2 \sin\theta = -\left(1 + \frac{\mathring{r}}{R}\right)^2 \sin\theta$$
$$= -\left(1 + \varsigma\,\mathring{r}^\star\right)^2 \sin(\gamma\theta^\star) \simeq -\sin(\gamma\theta^\star).$$

Above we neglected the term $\varsigma\,\mathring{r}^\star = \mathcal{O}(\varsigma)$. By making similar transformation with covariant metric tensor components, we obtain

$$g^\star_{11} = \sin^2(\gamma\theta^\star), \qquad g^\star_{22} = 1, \qquad g^\star_{01} = \Omega^\star \sin^2(\gamma\theta^\star).$$

For the sake of completeness we provide also the modified quantities in dimensional variables as well:

$$\mathcal{J} = -R^2 \sin\theta,$$
$$g_{11} = R^2\sin^2\theta, \qquad g_{22} = R^2, \qquad g_{01} = \Omega R^2 \sin^2\theta. \qquad (3.32)$$

After these modification we obtain the modified EULER equations, which have the same form as (3.27), (3.28), but the JACOBIAN $\mathcal{J}$ and metric tensor $\{g_{ij}\}$ are modified as we explained hereinabove. The right-hand side in momentum equation (3.28) is modified as well $\mathcal{S}_\beta \rightsquigarrow \grave{\mathcal{S}}_\beta$, since the quantities g_{11} and g_{22} do not depend on r anymore:

$$\grave{\mathcal{S}}_\beta = \begin{cases} 0, & \beta = 1, \\ (\Omega + \mathcal{V}^1)^2 R^2 \sin\theta\cos\theta, & \beta = 2, \\ -g, & \beta = 3. \end{cases} \qquad (3.33)$$

The simplification presented in this section can be also performed in the steady free surface profile which is thoroughly done in the following section.

Modified Stationary Free Surface Profile

After the modifications we introduced into EULER equations, we have to reconsider accordingly the question of the stationary water profile, which will serve us as the unperturbed water level. The modified EULER equations for the quiescent fluid are

$$\frac{\partial p}{\partial \lambda} = 0,$$
$$\frac{\partial p}{\partial \theta} = \rho\,\Omega^2 R^2 \sin\theta\,\cos\theta,$$
$$\frac{\partial p}{\partial r} = -\rho g.$$

The last equations can be similarly integrated and the most general solution is

$$p = -\rho g r + \frac{\rho\,\Omega^2 R^2}{2}\sin^2\theta + C,$$

where $C = \mathrm{const} \in \mathbb{R}$. Imposing the dynamic boundary condition $p(\lambda, \theta, r) = 0$ at the free surface $r = R + \eta_{00}(\lambda, \theta)$ and the additional condition on the North pole $\eta_{00}(\lambda, 0) = 0$, we obtain the following expression for the required function $\eta_{00}(\lambda, \theta)$:

$$\eta_{00}(\lambda, \theta) \equiv \eta_{00}(\theta) = \frac{\Omega^2 R^2}{2g}\sin^2\theta, \qquad 0 \leqslant \theta \leqslant \pi. \tag{3.34}$$

It can be readily seen that the equilibrium free surface profile $\eta_{00}(\theta)$ predicted by modified EULER equations is different from the expression (3.29) derived above. The modified expression (3.34) is depicted in Fig. 3.2(2). However, it is not difficult to show that the modified expression (3.34) can be obtained by neglecting the terms of order $\mathcal{O}(\varsigma)$ in the dimensionless counterpart of Eq. (3.29).

In case of modified EULER equations the bottom $r = \check{r}$ and free surface $r = \tilde{r}$ are described using corresponding deviations $-h(\lambda, \theta, t)$ and $\eta(\lambda, \theta, t)$ from the unperturbed water level $r = R + \eta_{00}(\theta)$. The kinematic boundary conditions on the free surface (3.30) and bottom (3.31) for modified EULER equations remain unchanged.

Modified Euler Equations in Dimensionless Variables

In this section we summarize the developments made so far and, thus, we provide explicitly the modified EULER equations in scaled variables. For the sake of simplicity, we drop the superscript $\star$, which denotes dimensionless quantities. The continuity equation (3.27) becomes:

$$[\mathfrak{J}\mathcal{V}^1]_\lambda + [\mathfrak{J}\mathcal{V}^2]_\theta + [\mathfrak{J}\mathcal{V}^3]_{\mathring{r}} = 0 .$$

Non-conservative momentum equations (3.28) are

$$\big(1 + (\mu^2 - 1)\,\delta_\beta^3\big)\cdot\big[\mathcal{V}_{\beta,t} + \mathcal{V}^1\cdot\mathcal{V}_{\beta,\lambda} + \mathcal{V}^2\cdot\mathcal{V}_{\beta,\theta} + \mathcal{V}^3\cdot\mathcal{V}_{\beta,\mathring{r}}\big] + p_{q^\beta} = \mathcal{S}_\beta ,$$

where $\beta = 1, 2, 3$ and as usually $q^1 \equiv \lambda$, $q^2 \equiv \theta$, $q^3 \equiv \mathring{r}$. The second index denotes the partial derivative operation, i.e. $\mathcal{V}_{\beta,q^\alpha} \overset{\text{def}}{:=} \frac{\partial \mathcal{V}_\beta}{\partial q^\alpha}$. Covariant and contravariant components of the velocity are related in the following way:

$$\mathcal{V}_1 = g_{01} + g_{11}\cdot\mathcal{V}^1 , \qquad \mathcal{V}_2 = \mathcal{V}^2 , \qquad \mathcal{V}_3 = \mathcal{V}^3 .$$

Dimensionless right-hand side $\mathcal{S}_\beta$ has the following components:

$$\mathcal{S}_1 = 0 , \qquad \mathcal{S}_2 = \frac{\varsigma}{\mu}\,(\Omega + \mathcal{V}^1)^2 \sin(\gamma\theta)\cos(\gamma\theta) , \qquad \mathcal{S}_3 = -1 . \tag{3.35}$$

Dimensionless kinematic and dynamic boundary conditions take the following form:

$$\begin{aligned}
\varepsilon\,\eta_t + \varepsilon\,\mathcal{V}^1\cdot\eta_\lambda + \mathcal{V}^2\cdot(\varepsilon\,\eta + \eta_{00})_\theta - \mathcal{V}^3 &= 0 , & \mathring{r} &= \mathring{r}^{\,\mathrm{s}} , \\
h_t + \mathcal{V}^1\cdot h_\lambda + \mathcal{V}^2\cdot(h - \eta_{00})_\theta + \mathcal{V}^3 &= 0 , & \mathring{r} &= \mathring{r}_{\mathrm{b}} , \\
p &= 0 , & \mathring{r} &= \mathring{r}^{\,\mathrm{s}} ,
\end{aligned}$$

where we introduced the traces of the shifted (dimensionless) radial variable $\mathring{r}$ at the bottom and free surface correspondingly:

$$\mathring{r}_{\mathrm{b}} \overset{\text{def}}{:=} \eta_{00} - h , \qquad \mathring{r}^{\,\mathrm{s}} \overset{\text{def}}{:=} \eta_{00} + \varepsilon\,\eta .$$

Finally, the dimensionless ‘still’ water level is given by the following formula:

$$\eta_{00}(\theta) = \frac{\Omega^2}{2}\,\sin^2(\gamma\theta) .$$

Below we shall need also the gradients of this profile:

$$\partial_t\,\eta_{00} = 0 , \qquad \partial_\lambda\,\eta_{00} = 0 , \qquad \partial_\theta\,\eta_{00} = \frac{\varsigma}{\mu}\,\Omega^2 \sin(\gamma\theta)\cos(\gamma\theta) . \tag{3.36}$$

Based on the modified EULER equations presented above, we shall develop a new approximate long wave model in the sections below by using asymptotic methods.

3.3 Nonlinear Dispersive Shallow Water Wave Model

In order to derive an approximate long wave model, we shall work with dimensionless modified EULER equations summarized in the preceding section. Moreover, by analogy with the globally flat case (see Chap. 1), we would like to separate momentum equations into 'horizontal', i.e. tangential and 'vertical', i.e. radial components. The complete set of equations is given here:

$$\bar{\nabla}\cdot[\mathfrak{J}\mathscr{U}] + [\mathfrak{J}\mathscr{W}]_{\mathring{r}} = 0, \tag{3.37}$$

$$\mathscr{V}_t + (\mathscr{U}\cdot\nabla)\mathscr{V} + \mathscr{W}\mathscr{V}_{\mathring{r}} + \nabla p = \mathscr{S}, \tag{3.38}$$

$$\mu^2(\mathscr{W}_t + \mathscr{U}\cdot\nabla\mathscr{W} + \mathscr{W}\mathscr{W}_{\mathring{r}}) + p_{\mathring{r}} = -1, \tag{3.39}$$

where vectors $\mathscr{U} \stackrel{\text{def}}{:=} (\mathcal{V}^1, \mathcal{V}^2)$ and $\mathscr{V} \stackrel{\text{def}}{:=} (\mathcal{V}_1, \mathcal{V}_2)^\top$ are contravariant and covariant components of the 'horizontal' velocity. Moreover, we have the following relation among them

$$\mathscr{V} = \mathscr{G} + \mathbb{G}\cdot\mathscr{U}, \qquad \mathscr{G} \stackrel{\text{def}}{:=} \begin{pmatrix} g_{01} \\ 0 \end{pmatrix}, \qquad \mathbb{G} \stackrel{\text{def}}{:=} \begin{pmatrix} g_{11} & 0 \\ 0 & 1 \end{pmatrix}. \tag{3.40}$$

The 'vertical' component of velocity was denoted by $\mathscr{W} \stackrel{\text{def}}{:=} \mathcal{V}^3$. On the right-hand side we have vector[5] $\mathscr{S} \stackrel{\text{def}}{:=} (\mathcal{S}_1, \mathcal{S}_2)^\top$. The 'horizontal' gradient operator $\nabla \stackrel{\text{def}}{:=} (\partial_\lambda, \partial_\theta)$ and the associated 'flat' divergence operator is:

$$\bar{\nabla}\cdot\mathscr{U} \stackrel{\text{def}}{:=} \frac{\partial \mathcal{V}^1}{\partial\lambda} + \frac{\partial \mathcal{V}^2}{\partial\theta}.$$

We have the following relations similar to the flat case:

$$\mathscr{U}\cdot\nabla\mathscr{W} = \mathcal{V}^1\cdot\mathscr{W}_\lambda + \mathcal{V}^2\cdot\mathscr{W}_\theta, \qquad (\mathscr{U}\cdot\nabla)\mathscr{V} = (\mathscr{U}\cdot\nabla\mathcal{V}_1, \mathscr{U}\cdot\nabla\mathcal{V}_2)^\top.$$

In order to write boundary conditions in a compact form similar to the flat case, we introduce two new functions:

$$\mathring{h} \stackrel{\text{def}}{:=} h - \eta_{00}, \qquad \varepsilon\mathring{\eta} \stackrel{\text{def}}{:=} \varepsilon\eta + \eta_{00}.$$

Finally, the boundary conditions in new variables become

[5] In the sequel by $^\top$ we denote the transposition operator of linear objects such as vectors and matrices.

$$\varepsilon\,\mathring{\eta}_t + \varepsilon\,\mathscr{U}\cdot\nabla\mathring{\eta} - \mathscr{W} = 0, \qquad \mathring{r} = \varepsilon\,\mathring{\eta}, \tag{3.41}$$

$$\mathring{h}_t + \mathscr{U}\cdot\nabla\mathring{h} + \mathscr{W} = 0, \qquad \mathring{r} = -\mathring{h}, \tag{3.42}$$

$$p = 0, \qquad \mathring{r} = \varepsilon\,\mathring{\eta}. \tag{3.43}$$

3.3.1 Horizontal Velocity Approximation

In long wave models we describe traditionally the flow using the total water depth $\mathcal{H}$ and some velocity $\mathbf{u}(t, \lambda, \theta) = \bigl(u^1(t, \lambda, \theta), u^2(t, \lambda, \theta)\bigr)$ variable, which is supposed to approximate the 'horizontal' velocity $\mathscr{U}(t, \lambda, \theta, \mathring{r})$. In weakly dispersive models we can assume that $\mathbf{u}$ approximates $\mathscr{U}$ up to the order $\mathcal{O}(\mu^2)$ in the dispersion parameter. Mathematically it can be written as

$$\mathscr{U}(t, \lambda, \theta, \mathring{r}) = \mathbf{u}(t, \lambda, \theta) + \mu^2\,\mathbf{U}_d(t, \lambda, \theta, \mathring{r}). \tag{3.44}$$

Here by $\mathbf{U}_d(t, \lambda, \theta, \mathring{r}) = \bigl(\mathcal{U}_d^1(t, \lambda, \theta, \mathring{r}), \mathcal{U}_d^2(t, \lambda, \theta, \mathring{r})\bigr)$ we denote the dispersive[6] component of the velocity field. For potential flows one can compute explicitly an approximation to $\mathbf{U}_d(t, \lambda, \theta, \mathring{r})$. However, in the present derivation we do not adopt this simplifying assumption.

For example, in [197] the authors choose $\mathbf{u}(t, \lambda, \theta)$ to be the 'horizontal' flow velocity computed on a certain surface $\mathring{r} = \mathring{r}(t, \lambda, \theta)$ which lies between[7] the bottom and free surface so that the following expression makes sense:

$$\mathbf{u}(t, \lambda, \theta) \stackrel{\text{def}}{:=} \mathscr{U}\bigl(t, \lambda, \theta, \mathring{r}(t, \lambda, \theta)\bigr).$$

In other works (see e.g. [123, 333]) $\mathbf{u}(t, \lambda, \theta)$ is taken to be the depth-averaged 'horizontal' velocity $\mathscr{U}(t, \lambda, \theta, \mathring{r})$ of (modified) Euler equations.

Using relation (3.40) we can similarly write the following decomposition for the covariant velocity vector $\mathscr{V}(t, \lambda, \theta, \mathring{r})$ as a sum of a component $\mathbf{v}(t, \lambda, \theta)$ independent from the 'vertical' coordinate and a dispersive addition $\mathbf{V}_d = \bigl(\mathcal{V}_{d,1}(t, \lambda, \theta, \mathring{r}), \mathcal{V}_{d,2}(t, \lambda, \theta, \mathring{r})\bigr)^{\top}$:

$$\mathscr{V}(t, \lambda, \theta, \mathring{r}) = \mathbf{v}(t, \lambda, \theta) + \mu^2\,\mathbf{V}_d(t, \lambda, \theta, \mathring{r}), \tag{3.45}$$

where as in (3.40) we have the following relations:

$$\mathbf{v} = \mathscr{G} + \mathbb{G}\cdot\mathbf{u}, \qquad \mathbf{V}_d = \mathbb{G}\cdot\mathbf{U}_d. \tag{3.46}$$

[6] We call this component dispersive, since it disappears from the equations if we take the dispersionless limit $\mu \to 0$.

[7] The moving boundaries can be included.

By integrating the representation (3.44) over the water depth, we trivially obtain:

$$\frac{1}{\mathcal{H}} \int_{-\mathring{h}}^{\varepsilon \mathring{\eta}} \mathscr{U}\,(t,\, \lambda,\, \theta,\, \mathring{r})\; \mathrm{d}\mathring{r} \;=\; \mathbf{u}\,(t,\, \lambda,\, \theta) \,+\, \mu^{2}\,\mathcal{U}\,(t,\, \lambda,\, \theta)\,, \tag{3.47}$$

where the total water depth $\mathcal{H}\,(t,\, \lambda,\, \theta)$ is defined as

$$\mathcal{H} \stackrel{\text{def}}{:=} \varepsilon\,\mathring{\eta} \,+\, \mathring{h} \;\equiv\; \varepsilon\,\eta \,+\, h\,.$$

We introduced another depth-averaged contravariant velocity variable:

$$\mathcal{U}\,(t,\, \lambda,\, \theta) \;=\; \bigl(\mathcal{U}^{1}\,(t,\, \lambda,\, \theta),\, \mathcal{U}^{2}\,(t,\, \lambda,\, \theta)\bigr)^{\top} \stackrel{\text{def}}{:=} \frac{1}{\mathcal{H}} \int_{-\mathring{h}}^{\varepsilon \mathring{\eta}} \mathbf{U}_{d}\;(t,\, \lambda,\, \theta,\, \mathring{r})\; \mathrm{d}\mathring{r}\,.$$

Similarly, one can introduce the depth-averaged covariant velocity component:

$$\mathcal{V}\,(t,\, \lambda,\, \theta) \;=\; \bigl(\mathcal{V}_{1}\,(t,\, \lambda,\, \theta),\, \mathcal{V}_{2}\,(t,\, \lambda,\, \theta)\bigr)^{\top} \stackrel{\text{def}}{:=}$$
$$\frac{1}{\mathcal{H}} \int_{-\mathring{h}}^{\varepsilon \mathring{\eta}} \mathbf{V}_{d}\;(t,\, \lambda,\, \theta,\, \mathring{r})\; \mathrm{d}\mathring{r} \;\equiv\; \mathbb{G}\boldsymbol{\cdot}\mathcal{U}\,.$$

The last identity comes from the independence of metric tensor components from the radial ('vertical') coordinate $\mathring{r}$ in modified EULER equations.

In general, the vector field $\mathcal{U}\,(t,\, \lambda,\, \theta)$ is not uniquely defined, unless some additional simplifying assumptions are adopted. For the moment we shall keep this function arbitrary (as we did in the globally plane case [189]) to derive the most general base long wave model. However, before this model can be applied to any particular situation, one has to express $\mathcal{U}$ in terms of other variables $\mathcal{H}\,(t,\, \lambda,\, \theta)$ and $\mathbf{u}\,(t,\, \lambda,\, \theta)$. In Physics such relations are usually called the *closures* (see e.g. [31, 313]).

3.3.2 Continuity Equation and the Radial Velocity Component

Let us integrate the continuity equation (3.37) in variable $\mathring{r}$ over the total water depth:

$$\bar{\boldsymbol{\nabla}}\boldsymbol{\cdot}\Bigl[\,\mathcal{J} \int_{-\mathring{h}}^{\varepsilon \mathring{\eta}} \mathscr{U}\; \mathrm{d}\mathring{r}\,\Bigr] - \mathcal{J}\,\bigl(\varepsilon\,\mathscr{U}\boldsymbol{\cdot}\boldsymbol{\nabla}\mathring{\eta} - \mathscr{W}\bigr)\,\bigr|^{\mathring{r}=\varepsilon\,\mathring{\eta}} - \mathcal{J}\,\bigl(\mathscr{U}\boldsymbol{\cdot}\boldsymbol{\nabla}\mathring{h} + \mathscr{W}\bigr)\Bigr|_{\mathring{r}=-\mathring{h}} = 0\,.$$

From the last identity using boundary conditions (3.41), (3.42) along with Eq. (3.47), we obtain the continuity equation on a sphere:

$$(\mathcal{J}\mathcal{H})_t + \bar{\nabla}\cdot\bigl[\mathcal{J}\mathcal{H}\mathbf{u}\bigr] = -\mu^2\,\bar{\nabla}\cdot\bigl[\mathcal{J}\mathcal{H}\mathcal{U}\bigr].$$

Finally, by using the curvilinear divergence operator definition (3.12) we can rewrite the continuity equation in a more familiar form (to be compared with the flat case from the Chap. 1):

$$\mathcal{H}_t + \nabla\cdot\bigl[\mathcal{H}\mathbf{u}\bigr] = -\mu^2\,\nabla\cdot\bigl[\mathcal{H}\mathcal{U}\bigr], \tag{3.48}$$

where we remind that

$$\nabla\cdot\bigl[\mathcal{H}\mathbf{u}\bigr] \equiv \frac{(\mathcal{J}\mathcal{H}\mathrm{u}^1)_\lambda + (\mathcal{J}\mathcal{H}\mathrm{u}^2)_\theta}{\mathcal{J}}, \qquad \nabla\cdot\bigl[\mathcal{H}\mathcal{U}\bigr] \equiv \frac{(\mathcal{J}\mathcal{H}\mathcal{U}^1)_\lambda + (\mathcal{J}\mathcal{H}\mathcal{U}^2)_\theta}{\mathcal{J}}.$$

By integrating the same continuity equation (3.37) in the radial coordinate from $-\mathring{h}$ to $\mathring{r}$ we obtain

$$\mathcal{W} = \underbrace{-\mathcal{D}\mathring{h} - (\mathring{r} + \mathring{h})\,\nabla\cdot\mathbf{u}}_{(\blacklozenge)} + \mathcal{O}(\mu^2). \tag{3.49}$$

It is natural to adopt the term ($\blacklozenge$) as the 'vertical' velocity in the nonlinear dispersive model:

$$\varpi \stackrel{\mathrm{def}}{:=} -\mathcal{D}\mathring{h} - (\mathring{r} + \mathring{h})\,\nabla\cdot\mathbf{u}, \qquad \varpi \preccurlyeq \mathcal{W} + \mathcal{O}(\mu^2).$$

The last relation $\varpi \preccurlyeq \mathcal{W}$ denotes the fact that a quantity ϖ is obtained from $\mathcal{W}$ by an asymptotic truncation. Thus, we can say informally that ϖ contains less information than $\mathcal{W}$. The operator $\mathcal{D}$ is the 'horizontal' material derivative defined traditionally as

$$\mathcal{D} \stackrel{\mathrm{def}}{:=} \frac{\partial}{\partial t} + \mathbf{u}\cdot\nabla.$$

3.3.3 Pressure Representation

In order to derive the pressure field approximation in terms of variables $\mathcal{H}$ and $\mathbf{u}$, we integrate the 'vertical' momentum equation (3.39) over the radial coordinate from $\mathring{r}$ to $\varepsilon\,\mathring{\eta}$. By taking into account the 'horizontal' velocity ansatz (3.44) we obtain

$$p = \mu^2 \int_{\mathring{r}}^{\varepsilon\mathring{\eta}} \underbrace{\bigl(\mathcal{D}\mathcal{W} + \mathcal{W}\mathcal{W}_\rho + \mathcal{O}(\mu^2)\bigr)}_{(\lozenge)}\,\mathrm{d}\rho - \mathring{r} + \varepsilon\,\mathring{\eta}. \tag{3.50}$$

We transform the expression ($\Diamond$) under the integral by employing the approximation (3.49) for the 'vertical' velocity component $\mathscr{W} \succcurlyeq \varpi$:

$$\mathcal{D}\mathscr{W} + \mathscr{W}\mathscr{W}_{\mathring{r}} = \mathcal{D}\varpi + \varpi\frac{\partial\varpi}{\partial\mathring{r}} + \mathcal{O}(\mu^2) =$$
$$-\mathcal{D}^2\mathring{h} - (\mathring{r} + \mathring{h})\cdot\mathcal{D}(\nabla\cdot\mathbf{u}) + (\mathring{r} + \mathring{h})\cdot(\nabla\cdot\mathbf{u})^2 + \mathcal{O}(\mu^2) \equiv$$
$$-(\mathring{r} + \mathring{h})\mathscr{R}_1 - \mathscr{R}_2 + \mathcal{O}(\mu^2),$$

where we introduced

$$\mathscr{R}_1 \stackrel{\text{def}}{:=} \mathcal{D}(\nabla\cdot\mathbf{u}) - (\nabla\cdot\mathbf{u})^2, \qquad \mathscr{R}_2 \stackrel{\text{def}}{:=} \mathcal{D}^2\mathring{h}.$$

By substituting the last asymptotic approximation for ($\Diamond$) into (3.50), we obtain the approximate pressure distribution over the fluid layer:

$$p = \underbrace{\mathcal{H} - (\mathring{r} + \mathring{h}) - \mu^2\Big[\Big(\mathcal{H} - (\mathring{r} + \mathring{h})\Big)\mathscr{R}_2 + \Big(\frac{\mathcal{H}^2}{2} - \frac{(\mathring{r} + \mathring{h})^2}{2}\Big)\mathscr{R}_1\Big]}_{(\square)} + \mathcal{O}(\mu^4).$$

The term ($\square$) will serve us as the pressure in the nonlinear long wave model:

$$\Pi \stackrel{\text{def}}{:=} \mathcal{H} - (\mathring{r} + \mathring{h}) - \mu^2\Big[\Big(\mathcal{H} - (\mathring{r} + \mathring{h})\Big)\mathscr{R}_2 + \Big(\frac{\mathcal{H}^2}{2} - \frac{(\mathring{r} + \mathring{h})^2}{2}\Big)\mathscr{R}_1\Big]. \tag{3.51}$$

We would like to insist on two important facts that have been just shown:

- Pressure Π approximates the three-dimensional pressure distribution p to the asymptotic order $\mathcal{O}(\mu^4)$, i.e. $p = \Pi + \mathcal{O}(\mu^4)$. This fact we will be also denoted by $\Pi \preccurlyeq p + \mathcal{O}(\mu^4)$.
- The expression (3.51) does not depend on the variable $\mathcal{U}$, which is to be specified later.

Remark 3.2 One can notice that

$$\mathcal{D}^2\mathring{h} = \mathcal{D}^2 h - \mathcal{D}^2\eta_{00} = \mathcal{D}^2 h - \mathcal{D}\Big(\frac{\varsigma}{\mu}u^2\Omega^2\sin(\gamma\theta)\cos(\gamma\theta)\Big).$$

Thus, we can write:

$$\mathfrak{D}^2 \mathring{h} = \mathscr{D}^2 h + \mathcal{O}\left(\frac{\varsigma}{\mu}\right).$$

By noticing that terms $\mathscr{R}_{1,2}$ appear with the coefficient μ^2, one can equivalently define $\mathscr{R}_2$ as

$$\mathscr{R}_2^\star \stackrel{\text{def}}{:=} \mathscr{D}^2 h,$$

consistently with modified EULER equations. The difference between two quantities being asymptotically negligible, i.e.

$$\mu^2 \left(\mathscr{R}_2 - \mathscr{R}_2^\star\right) = \mathcal{O}(\varsigma\,\mu).$$

3.3.4 Momentum Equations

In order to derive the 'horizontal' momentum equations in the nonlinear dispersive wave model, we integrate the horizontal momentum equation (3.38) over the fluid layer depth. By taking into account the dynamic boundary condition (3.43), we obtain:

$$\int_{-\mathring{h}}^{\varepsilon\mathring{\eta}} \underbrace{\left(\mathscr{V}_t + (\mathscr{U}\cdot\nabla)\mathscr{V} + \mathscr{W}\,\mathscr{V}_{\mathring{r}}\right)}_{(\bigstar)} \mathrm{d}\mathring{r} + \nabla \int_{-\mathring{h}}^{\varepsilon\mathring{\eta}} p\,\mathrm{d}\mathring{r} - p\,|_{\mathring{r}=-\mathring{h}} \cdot \nabla\mathring{h} = \int_{-\mathring{h}}^{\varepsilon\mathring{\eta}} \mathscr{S}\,\mathrm{d}\mathring{r}. \qquad (3.52)$$

We underline the fact that the last identity is exact. In order to simplify this equation, we exploit first the approximation for p worked out above:

$$\nabla \int_{-\mathring{h}}^{\varepsilon\mathring{\eta}} p\,\mathrm{d}\mathring{r} = \nabla\mathscr{P} + \mathcal{O}(\mu^4), \qquad p\,|_{\mathring{r}=-\mathring{h}} = \check{p} + \mathcal{O}(\mu^4),$$

where we introduced the depth-integrated and bottom pressures defined, respectively, as

$$\mathscr{P} \stackrel{\text{def}}{:=} \int_{-\mathring{h}}^{\varepsilon\mathring{\eta}} \Pi\,\mathrm{d}\mathring{r} \equiv \frac{\mathcal{H}^2}{2} - \mu^2\wp,$$

$$\check{p} \stackrel{\text{def}}{:=} \Pi\,|_{\mathring{r}=-\mathring{h}} \equiv \mathcal{H} - \mu^2\check{\wp}.$$

Above we separated hydrostatic terms from non-hydrostatic ones summarized in functions $\wp$ and $\check{\wp}$ defined as

$$\wp \stackrel{\text{def}}{:=} \frac{\mathcal{H}^3}{3}\,\mathscr{R}_1 + \frac{\mathcal{H}^2}{2}\,\mathscr{R}_2^{\star}, \qquad \check{\wp} \stackrel{\text{def}}{:=} \frac{\mathcal{H}^2}{2}\,\mathscr{R}_1 + \mathcal{H}\mathscr{R}_2^{\star}.$$

Physically, in our dispersive wave model $\check{\wp}$ is the non-hydrostatic pressure trace at the bottom and $\wp$ is the depth-integrated non-hydrostatic pressure component. Then, we have

$$\check{\wp}\,\nabla\mathring{h} = \bigl(\mathcal{H} - \mu^2\check{\wp}\bigr)\cdot\bigl(\nabla h - \nabla\eta_{00}\bigr) = \bigl(\mathcal{H} - \mu^2\check{\wp}\bigr)\,\nabla h - \mathcal{H}\,\nabla\eta_{00} + \mu^2\,\check{\wp}\,\nabla\eta_{00}\,.$$

From (3.36) it follows that

$$\nabla\eta_{00} = \begin{pmatrix} 0 \\ \dfrac{\varsigma}{\mu}\,\Omega^2\,\sin(\gamma\theta)\,\cos(\gamma\theta)\end{pmatrix},$$

and consequently we have

$$\check{p}\,\nabla\mathring{h} = \bigl(\mathcal{H} - \mu^2\check{\wp}\bigr)\,\nabla h - \mathcal{H}\,\nabla\eta_{00} + \mathcal{O}\bigl(\mu^4 + \varsigma\mu + \varsigma^2\bigr).$$

Now we proceed to the approximation of remaining terms in the depth-integrated 'horizontal' momentum equation (3.52). The term involving the 'vertical' velocity component can be approximated as follows:

$$\int_{-\mathring{h}}^{\varepsilon\mathring{\eta}} \mathscr{W}\,\mathscr{V}_{\mathring{r}}\,\mathrm{d}\mathring{r} = \mu^2 \int_{-\mathring{h}}^{\varepsilon\mathring{\eta}} \varpi\,\frac{\partial\mathbf{V}_d}{\partial\mathring{r}}\,\mathrm{d}\mathring{r} + \mathcal{O}(\mu^4) =$$

$$\mu^2\Bigl\{ (\varpi\,\mathbf{V}_d)\bigr|_{\mathring{r}=-\mathring{h}}^{\mathring{r}=\varepsilon\mathring{\eta}} - \int_{-\mathring{h}}^{\varepsilon\mathring{\eta}} \mathbf{V}_d\,\frac{\partial\varpi}{\partial\mathring{r}}\,\mathrm{d}\mathring{r}\Bigr\} + \mathcal{O}(\mu^4)$$

$$= \mu^2\Bigl\{ -\bigl(\mathcal{D}\mathring{h} + \mathcal{H}(\nabla\cdot\mathbf{u})\bigr)\,\mathbf{V}_d\Bigr|^{\mathring{r}=\varepsilon\mathring{\eta}} +$$

$$\mathcal{D}\mathring{h}\,\mathbf{V}_d\Bigr|_{\mathring{r}=-\mathring{h}} + \mathcal{H}\mathcal{V}(\nabla\cdot\mathbf{u})\Bigr\} + \mathcal{O}(\mu^4).$$

The group of terms in (3.52) involving the 'horizontal' velocities are similarly transformed:

$$\int_{-\mathring{h}}^{\varepsilon\mathring{\eta}} \bigl[\mathscr{V}_t + (\mathscr{U}\cdot\nabla)\mathscr{V}\bigr]\,\mathrm{d}\mathring{r} =$$

$$\int_{-\mathring{h}}^{\varepsilon\mathring{\eta}} \Bigl[\bigl(\mathbf{v} + \mu^2\mathbf{V}_d\bigr)_t + \Bigl(\bigl(\mathbf{u} + \mu^2\mathbf{U}_d\bigr)\cdot\nabla\Bigr)\cdot\bigl(\mathbf{v} + \mu^2\mathbf{V}_d\bigr)\Bigr]\,\mathrm{d}\mathring{r}$$

$$= \int_{-\mathring{h}}^{\varepsilon\mathring{\eta}} \Big[\mathbf{v}_t + (\mathbf{u}\cdot\nabla)\,\mathbf{v}\Big]\,\mathrm{d}\mathring{r} +$$

$$\mu^2 \int_{-\mathring{h}}^{\varepsilon\mathring{\eta}} \Big[\mathbf{V}_{d\,t} + (\mathbf{u}\cdot\nabla)\,\mathbf{V}_d + (\mathbf{U}_d\cdot\nabla)\,\mathbf{v}\Big]\,\mathrm{d}\mathring{r} + \mathcal{O}(\mu^4)$$

$$= \mathcal{H}\big[\mathbf{v}_t + (\mathbf{u}\cdot\nabla)\,\mathbf{v}\big] + \mu^2\Big\{(\mathcal{H}\mathcal{V})_t - \varepsilon\,\mathring{\eta}_t\,\mathbf{V}_d\,\Big|^{\mathring{r}=\varepsilon\mathring{\eta}}$$

$$- \mathring{h}_t\,\mathbf{V}_d\,\Big|_{\mathring{r}=-\mathring{h}} + (\mathbf{u}\cdot\nabla)\cdot(\mathcal{H}\mathcal{V})$$

$$- \varepsilon\,(\mathbf{u}\cdot\nabla\mathring{\eta})\,\mathbf{V}_d\,\Big|^{\mathring{r}=\varepsilon\mathring{\eta}} - (\mathbf{u}\cdot\nabla\mathring{h})\,\mathbf{V}_d\,\Big|_{\mathring{r}=-\mathring{h}} + \mathcal{H}\,(\mathcal{U}\cdot\nabla)\,\mathbf{v}\Big\} + \mathcal{O}(\mu^4)$$

$$= \mathcal{H}\big[\mathbf{v}_t + (\mathbf{u}\cdot\nabla)\,\mathbf{v}\big] + \mu^2\Big\{(\mathcal{H}\mathcal{V})_t + (\mathbf{u}\cdot\nabla)\,(\mathcal{H}\mathcal{V})$$

$$+ \mathcal{H}\,(\mathcal{U}\cdot\nabla)\,\mathbf{v} - \varepsilon\,(\mathcal{D}\,\mathring{\eta})\,\mathbf{V}_d\,\Big|^{\mathring{r}=\varepsilon\mathring{\eta}} - (\mathcal{D}\,\mathring{h})\,\mathbf{V}_d\,\Big|_{\mathring{r}=-\mathring{h}}\Big\} + \mathcal{O}(\mu^4)\,.$$

Combining the last two results we obtain the following expression for the term (★):

$$\int_{-\mathring{h}}^{\varepsilon\mathring{\eta}} \underbrace{\big(\mathscr{V}_t + (\mathscr{U}\cdot\nabla)\mathscr{V} + \mathscr{W}\,\mathscr{V}_{\mathring{r}}\big)}_{(\bigstar)}\,\mathrm{d}\mathring{r} =$$

$$\mathcal{H}\big[\mathbf{v}_t + (\mathbf{u}\cdot\nabla)\,\mathbf{v}\big] - \mu^2\,\underbrace{\big[\mathcal{D}\mathcal{H} + \mathcal{H}\,(\nabla\cdot\mathbf{u})\big]}_{(\maltese)}\,\mathbf{V}_d\,\Big|^{\mathring{r}=\varepsilon\mathring{\eta}}$$

$$+ \mu^2\Big\{(\mathcal{H}\mathcal{V})_t + (\mathbf{u}\cdot\nabla)\cdot(\mathcal{H}\mathcal{V}) + \mathcal{H}\,(\mathcal{U}\cdot\nabla)\,\mathbf{v} + \mathcal{H}\mathcal{V}\,(\nabla\cdot\mathbf{u})\Big\} + \mathcal{O}(\mu^4)\,.$$

The term (✠) can be simplified by taking into account the mass conservation equation (3.48):

$$(\maltese) \equiv \mathcal{D}\mathcal{H} + \mathcal{H}\,(\nabla\cdot\mathbf{u}) \overset{(3.48)}{=} -\mu^2\,\nabla\cdot(\mathcal{H}\mathcal{U}) = \mathcal{O}(\mu^2) \succcurlyeq 0\,.$$

The last step consists in averaging the right-hand side of Eq. (3.38):

$$\int_{-\mathring{h}}^{\varepsilon\mathring{\eta}} \mathscr{S}\,\mathrm{d}\mathring{r} = \mathcal{H}\,\boldsymbol{S} + \mathcal{O}(\varsigma\,\mu)\,.$$

The term $\mathcal{O}(\varsigma\,\mu)$ can be consistently neglected in accordance with modified EULER equations. The components of $\mathscr{S} = (\mathcal{S}_1,\,\mathcal{S}_2)^\top$ are defined in (3.35). The first component $\mathcal{S}_1 \equiv 0$, so its averaging is rather a trivial task. Let us focus on $\mathcal{S}_2$ component:

$$\mathcal{S}_2 = \frac{\varsigma}{\mu}\,(\Omega + u^1 + \mu^2 U_d^1)^2 \sin(\gamma\theta)\cos(\gamma\theta) = \underbrace{\frac{\varsigma}{\mu}\,(\Omega + u^1)^2 \sin(\gamma\theta)\cos(\gamma\theta)}_{S_2} + \mathcal{O}(\varsigma\,\mu)\,.$$

Thus, $\boldsymbol{S} = (0,\ S_2)^\top$ with S_2 defined on the line above.

After combining all the developments made above and neglecting the terms of order $\mathcal{O}(\mu^4)$, we obtain the required momentum balance equation in dimensionless variables:

$$\mathcal{H}\big(\mathbf{v}_t + (\mathbf{u}\boldsymbol{\cdot}\nabla)\,\mathbf{v}\big) + \nabla\Big(\frac{\mathcal{H}^2}{2}\Big) = \mathcal{H}\,\nabla h + \mathcal{H}\big(\boldsymbol{S} - \nabla\eta_{00}\big) + \mu^2\big(\nabla\wp - \breve{\wp}\,\nabla h\big)$$
$$- \mu^2\Big[(\mathcal{H}\mathcal{V})_t + (\mathbf{u}\boldsymbol{\cdot}\nabla)(\mathcal{H}\mathcal{V}) + \mathcal{H}(\mathcal{U}\boldsymbol{\cdot}\nabla)\,\mathbf{v} + \mathcal{H}\mathcal{V}(\nabla\boldsymbol{\cdot}\mathbf{u})\Big]. \qquad (3.53)$$

3.3.5 Base Model in Conservative Form

For theoretical and numerical analysis of the governing equations, it can be beneficial to recast equations in the conservative form. In order to achieve this goal, we have to introduce the tensorial product operation $\otimes$. For our modest purposes it is sufficient to define this operation on vectors in $\mathbb{R}^2$. Let us take a covariant vector $\boldsymbol{\alpha} = (\alpha_1,\ \alpha_2)$ and a contravariant vector $\boldsymbol{\beta} = (\beta^1,\ \beta^2)$. Then, their tensorial product is defined as

$$\boldsymbol{\alpha}\otimes\boldsymbol{\beta} \stackrel{\text{def.}}{:=} \begin{pmatrix} \alpha_1\beta^1 & \alpha_1\beta^2 \\ \alpha_2\beta^1 & \alpha_2\beta^2 \end{pmatrix}.$$

The divergence of the 2–tensor $\boldsymbol{\alpha}\otimes\boldsymbol{\beta}$ is a vector defined as

$$\nabla\boldsymbol{\cdot}(\boldsymbol{\alpha}\otimes\boldsymbol{\beta}) \stackrel{\text{def}}{:=} \begin{pmatrix} \nabla\boldsymbol{\cdot}(\alpha_1\boldsymbol{\beta}) \\ \nabla\boldsymbol{\cdot}(\alpha_2\boldsymbol{\beta}) \end{pmatrix},$$

or in component-wise form as

$$\nabla\boldsymbol{\cdot}(\boldsymbol{\alpha}\otimes\boldsymbol{\beta}) \stackrel{\text{def}}{:=} \frac{1}{\mathcal{J}}\begin{pmatrix} \big(\mathcal{J}\alpha_1\beta^1\big)_\lambda + \big(\mathcal{J}\alpha_1\beta^2\big)_\theta \\ \big(\mathcal{J}\alpha_2\beta^1\big)_\lambda + \big(\mathcal{J}\alpha_2\beta^2\big)_\theta \end{pmatrix}. \qquad (3.54)$$

Then, by using this definition of the tensor product (in curvilinear coordinates), one can show the following formula remains true by analogy with the usual (i.e. ‘flat’) vector calculus:

$$\nabla \cdot (\boldsymbol{\alpha} \otimes \boldsymbol{\beta}) \equiv \boldsymbol{\alpha} \, (\nabla \cdot \boldsymbol{\beta}) + (\boldsymbol{\beta} \cdot \nabla) \, \boldsymbol{\alpha} \, .$$

We have set up all the tools to transform the momentum equation (3.53). By multiplying the continuity equation (3.48) by $\mathbf{v}$ and adding it to (3.53) we obtain the balance equation for the total 'horizontal' momentum:

$$(\mathcal{H} \mathbf{v})_t + \nabla \cdot (\mathcal{H} \mathbf{v} \otimes \mathbf{u}) + \nabla \Big(\frac{\mathcal{H}^2}{2} \Big) = \mathcal{H} \nabla h + \mathcal{H} \big(\boldsymbol{S} - \nabla \eta_{00} \big)$$
$$+ \mu^2 \big(\nabla \wp - \check{\wp} \nabla h \big) - \mu^2 \Big[(\mathcal{H} \mathcal{V})_t + \nabla \cdot (\mathcal{H} \mathbf{v} \otimes \mathcal{U}) + \nabla \cdot (\mathcal{H} \mathcal{V} \otimes \mathbf{u}) \Big] .$$

By using Formula (3.54), we can rewrite the last equation in partial derivatives:

$$(\mathcal{J} \mathcal{H} \mathbf{v})_t + \big[\mathcal{J} \mathcal{H} \mathbf{v} u^1 \big]_\lambda + \big[\mathcal{J} \mathcal{H} \mathbf{v} u^2 \big]_\theta + \nabla \Big(\mathcal{J} \frac{\mathcal{H}^2}{2} \Big)$$
$$= \mathcal{J} \mathcal{H} \nabla h + \frac{\mathcal{H}^2}{2} \nabla \mathcal{J} + \mathcal{J} \mathcal{H} \big(\boldsymbol{S} - \nabla \eta_{00} \big) + \mu^2 \mathcal{J} \big(\nabla \wp - \check{\wp} \nabla h \big)$$
$$- \mu^2 \Big\{ (\mathcal{J} \mathcal{H} \mathcal{V})_t + \big(\mathcal{J} \mathcal{H} (\mathcal{U}^1 \mathbf{v} + u^1 \mathcal{V}) \big)_\lambda + \big(\mathcal{J} \mathcal{H} (\mathcal{U}^2 \mathbf{v} + u^2 \mathcal{V}) \big)_\theta \Big\} .$$

The conservative structure of the governing equations will be exploited in the following chapter in order to construct a modern finite volume TVD scheme for the numerical simulation of nonlinear dispersive waves on a sphere.

Transformation of the Source Term

Some additional simplification can be achieved in the last equation if we analyse the expression $\mathcal{J} \mathcal{H} \big(\boldsymbol{S} - \nabla \eta_{00} \big)$. Indeed, this expression contains centrifugal force terms (proportional to) $\propto \Omega^2$. It is not difficult to see that these terms cancel each other due to the judicious choice of the 'still water' level $r = \eta_{00}(\theta)$. As a result, we obtain

$$\boldsymbol{S} - \nabla \eta_{00} = \begin{pmatrix} 0 \\ \frac{\varsigma}{\mu} \big(2 \Omega u^1 + (u^1)^2 \big) \sin(\gamma \theta) \cos(\gamma \theta) \end{pmatrix} \stackrel{\text{def}}{=:} \boldsymbol{S}^\star .$$

The momentum balance equation reads now

$$(\mathcal{J} \mathcal{H} \mathbf{v})_t + \big[\mathcal{J} \mathcal{H} \mathbf{v} u^1 \big]_\lambda + \big[\mathcal{J} \mathcal{H} \mathbf{v} u^2 \big]_\theta + \nabla \Big(\mathcal{J} \frac{\mathcal{H}^2}{2} \Big)$$

$$= \mathcal{J}\mathcal{H}\nabla h + \frac{\mathcal{H}^2}{2}\nabla\mathcal{J} + \mathcal{J}\mathcal{H}\boldsymbol{S}^\star + \mu^2\mathcal{J}(\nabla\wp - \check{\wp}\nabla h)$$
$$- \mu^2\Big\{(\mathcal{J}\mathcal{H}\mathcal{V})_t + (\mathcal{J}\mathcal{H}(\mathcal{U}^1\mathbf{v} + u^1\mathcal{V}))_\lambda + (\mathcal{J}\mathcal{H}(\mathcal{U}^2\mathbf{v} + u^2\mathcal{V}))_\theta\Big\}.$$

3.3.6 Base Model in Dimensional Variables

In applications it can be useful to have the governing equations in dimensional (unscaled) variables. In this way we obtain the following system of equations which constitute what we call the *base model* in the present study:

$$(\mathcal{J}\mathcal{H})_t + \bar{\nabla}\cdot\big[\mathcal{J}\mathcal{H}\mathbf{u}\big] = -\bar{\nabla}\cdot\big[\mathcal{J}\mathcal{H}\mathcal{U}\big], \tag{3.55}$$

$$(\mathcal{J}\mathcal{H}\mathbf{v})_t + \bar{\nabla}\cdot\big[\mathcal{J}\mathcal{H}\mathbf{v}\bar{\otimes}\mathbf{u}\big] + g\nabla\Big(\mathcal{J}\frac{\mathcal{H}^2}{2}\Big)$$
$$= g\mathcal{H}\mathcal{J}\nabla h + \mathcal{J}\mathcal{H}\boldsymbol{S}^\star + g\frac{\mathcal{H}^2}{2}\nabla\mathcal{J} + \mathcal{J}(\nabla\wp - \check{\wp}\nabla h)$$
$$- \Big\{(\mathcal{J}\mathcal{H}\mathcal{V})_t + (\mathcal{J}\mathcal{H}(\mathcal{U}^1\mathbf{v} + u^1\mathcal{V}))_\lambda + (\mathcal{J}\mathcal{H}(\mathcal{U}^2\mathbf{v} + u^2\mathcal{V}))_\theta\Big\}, \tag{3.56}$$

where $\bar{\otimes}$ is the usual 'flat' tensorial product of two vectors, $\mathcal{H} \stackrel{\text{def}}{:=} \eta + h$ and covariant $\mathbf{v} = (v_1, v_2)^\top$ / contravariant $\mathbf{u} = (u^1, u^2)^\top$ components of the velocity vectors are related as

$$\mathbf{v} = \mathscr{G} + \mathbb{G}\cdot\mathbf{u}, \qquad \mathscr{G} \stackrel{\text{def}}{:=} \begin{pmatrix} g_{01} \\ 0 \end{pmatrix}, \qquad \mathbb{G} \stackrel{\text{def}}{:=} \begin{pmatrix} g_{11} & 0 \\ 0 & g_{22} \end{pmatrix}.$$

The JACOBIAN $\mathcal{J}$ and the metric tensor components $\{g_{ij}\}_{0 \leqslant i, j \leqslant 2}$ are computed as specified in (3.32). The velocity variable $\mathcal{U} = (\mathcal{U}^1, \mathcal{U}^2)^\top$ has to be specified by a closure relation. Then, the vector function $\mathcal{V} = (\mathcal{V}_1, \mathcal{V}_2)^\top$ is recomputed using relation $\mathcal{V} = \mathbb{G}\cdot\mathcal{U}$.

Physically, the vector function $\mathcal{U}$ describes the deviation of the chosen 'horizontal' velocity variable $\mathbf{u}$ from the depth-averaged profile. Below we shall consider two particular (and also popular) choices of $\mathcal{U}$ leading to different models which can be already used in practical applications.

3.4 Depth-Averaged Velocity Variable

One natural way to choose the approximate velocity variable $\mathbf{u}$ in the dispersive wave model is to take the three-dimensional 'horizontal' velocity field $\mathscr{U}(t, \lambda, \theta, \mathring{r})$ and replace it by its average over the total fluid depth, i.e.

$$\mathbf{u} = \frac{1}{\mathcal{H}} \int_{-\mathring{h}}^{\mathring{\eta}} \mathscr{U}(t, \lambda, \theta, \mathring{r})\, \mathrm{d}\mathring{r} .$$

Then, from formula (3.47) it follows that $\mathfrak{U} \equiv \mathbf{0}$. It is probably the simplest possible closure relation that one can find. By substituting $\mathfrak{U} = \mathbf{0}$ into governing equations of the base model we obtain:

$$(\mathcal{J}\mathcal{H})_t + \bar{\nabla}\cdot\big[\mathcal{J}\mathcal{H}\mathbf{u}\big] = 0, \tag{3.57}$$

$$\begin{aligned}(\mathcal{J}\mathcal{H}\mathbf{v})_t + \bar{\nabla}\cdot\big[\mathcal{J}\mathcal{H}\mathbf{v}\bar{\otimes}\mathbf{u}\big] + g\,\nabla\Big(\mathcal{J}\,\frac{\mathcal{H}^2}{2}\Big) \\ = g\,\mathcal{H}\mathcal{J}\nabla h + \mathcal{J}\mathcal{H}\boldsymbol{S}^\star + g\,\frac{\mathcal{H}^2}{2}\,\nabla\mathcal{J} + \mathcal{J}\big(\nabla\wp - \check{\wp}\,\nabla h\big).\end{aligned} \tag{3.58}$$

3.4.1 *Base Model in Terms of the Linear Velocity Components*

The last system of equations describes the evolution of the total water depth $\mathcal{H}$ and two contravariant components u^1, u^2 of the velocity vector $\mathbf{u}$. However, for the numerical modelling and the interpretation of obtained results it can be more convenient to use directly the linear components u, v of this vector:

$$u \overset{\text{def}}{:=} \sqrt{g_{11}}\,u^1 \equiv R\,u^1 \sin\theta, \qquad v \overset{\text{def}}{:=} \sqrt{g_{22}}\,u^2 \equiv R\,u^2 .$$

By using Formulas (3.32) and relation (3.46), Eqs. (3.57) and (3.58) become:

$$(\mathcal{H}R\sin\theta)_t + \big[\mathcal{H}u\big]_\lambda + \big[\mathcal{H}v\sin\theta\big]_\theta = 0, \tag{3.59}$$

$$\begin{aligned}(\mathcal{H}u R\sin\theta)_t + \Big[\mathcal{H}u^2 + g\,\frac{\mathcal{H}^2}{2}\Big]_\lambda + \big[\mathcal{H}u v\sin\theta\big]_\theta \\ = g\,\mathcal{H}h_\lambda - \mathcal{H}u v\cos\theta - \digamma\,\mathcal{H}v R\sin\theta + \wp_\lambda - \check{\wp}\,h_\lambda,\end{aligned} \tag{3.60}$$

$$(\mathcal{H} v R \sin\theta)_t + \bigl[\mathcal{H} u v\bigr]_\lambda + \Bigl[\Bigl(\mathcal{H} v^2 + g\,\frac{\mathcal{H}^2}{2}\Bigr)\sin\theta\Bigr]_\theta$$
$$= g\,\mathcal{H} h_\theta \sin\theta + g\,\frac{\mathcal{H}^2}{2}\cos\theta + \mathcal{H} u^2\cos\theta +$$
$$\digamma\,\mathcal{H} u R \sin\theta + \bigl(\wp_\theta - \check{\wp}\, h_\theta\bigr)\sin\theta\,, \qquad (3.61)$$

where $\digamma \stackrel{\text{def}}{:=} 2\,\Omega\,\cos\,\theta$ is the CORIOLIS parameter defined in terms of the colatitude θ. In equations above u^2 and v^2 denote $u \cdot u$ and $v \cdot v$, respectively. The quantities $\mathscr{R}_\alpha, \alpha = 1, 2$ arising in $\wp$ and $\check{\wp}$ can be computed by the following formulas:

$$\mathscr{R}_1 \stackrel{\text{def}}{:=} (\nabla \cdot \mathbf{u})_t + \frac{1}{R\,\sin\,\theta}\Bigl\{u\,(\nabla \cdot \mathbf{u})_\lambda + v\,(\nabla \cdot \mathbf{u})_\theta\,\sin\,\theta\Bigr\} - (\nabla \cdot \mathbf{u})^2\,,$$
$$\mathscr{R}_2^\star \stackrel{\text{def}}{:=} (\mathcal{D}\,h)_t + \frac{1}{R\,\sin\,\theta}\Bigl\{u\,(\mathcal{D}\,h)_\lambda + v\,(\mathcal{D}\,h)_\theta\,\sin\,\theta\Bigr\}\,,$$

where the divergence operator $\nabla \cdot \mathbf{u}$ and the material derivative $\mathcal{D}\,h$ can be computed in linear velocity components u, v as

$$\nabla \cdot \mathbf{u} \stackrel{\text{def}}{:=} \frac{1}{R\,\sin\,\theta}\Bigl\{u_\lambda + (v\,\sin\,\theta)_\theta\Bigr\}\,,$$
$$\mathcal{D}\,h \stackrel{\text{def}}{:=} h_t + \frac{1}{R\,\sin\,\theta}\Bigl\{u\,h_\lambda + v\,h_\theta\,\sin\,\theta\Bigr\}\,.$$

The hydrodynamic model (3.59)–(3.61) presented in this section is the base fully nonlinear weakly dispersive model with depth-averaged velocity written in dimensional variables. We stress out that in the derivation above the flow irrotationality has never been assumed. It plays the same rôle in the spherical geometry as the so popular nowadays SERRE–GREEN–NAGHDI equations [147, 148, 283, 297] in the flat case (see Chap. 1 for the derivation of the base model on a globally flat space). By applying further simplifications to these equations we can obtain weakly nonlinear dispersive and fully nonlinear dispersionless equations.

3.4.2 *Weakly Nonlinear Model*

Above we considered the fully nonlinear base model with depth-averaged velocity variable. During the derivation of this model we have not assumed that the wave scaled amplitude $\varepsilon = \mathcal{O}(1)$ is a small parameter. In the present section we propose a weakly nonlinear weakly dispersive model. Analogues of this model have been used for the numerical modelling of tsunami propagation in the ocean [80, 223, 224].

Weakly nonlinear models can be easily obtained from their fully nonlinear counterparts by adopting the simplifying assumption $\varepsilon \ll 1$. It is quite common to work in the so-called BOUSSINESQ regime [43, 46, 93] where we relate the nonlinear parameter to the magnitude of the dispersion:

$$\varepsilon = \mathcal{O}(\mu^2).$$

Thus, the terms of order $\mathcal{O}(\varepsilon^2 + \varepsilon\mu^2)$ can be neglected in the governing equations. As a result we obtain the same governing equations (3.59)–(3.61) with one important modification—in the computation of dispersive terms we use the following linearized formulas:[8]

$$\wp \stackrel{\text{def}}{:=} \frac{h^3}{3}\mathscr{R}_1^{\text{wnl}} + \frac{h^2}{2}\mathscr{R}_2^{\text{wnl}},$$

$$\check{\wp} \stackrel{\text{def}}{:=} \frac{h^2}{2}\mathscr{R}_1^{\text{wnl}} + h\,\mathscr{R}_2^{\text{wnl}},$$

and $\mathscr{R}_\alpha^{\text{wnl}}$, $\alpha = 1, 2$ are defined as

$$\mathscr{R}_1^{\text{wnl}} \stackrel{\text{def}}{:=} (\boldsymbol{\nabla}\cdot\mathbf{u})_t, \qquad \mathscr{R}_2^{\text{wnl}} \stackrel{\text{def}}{:=} (\mathcal{D}h)_t + \frac{1}{R\sin\theta}\Big\{u\,h_{t\lambda} + v\,h_{t\theta}\sin\theta\Big\}.$$

3.4.3 Dispersionless Shallow Water Equations

Another important particular system can be trivially obtained from the base model (3.59)–(3.61) with the depth-averaged velocity by neglecting the non-hydrostatic terms $\wp$, $\check{\wp}$, which have the asymptotic order $\mathcal{O}(\mu^2)$. We reiterate on the fact that the three-dimensional flow irrotationality is not needed as a simplifying assumption. In this way we obtain a dispersionless model similar to nonlinear shallow water (or SAINT-VENANT) equations in the globally flat space [84]. This system of equations (3.59)–(3.61) (without non-hydrostatic terms) has the hyperbolic type. Consequently, it is natural to use finite volume methods for the numerical discretization of these equations [19, 215]. This method was proven to be very successful in solving hydrodynamic problems in coastal areas (see e.g. [113, 187]). The unique form of Eqs. (3.59)–(3.61) for the entire hierarchy of asymptotic hydrodynamic models is very beneficial for the development of efficient numerical algorithms. Namely, we expect that neglecting non-hydrostatic terms in the discrete equations will result in a robust finite volume scheme for the remaining hyperbolic part of the equations. Numerical discretizations respecting the hierarchy of hydrodynamic models will be developed in the following Chap. 4 by analogy with the globally flat space.

[8]In the superscripts we use the abbreviation 'wnl' which stands for 'weakly nonlinear'.

3.4.4 State-of-the-Art

In this section we make a review of published literature devoted to the derivation and/or application of nonlinear dispersive wave models with the depth-averaged velocity variable. The case of the velocity defined on an arbitrary surface in the fluid bulk will be discussed below in Sect. 3.6 (see Sect. 3.6.4 for the corresponding literature review).

We can report a BOUSSINESQ-type model in spherical coordinates which uses the depth-averaged velocity variable [80]. The equations presented in that study have the advantage of being written in the conservative form. The bathymetry is assumed to be stationary. The CORIOLIS force is taken into account. However, some nonlinear terms in the right-hand side (such as $\mathcal{H}\, u\, v\, \cos\theta$ and $\mathcal{H}\, u^2\, \cos\theta$) are omitted. Finally, the dispersive terms are written as if the bottom were flat. In the notation of our study, Dao and Tkalich (2007) take the dispersive terms as

$$\wp = \frac{\mathcal{H}\, h^2}{3}\, \mathscr{R}_1^{\,\mathrm{wnl}}, \qquad \check{\wp} \equiv 0, \qquad \mathscr{R}_2^{\,\mathrm{wnl}} \equiv 0.$$

Moreover, $\mathscr{R}_1^{\,\mathrm{wnl}}$ is simplified by assuming that $\cos\theta \ll 1$. For the justification of this form of dispersive terms Dao and Tkalich refer to [168]. Then, this weakly nonlinear and weakly dispersive model was incorporated into TUNAMI-N2 code, which was used to study SUMATRA 2004 event. The authors came to the conclusion that the inclusion of dispersive effects and EARTH's sphericity are needed to reproduce the observed data. In the aforementioned paper [168] the authors used CARTESIAN coordinates only. Their dispersive terms were directly transformed into spherical coordinates by Dao and Tkalich [80]. In the following publication Horrillo et al. [169] used the spherical coordinates and depth-integrated equations with non-hydrostatic pressure (similar non-hydrostatic barotropic models are well-known for the flat space, see e.g. [50, 82]). However, in contrast to [168], in [169] Horrillo et al. do not write explicitly non-hydrostatic terms.

Another dispersive model with depth-averaged velocity variable in spherical geometry was published in [223, 224]. This weakly nonlinear and weakly dispersive model is presented in a non-conservative form. Their model is the spherical counterpart of the classical PEREGRINE model well-known in flat space [264]. However, some additional dispersive terms are added in order to improve the linear dispersion properties. The new terms come with a coefficient γ. If we set $\gamma = 0$ in their model, one obtains weakly dispersive model presented above with all nonlinear dispersive and CORIOLIS terms neglected. Strictly speaking, only with $\gamma = 0$ their velocity variable can be interpreted as the depth-integrated one. For $\gamma \neq 0$ we rather have a spherical analogue of BEJI–NADAOKA system [27]. The derivation of this model can be found only in a technical report [259]. Later this model was called GLOBOUSS. This model was applied to model the propagation of a trans-Atlantic hypothetic tsunami resulting from an eventual landslide at the LA PALMA island [83, 164, 223, 322].

A linear dispersive model in spherical coordinates was used in [245]. The linearization was justified by the need to produce a 'fast' solution for the operational real time tsunami hazard forecast. The spherical BOUSSINESQ system was borrowed from [302]. Recently a parallel implementation of spherical weakly nonlinear BOUSSINESQ equations was reported in [15]. The authors used a conservative form of equations along with conservative variables. The authors came to the following conclusions:

> [...] A clear discrepancy was apparent from comparison of tsunami waveforms derived from dispersive and non-dispersive simulations at the DART21418 buoy located in the deep ocean. Tsunami soliton fission near the coast recorded by helicopter observations was accurately reproduced by the dispersive model with the high-resolution grids [...]

We are not aware of any fully nonlinear models using the depth-averaged velocity variable. In this respect the present work fills in this gap. In the following Chap. 4 we shall describe an efficient splitting[9]-type approach to solve an important representative of this class of dispersive wave models.

3.5 Conservation of Energy

The mathematical study of approximate models can be greatly simplified if a positive definite conserved energy functional can be found. The physical soundness of an approximation requires that the approximate model possesses at least the conserved mass, momentum and energy properties. Moreover, the conserved densities should be consistent with the conservation laws of the parent model. The presence of these properties greatly increases the value of a model. Below, in the spherical case, we provide the equation of the total energy balance in the spherical EULER and FNWD models with the depth-averaged velocity variable. In the incompressible case, this equation is a differential consequence of the mass conservation and momentum balance equations. Moreover, we show the consistency of the approximate energy density with respect to the spherical EULER equations to the same approximation order as used in the long wave approximation.

3.5.1 Energy Equation in the Spherical Euler Equations

We begin the consideration of the energy conservation property in the framework of the full (spherical) EULER equations (3.27) and (3.28) since they occupy the highest position in the hierarchy of mathematical models considered in this book. Moreover, this result will serve us as a reference expression for the sake of validation

[9]The splitting is naturally performed in the hyperbolic and elliptic parts of the governing equations. See [167] for a recent general introduction to the splitting methodology and [201] for the origins.

of the differential consequences of approximate models which qualify for the total mechanical energy approximation. The consistency of the approximation can be established by applying the same asymptotic expansions (and/or some other approximations) to the reference expression.

Taking into account the physical effects present in the model, the total energy $\mathcal{E}$ of the fluid flow described by the full EULER equations is a sum of the kinetic and potential energies. The latter includes the contributions of the gravity and centrifugal forces. Hence, the total energy density can be written as [131]:

$$\mathcal{E} = \frac{\rho}{2}\left[(r \sin\theta\, \mathcal{V}^1)^2 + (r\,\mathcal{V}^2)^2 + (\mathcal{V}^3)^2\right] + \rho\, g\,(r - R) - \frac{\rho}{2}\,\Omega^2 r^2 \sin^2\theta\,. \tag{3.62}$$

By multiplying each (scalar) Eq. (3.28) by the corresponding contravariant component of the velocity $\{\mathcal{V}^\beta\}_{\beta=1}^{\beta=3}$ and summing up the results, we obtain the energy conservation equation for the full (spherical) EULER equations:

$$(\mathcal{J}\,\mathcal{E})_t + \left[\mathcal{J}(\mathcal{E} + p)\,\mathcal{V}^1\right]_\lambda + \left[\mathcal{J}(\mathcal{E} + p)\,\mathcal{V}^2\right]_\theta + \left[\mathcal{J}(\mathcal{E} + p)\,\mathcal{V}^3\right]_r = 0\,. \tag{3.63}$$

We underline that the last equation is a conservation law and the quantity $\mathcal{J}\,\mathcal{E}$ is locally conserved as it can be easily seen.

3.5.2 *Energy Equation in the Modified Spherical Euler Equations*

Slightly below in the hierarchy of the mathematical models we have the modified EULER equations. Equations of this model are obtained under the assumption that the fluid layer depth is much smaller than the radius of the sphere. The modification consists in neglecting in the full EULER equations the terms which contain the relative fluid layer thickness ς as a multiplier of other finite quantities (cf. also Sect. 3.2.3). A similar method can be applied to derive also the energy conservation equation for the modified EULER equations. The first step consists in writing the total energy expression (3.62) in dimensionless variables:

$$\mathcal{E}^\star = \frac{1}{2}\Big[\big(\mathcal{V}^{1,\star}(\mathring{r}^\star\varsigma + 1)\sin(\gamma\theta^\star)\big)^2 + \big(\mathcal{V}^{2,\star}(\mathring{r}^\star\varsigma + 1)\big)^2 + \mu^2\big(\mathcal{V}^{3,\star}\big)^2\Big] \\ + \mathring{r}^\star - \frac{1}{2}(\mathring{r}^\star\varsigma + 1)^2(\Omega^\star)^2\sin^2(\gamma\theta^\star)\,,$$

where we wish to neglect the terms of the orders $\mathcal{O}(\varsigma)$ and $\mathcal{O}(\varsigma^2)$. As a result of this operation, we obtain the expression for the energy in modified EULER equations:

$$\mathcal{E}_{\mathrm{m}}^\star = \frac{1}{2}\Big[\big(\mathcal{V}^{1,\star}\sin(\gamma\theta^\star)\big)^2 + \big(\mathcal{V}^{2,\star}\big)^2 + \mu^2\big(\mathcal{V}^{3,\star}\big)^2\Big] + \mathring{r}^\star - \eta_{00}^\star(\theta^\star)\,. \tag{3.64}$$

Then, the energy conservation equation in the modified model can be consistently obtained by applying the same simplification in Eq. (3.63). It is not difficult to see that the modified equation in dimensional variables has the same structure as (3.63):

$$(\mathcal{J}\,\mathcal{E}_{\mathrm{m}})_t + \left[\mathcal{J}\,(\mathcal{E}_{\mathrm{m}} + p)\,\mathcal{V}^1\right]_\lambda + \left[\mathcal{J}\,(\mathcal{E}_{\mathrm{m}} + p)\,\mathcal{V}^2\right]_\theta + \left[\mathcal{J}\,(\mathcal{E}_{\mathrm{m}} + p)\,\mathcal{V}^3\right]_{\mathring{r}} = 0\,, \qquad (3.65)$$

where the energy density in dimensional variables reads:

$$\mathcal{E}_{\mathrm{m}} = \frac{\rho}{2}\left[\left(\mathcal{V}^1 R \sin\theta\right)^2 + \left(R\,\mathcal{V}^2\right)^2 + \left(\mathcal{V}^3\right)^2\right] + \rho\, g\, \mathring{r} - \rho\, g\, \eta_{00}(\theta)\,. \qquad (3.66)$$

We remind that the JACOBIAN $\mathcal{J}$ can be computed according to Formula (3.32) and the function $\eta_{00}(\theta)$ was provided in Eq. (3.34). We would like to mention also that Eq. (3.65) in dimensionless variables remains the same modulo the fact that the dimensionless JACOBIAN function can be computed as

$$\mathcal{J}^{\star} = -\sin(\gamma\,\theta^{\star})\,.$$

The derivation of this formula can be found above in Sect. 3.2.3.

Energy Equation as a Differential Consequence

In the previous Sect. 3.5.2, the energy equation for the modified EULER equations was obtained by taking the limit $\varsigma \rightarrow 0$ in the energy equation (3.63) for the complete (spherical) EULER equations. However, it turns out that Eq. (3.65) can be obtained directly from the modified EULER equations without reverting to the parent model. Indeed, by multiplying each equation of motion (3.28) by the corresponding contravariant component of velocity $\mathcal{V}^{\beta}$ ($\beta = 1, 2, 3$), summing up the results, taking into account formulas (3.32) for the covariant components of the metric tensor $\{g_{ij}\}$ and the modified expressions (3.33) for the right-hand side member $\grave{\mathcal{S}}_{\beta}$ in the momentum equation, we arrive to the following equality:

$$\begin{aligned}
&\frac{1}{2}\frac{\partial}{\partial t}\left[\left(\mathcal{V}^1 R \sin\theta\right)^2 + \left(R\,\mathcal{V}^2\right)^2 + \left(\mathcal{V}^3\right)^2\right] + \\
&\qquad \frac{1}{2}\,\mathcal{V}^{\beta}\,\frac{\partial}{\partial q^{\beta}}\left[\left(\mathcal{V}^1 R \sin\theta\right)^2 + \left(R\,\mathcal{V}^2\right)^2 + \left(\mathcal{V}^3\right)^2\right] + \\
&\qquad \frac{1}{\rho}\,\mathcal{V}^{\beta}\,\frac{\partial p}{\partial q^{\beta}} + g\,\mathcal{V}^3 + 2\,\mathcal{V}^1\mathcal{V}^2\,\Omega\, R^2 \sin\theta\cos\theta + \\
&(\mathcal{V}^1)^2\,\mathcal{V}^2 R^2 \sin\theta\cos\theta - (\Omega + \mathcal{V}^1)^2\,\mathcal{V}^2 R^2 \sin\theta\cos\theta = 0\,, \qquad (3.67)
\end{aligned}$$

where $q^1 = \lambda, q^2 = \theta, q^3 = \mathring{r}$ and the implicit summation of the repeated index $\beta \in \{1, 2, 3\}$ is performed. Taking into account the analogue of equalities (3.34) for dimensional variables, we obtain the following expression for the source terms involving the trigonometric functions:

$$-\mathcal{V}^2\,\Omega^2\,R^2\,\sin\theta\,\cos\theta \overset{(3.34)}{=} -\mathcal{V}^2\,\frac{\partial\big(g\,\eta_{00}\,(\theta)\big)}{\partial\theta} = -\frac{\partial\big(g\,\eta_{00}\big)}{\partial t} - \mathcal{V}^\beta\,\frac{\partial\big(g\,\eta_{00}\big)}{\partial q^\beta}\,.$$

Moreover, we may write the following trivial identity, which plays a certain rôle in our derivation:

$$g\,\mathcal{V}^3 = \mathcal{V}^3\,\frac{\partial(g\,\mathring{r})}{\partial\mathring{r}} = \frac{\partial(g\,\mathring{r})}{\partial t} + \mathcal{V}^\beta\,\frac{\partial(g\,\mathring{r})}{\partial q^\beta}\,.$$

In light of these elements, Eq. (3.67) can be rewritten in a much more compact form:

$$\frac{\partial\mathcal{E}_{\mathrm{m}}}{\partial t} + \mathcal{V}^\beta\,\frac{\partial\mathcal{E}_{\mathrm{m}}}{\partial q^\beta} = 0\,, \tag{3.68}$$

with $\mathcal{E}_{\mathrm{m}}$ being defined in Eq. (3.66). If one multiplies Eq. (3.66) by the JACOBIAN $\mathcal{J}$ and the continuity equation (3.22) by $\mathcal{E}_{\mathrm{m}}$, then sums up the results, we obtain the total energy conservation equation (3.68) in the conservative form.

In this section we showed that the energy conservation equation (3.65) can be obtained as a formal limit $\varsigma \to 0$ from the full EULER energy equation (3.63). Moreover, the same equation can be obtained also as a differential consequence from the modified EULER equations. Consequently, we can say that Eq. (3.65) is asymptotically consistent in the parameter ς with the total energy conservation equation (3.63) of the parent model. This situation is similar to what we already saw in Sect. 1.1.5 for the SGN model with respect to the small parameter μ and the full EULER equations (in the flat case). The persistence of these properties through the mathematical models presented in this book is a direct consequence of good modelling practices to which the authors subscribe.

3.5.3 Energy Conservation in the FNWD Model with the Depth-Averaged Velocity

The methodology of the total mechanical energy equation derivation for FNWD models in the globally flat and globally spherical geometries is essentially the same. Only small technical details differ. In this section we shall follow the well-known methodology that the reader already got familiarized with in Chap. 1. First of all, we shall obtain the expression for the total energy in the FNWD model with the depth-averaged velocity variable. Then, the energy balance equation will be derived by two different approaches. The fact that both approaches give the same result allows us to declare the asymptotic consistency property with the modified EULER equations on a rotating sphere.

In this section we shall use the usual notations. The computations will be performed in dimensional and, especially, dimensionless variables. For the sake of notation compactness, we shall omit here the superscript $\star$ over dimensionless variables. There will be no ambiguity here since dimensionless equations contain dimensionless coefficients such as μ^2, γ and, in some cases, the BACHMANN–LANDAU notation $\mathcal{O}(\cdot)$. For example, the expression (3.64) of the total mechanical energy for the modified EULER equations in this notation reads:

$$\mathcal{E}_{\mathrm{m}} = \frac{1}{2}\Big[\big(\mathcal{V}^1 \sin(\gamma\theta)\big)^2 + (\mathcal{V}^2)^2 + \mu^2 \mathcal{W}^2\Big] + \mathring{r} - \eta_{00}(\theta), \qquad (3.69)$$

where the steady free surface profile becomes:

$$\eta_{00}(\theta) = \frac{\Omega^2}{2}\sin^2(\gamma\theta).$$

With these notations in mind, we can begin the derivation.

Energy Expression in the FNWD Model with the Depth-Averaged Velocity

Let us integrate the expression (3.69) for the total energy in the modified EULER equations in dimensionless variables. If we take into account the decomposition (3.44) with the following closure relation valid for the depth-averaged velocity:

$$\int_{-\mathring{h}}^{\varepsilon\mathring{\eta}} \mathbf{U}_d\,(t, \lambda, \theta, \mathring{r})\,\mathrm{d}\mathring{r} = \mathbf{0}, \qquad (3.70)$$

along with the formula $\varpi \preccurlyeq \mathcal{W} + \mathcal{O}(\mu^2)$ for the radial component of the velocity (see Sect. 3.3.2), we obtain:

$$\frac{1}{\mathcal{H}}\int_{-\mathring{h}}^{\varepsilon\mathring{\eta}} \mathcal{E}_{\mathrm{m}}\,\mathrm{d}\mathring{r} = \frac{1}{\mathcal{H}}\int_{-\mathring{h}}^{\varepsilon\mathring{\eta}} \frac{1}{2}\Big[(u^1 + \mu^2\mathcal{U}_d^1)^2 \sin^2(\gamma\theta) + (u^2 + \mu^2\mathcal{U}_d^2)^2$$
$$+ \mu^2\big(\mathscr{D}\mathring{h} + (\mathring{r} + \mathring{h})(\boldsymbol{\nabla}\cdot\mathbf{u}) + \mathcal{O}(\mu^2)\big)^2\Big]\mathrm{d}\mathring{r} + \frac{1}{\mathcal{H}}\int_{-\mathring{h}}^{\varepsilon\mathring{\eta}} (\mathring{r} - \eta_{00})\,\mathrm{d}\mathring{r} =$$
$$\frac{1}{2}\Big[\big(u^1 \sin(\gamma\theta)\big)^2 + (u^2)^2 + \mu^2\Big(\frac{\mathcal{H}^2}{3}(\boldsymbol{\nabla}\cdot\mathbf{u})^2 + \mathcal{H}(\mathscr{D}\mathring{h})\cdot(\boldsymbol{\nabla}\cdot\mathbf{u}) + (\mathscr{D}\mathring{h})^2\Big)\Big]$$
$$+ \frac{1}{2\mathcal{H}}(\mathring{r} - \eta_{00})^2\Big|_{\mathring{r}=-\mathring{h}}^{\mathring{r}=\varepsilon\mathring{\eta}} + \mathcal{O}(\mu^4).$$

According to Remark 3.2, the quantity $\mathscr{D}\mathring{h}$ can be replaced by $\mathscr{D}h$. Consequently,

$$\frac{1}{\mathcal{H}}\int_{-\mathring{h}}^{\varepsilon\mathring{\eta}} \mathcal{E}_{\mathrm{m}}\,\mathrm{d}\mathring{r} = \mathscr{E} + \mathcal{O}(\mu^4), \tag{3.71}$$

where the quantity $\mathscr{E}$ is defined as

$$\mathscr{E} = \frac{\left(u^1 \sin(\gamma\theta)\right)^2 + (u^2)^2}{2} + \\ \mu^2\left[\frac{\mathcal{H}^2}{6}(\nabla\cdot\mathbf{u})^2 + \frac{\mathcal{H}}{2}(\mathscr{D}h)\cdot(\nabla\cdot\mathbf{u}) + \frac{(\mathscr{D}h)^2}{2}\right] + \frac{\mathcal{H} - 2h}{2}. \tag{3.72}$$

We can see that the quantity $\mathscr{E}$ approximates the depth-averaged total energy $\mathcal{E}_{\mathrm{m}}$ of the 3D flow with the same asymptotic order $\mathcal{O}(\mu^4)$, which were retained in the derivation of the spherical FNWD model with the depth-averaged velocity variable. Consequently, we may identify the quantity $\mathscr{E}$ defined in (3.72) with the total energy of the flow described by FNWD model.

Consistency of the Energy Balance Equation with the Modified Euler Energy Conservation Equation

In the beginning we perform the derivation of the energy conservation equation for the FNWD model starting from the corresponding Eq. (3.65) of the modified EULER equations. To achieve this goal, we shall integrate Eq. (3.65) in dimensionless variables over the total water depth. The derivatives will be taken out from the integral signs with the help of boundary conditions (3.41)–(3.43). Moreover, the use of the velocity decomposition (3.44) along with the formula $\varpi \preccurlyeq \mathscr{W} + \mathcal{O}(\mu^2)$ and Definition (3.51) allows us to perform the following sequence of transformations:

$$\int_{-\mathring{h}}^{\varepsilon\mathring{\eta}} \Big[(\mathcal{J}\mathcal{E}_{\mathrm{m}})_t + \left[\mathcal{J}(\mathcal{E}_{\mathrm{m}} + p)\mathcal{V}^1\right]_\lambda + \left[\mathcal{J}(\mathcal{E}_{\mathrm{m}} + p)\mathcal{V}^2\right]_\theta + \\ \left[\mathcal{J}(\mathcal{E}_{\mathrm{m}} + p)\mathscr{W}\right]_{\mathring{r}}\Big]\mathrm{d}\mathring{r} = \left[\mathcal{J}\int_{-\mathring{h}}^{\varepsilon\mathring{\eta}} \mathcal{E}_{\mathrm{m}}\,\mathrm{d}\mathring{r}\right]_t \\ - \varepsilon\mathring{\eta}_t\,\mathcal{J}\mathcal{E}_{\mathrm{m}}\big|^{\mathring{r}=\varepsilon\mathring{\eta}} - \mathring{h}_t\,\mathcal{J}\mathcal{E}_{\mathrm{m}}\big|_{\mathring{r}=-\mathring{h}} + \\ \left[\mathcal{J}\int_{-\mathring{h}}^{\varepsilon\mathring{\eta}} (\mathcal{E}_{\mathrm{m}} + p)\mathcal{V}^1\,\mathrm{d}\mathring{r}\right]_\lambda - \varepsilon\mathring{\eta}_\lambda\,\mathcal{J}\big((\mathcal{E}_{\mathrm{m}} + p)\mathcal{V}^1\big)\big|^{\mathring{r}=\varepsilon\mathring{\eta}} \\ - \mathring{h}_\lambda\,\mathcal{J}\big((\mathcal{E}_{\mathrm{m}} + p)\mathcal{V}^1\big)\big|_{\mathring{r}=-\mathring{h}} + \left[\mathcal{J}\int_{-\mathring{h}}^{\varepsilon\mathring{\eta}} (\mathcal{E}_{\mathrm{m}} + p)\mathcal{V}^2\,\mathrm{d}\mathring{r}\right]_\theta$$

$$- \varepsilon \mathring{\eta}_\theta \mathcal{J}\big((\mathcal{E}_{\mathrm{m}} + p)\mathcal{V}^2\big)\big|^{\mathring{r}=\varepsilon\mathring{\eta}} - \mathring{h}_\theta \mathcal{J}\big((\mathcal{E}_{\mathrm{m}} + p)\mathcal{V}^2\big)\big|_{\mathring{r}=-\mathring{h}} +$$

$$\mathcal{J}\big((\mathcal{E}_{\mathrm{m}} + p)\mathcal{W}\big)\big|^{\mathring{r}=\varepsilon\mathring{\eta}} - \mathcal{J}\big((\mathcal{E}_{\mathrm{m}} + p)\mathcal{W}\big)\big|_{\mathring{r}=-\mathring{h}} = (\mathcal{J}\mathcal{H}\mathscr{E})_t +$$

$$\Big[\underbrace{\mathcal{J}\int_{-\mathring{h}}^{\varepsilon\mathring{\eta}} (\mathcal{E}_{\mathrm{m}} + \Pi)\cdot(u^1 + \mu^2\mathcal{U}_d^1)\,\mathrm{d}\mathring{r}}_{(\triangle_1)}\Big]_\lambda +$$

$$\Big[\underbrace{\mathcal{J}\int_{-\mathring{h}}^{\varepsilon\mathring{\eta}} (\mathcal{E}_{\mathrm{m}} + \Pi)\cdot(u^2 + \mu^2\mathcal{U}_d^2)\,\mathrm{d}\mathring{r}}_{(\triangle_2)}\Big]_\theta + \mathcal{J}\mathring{h}_t\, p\big|_{\mathring{r}=-\mathring{h}} + \mathcal{O}(\mu^4).$$

The sequel of transformations of these integral relations uses identities (3.70) and (3.71). For example, the first integral $(\triangle_1)$ can be transformed as follows:

$$\int_{-\mathring{h}}^{\varepsilon\mathring{\eta}} (\mathcal{E}_{\mathrm{m}} + \Pi)\cdot(u^1 + \mu^2\mathcal{U}_d^1)\,\mathrm{d}\mathring{r} =$$

$$\mathcal{H}\mathscr{E}u^1 + u^1\int_{-\mathring{h}}^{\varepsilon\mathring{\eta}} \Pi\,\mathrm{d}\mathring{r} + \mu^2\int_{-\mathring{h}}^{\varepsilon\mathring{\eta}} (\mathcal{E}_{\mathrm{m}} + \Pi)\,\mathcal{U}_d^1\,\mathrm{d}\mathring{r} =$$

$$\mathcal{H}\mathscr{E}u^1 + \mathscr{P}u^1 + \mu^2\int_{-\mathring{h}}^{\varepsilon\mathring{\eta}} \Big[\tfrac{1}{2}\big((u^1\sin(\gamma\theta))^2 + (u^2)^2\big) +$$

$$\mathring{r} - \eta_{00} + \mathcal{H} - (\mathring{r} + \mathring{h}) + \mathcal{O}(\mu^2)\Big]\mathcal{U}_d^1\,\mathrm{d}\mathring{r} = \mathcal{H}\mathscr{E}u^1 + \mathscr{P}u^1 +$$

$$\mu^2\Big[\tfrac{1}{2}\big((u^1\sin(\gamma\theta))^2 + (u^2)^2\big) + \varepsilon\eta\Big]\underbrace{\int_{-\mathring{h}}^{\varepsilon\mathring{\eta}} \mathcal{U}_d^1\,\mathrm{d}\mathring{r}}_{\equiv 0 \text{ by } (3.70)} + \mathcal{O}(\mu^4) =$$

$$\mathcal{H}\mathscr{E}u^1 + \mathscr{P}u^1 + \mathcal{O}(\mu^4).$$

The second integral $(\triangle_2)$ can be transformed in a similar way. Henceforth, the following asymptotic identity follows from Eq. (3.65):

$$(\mathcal{J}\mathcal{H}\mathscr{E})_t + \big(\mathcal{J}(u^1\mathcal{H}\mathscr{E} + u^1\mathscr{P})\big)_\lambda + \big(\mathcal{J}(u^2\mathcal{H}\mathscr{E} + u^2\mathscr{P})\big)_\theta$$
$$+ \mathcal{J}\check{p}h_t + \mathcal{O}(\mu^4) = 0.$$

By neglecting in the last equation the asymptotic terms of the order $\mathcal{O}(\mu^4)$ and turning back to dimensional variables, we obtain:

$$(\mathcal{J}\mathcal{H}\mathscr{E})_t + \bar{\nabla}\cdot\Big[\mathcal{J}\mathcal{H}\Big(\mathscr{E} + \frac{\mathscr{P}}{\mathcal{H}}\Big)\mathbf{u}\Big] = -\mathcal{J}\check{p}h_t, \tag{3.73}$$

where $\mathcal{J} = -R^2 \sin\theta$. Consequently, we can say now that the energy conservation equation (3.73) of the FNWD spherical model is asymptotically consistent with the energy equation (3.65) of the 3D fluid flow described by the modified EULER equations on the rotating sphere in the long wave limit.

Energy Balance Equation as a Differential Consequence

Let us derive now the energy balance equation for the FNWD model by manipulating only the governing equations of this model. In this section we shall demonstrate that the energy balance property[10] is a differential consequence of the evolution equations of the FNWD model with the depth-averaged velocity variable. In this derivation we shall work with dimensional variables. In particular, the first step will consist in returning to dimensional variables in the total energy expression of the FNWD model with the depth-averaged velocity given in Eq. (3.72). After this simple operation we obtain:

$$\mathscr{E} = \frac{|\mathbf{u}|^2}{2} + \frac{\mathcal{H}^2}{6}(\nabla\cdot\mathbf{u})^2 + \frac{\mathcal{H}}{2}(\mathscr{D}h)\cdot(\nabla\cdot\mathbf{u}) + \frac{(\mathscr{D}h)^2}{2} + g\,\frac{\mathcal{H} - 2h}{2},$$

where $|\mathbf{u}|^2 = \left(u^1 R\sin\theta\right)^2 + (u^2)^2$ and we set $\rho \equiv 1$ for the sake for simplicity. Moreover, in our derivation we shall need also compact forms of Eqs. (3.57) and (3.58). Using the curvilinear divergence operator defined in Eq. (3.12), we may rewrite the mass conservation equation (3.57) in the following short-hand form:

$$\mathcal{H}_t + \nabla\cdot[\mathcal{H}\mathbf{u}] = 0, \tag{3.74}$$

from which we obtain:

$$\nabla\cdot\mathbf{u} = -\frac{\mathscr{D}\mathcal{H}}{\mathcal{H}}. \tag{3.75}$$

Similarly, taking into account the definitions of the pressure-related quantities $\mathscr{P}$ and $\check{p}$, we may rewrite also the momentum balance equation (3.58) in a more compact (non-conservative) form:

$$\mathbf{v}_t + (\mathbf{u}\cdot\nabla)\mathbf{v} + \frac{\nabla\mathscr{P}}{\mathcal{H}} = \frac{\check{p}\,\nabla h}{\mathcal{H}} + \mathring{\mathbf{S}}, \tag{3.76}$$

where

[10]This property becomes the energy conservation property in the case of the steady bottom.

$$\mathring{\mathbf{S}} \stackrel{\text{def}}{:=} \begin{pmatrix} 0 \\ \left(2\,\Omega\, u^1 + (u^1)^2\right) R^2 \sin\theta\cdot\cos\theta \end{pmatrix}.$$

In order to obtain the energy balance equation for the FNWD model we consider in this section, we multiply Eq. (3.76) by $\mathbf{u}$ in the sense of the scalar product:

$$\mathbf{u}\cdot\mathbf{v}_t + \mathbf{u}\cdot(\mathbf{u}\cdot\nabla)\,\mathbf{v} + \frac{1}{\mathcal{H}}\,\mathbf{u}\cdot\nabla\,\mathcal{P} = \frac{\check{p}}{\mathcal{H}}\,\mathbf{u}\cdot\nabla h + \mathbf{u}\cdot\mathring{\mathbf{S}}. \tag{3.77}$$

The three pressureless terms can be rewritten in a much more compact form:

$$\mathbf{u}\cdot\mathbf{v}_t + \mathbf{u}\cdot(\mathbf{u}\cdot\nabla)\,\mathbf{v} - \mathbf{u}\cdot\mathring{\mathbf{S}} \equiv \mathcal{D}\Big(\frac{|\mathbf{u}|^2}{2}\Big). \tag{3.78}$$

For two other pressure-related terms we have the following equality:

$$\frac{1}{\mathcal{H}}\,\mathbf{u}\cdot\nabla\,\mathcal{P} - \frac{\check{p}}{\mathcal{H}}\,\mathbf{u}\cdot\nabla h = \frac{\nabla\cdot(\mathcal{P}\,\mathbf{u})}{\mathcal{H}} - \underbrace{\Big(\frac{\mathcal{P}\,\nabla\cdot\mathbf{u}}{\mathcal{H}} + \frac{\check{p}}{\mathcal{H}}\,\mathcal{D}h\Big)}_{(\mathfrak{h})} + \frac{\check{p}}{\mathcal{H}}\,h_t.$$

The two terms $(\mathfrak{h})$ can be transformed using Eq. (3.75) as follows:

$$\begin{aligned}
(\mathfrak{h}) &\equiv \frac{\mathcal{P}\,\nabla\cdot\mathbf{u}}{\mathcal{H}} + \frac{\check{p}}{\mathcal{H}}\,\mathcal{D}h = -g\,\frac{\mathcal{D}\mathcal{H}}{2} + g\,\mathcal{D}h \\
&- \frac{\mathcal{H}}{3}\,\mathcal{D}\mathcal{H}\cdot(\nabla\cdot\mathbf{u})^2 - \frac{\mathcal{H}^2}{3}\,\mathcal{D}\Big(\frac{(\nabla\cdot\mathbf{u})^2}{2}\Big) - \frac{\mathcal{H}}{2}\,(\mathcal{D}^2 h)\cdot(\nabla\cdot\mathbf{u}) \\
&- \frac{\mathcal{H}}{2}\,(\mathcal{D}h)\cdot\mathcal{D}(\nabla\cdot\mathbf{u}) - \frac{\mathcal{D}h}{2}\,(\mathcal{D}\mathcal{H})\cdot(\nabla\cdot\mathbf{u}) - (\mathcal{D}h)\cdot(\mathcal{D}^2 h) = \\
&- g\,\mathcal{D}\Big(\frac{\mathcal{H} - 2h}{2}\Big) - \mathcal{D}\Big(\frac{\mathcal{H}^2}{6}\,(\nabla\cdot\mathbf{u})^2\Big) - \mathcal{D}\Big(\frac{\mathcal{H}}{2}\,(\mathcal{D}h)\cdot(\nabla\cdot\mathbf{u})\Big) \\
&- \mathcal{D}\Big(\frac{(\mathcal{D}h)^2}{2}\Big) = -\mathcal{D}\Big(\mathcal{E} - \frac{|\mathbf{u}|^2}{2}\Big).
\end{aligned}$$

Taking into account the last established identity along with Eq. (3.78), we may rewrite Eq. (3.77) as an evolution equation for the FNWD model energy $\mathcal{E}$:

$$\mathcal{E}_t + \mathbf{u}\cdot\nabla\,\mathcal{E} + \frac{1}{\mathcal{H}}\,\nabla\cdot(\mathcal{P}\,\mathbf{u}) = -\frac{1}{\mathcal{H}}\,\check{p}\,h_t.$$

After multiplying the last equation by $\mathcal{H}$ and summing it up with the continuity equation (3.74) multiplied in advance by $\mathcal{E}$, we obtain the energy balance equation in the conservative form (with a source term due to the bottom motion effect):

$$(\mathcal{H}\mathscr{E})_t + \nabla \cdot \left[\mathcal{H}\left(\mathscr{E} + \frac{\mathscr{P}}{\mathcal{H}} \right) \mathbf{u} \right] = -\check{p}\, h_t\,. \tag{3.79}$$

It is not difficult to see that Eq. (3.79) can be written also in form (3.73).

To make an intermediate conclusion, in this section we derived the total mechanical energy balance equation (3.79) for the FNWD model on a rotating sphere. The form of this equation is very similar to its counterpart on the 3D globally flat space (cf. see Sect. 1.1.5). This observation is probably one of the main arguments towards the choice of the FNWD model with the depth-averaged velocity comparing to its 'cousins'.

Remark 3.3 If in the expression (3.72) of the total energy $\mathscr{E}$ and in formulas to compute pressure-related quantities $\mathscr{P}$ and $\check{p}$ we neglect dispersive terms of the asymptotic order $\mathcal{O}(\mu^2)$, then Eq. (3.73) will correspond to the energy balance equation to the classical non-dispersive and hydrostatic[11] NSWE on a rotating sphere.

3.6 Velocity Variable Defined on a Given Surface

Another hierarchy of nonlinear dispersive wave models can be obtained by making a different choice of the velocity variable. A practically important choice consists in taking the trace of the three-dimensional 'horizontal' velocity field at a given surface lying in the fluid bulk, i.e.

$$\mathbf{u}(\lambda, \theta, t) \stackrel{\text{def}}{:=} \mathscr{U}\big(\lambda, \theta, \mathring{r}_\sigma(\lambda, \theta, t), t\big)\,. \tag{3.80}$$

This choice was hinted in a pioneering paper by Bona and Smith [35] and developed later by Nwogu [254] in the flat case. In order to close the system, one has to specify also the 'dispersive' component of the velocity field $\mathbf{U}_d = \mathbf{U}_d(\mathcal{H}, \mathbf{u})$ in terms of other dynamic variables $\mathcal{H}$ and $\mathbf{u}$. With the choice of the velocity $\mathbf{u}$ as specified above in (3.80), $\mathbf{U}_d \neq \mathbf{0}$ in general (similar to the case described in Sect. 3.4, where only the integral of $\mathbf{U}_d$ over the water column height has to vanish due to the choice of $\mathbf{u}$) and we need an additional assumption to close completely the system. Consequently, in this case we proceed as follows: first, we construct $\mathbf{U}_d$ to the required accuracy and then, we apply the depth-averaging operator to determine $\mathscr{U}$. For example, in [197] the authors assumed the 3D flow to be irrotational. Here we shall give a derivation under weaker assumptions. Namely, we assume that only first two components $\omega^{1,2}$ of the vorticity field (3.16) vanish. The 'vertical' vorticity component ω^3 can take any values. In dimensionless variables this assumption can

[11] The terms 'hydrostatic' and 'non-dispersive' are often used as synonyms. However, it is not correct. An example of a non-hydrostatic and non-dispersive model can be found in [101].

be expressed as

$$\mathscr{V}_{\mathring{r}} = \mu^2 \nabla \mathscr{W}. \tag{3.81}$$

By differentiating (3.45) with respect to $\mathring{r}$ and substituting the last identity into asymptotic expansion (3.49) we obtain

$$\frac{\partial \mathbf{V}_d}{\partial \mathring{r}} = \nabla w + \mathcal{O}(\mu^2) = -\nabla(\mathcal{D}\mathring{h}) - (\nabla \cdot \mathbf{u})\nabla \mathring{h} \\ - (\mathring{r} + \mathring{h})\nabla(\nabla \cdot \mathbf{u}) + \mathcal{O}(\mu^2).$$

By integrating this identity over the vertical coordinate $\mathring{r}$ one obtains the following expression for $\mathbf{V}_d$:

$$\mathbf{V}_d = -(\mathring{r} + \mathring{h})\Big[\nabla(\mathcal{D}\mathring{h}) + (\nabla \cdot \mathbf{u})\nabla \mathring{h}\Big] \\ - \frac{(\mathring{r} + \mathring{h})^2}{2}\nabla(\nabla \cdot \mathbf{u}) + \mathbf{V}_d\big|_{\mathring{r} = -\mathring{h}} + \mathcal{O}(\mu^2).$$

By using the connection between covariant and contravariant components $\mathbf{U}_d = \mathbb{G}^{-1} \cdot \mathbf{V}_d$, which follows from (3.46), we obtain an asymptotic approximation to the 3D 'horizontal' velocity field:

$$\mathscr{U}(\lambda, \theta, \mathring{r}, t) = \mathbf{u}(\lambda, \theta, t) + \\ \mu^2 \mathbb{G}^{-1} \cdot \underbrace{\left\{(\mathring{r} + \mathring{h})\mathscr{A} + \frac{(\mathring{r} + \mathring{h})^2}{2}\mathscr{B} + \mathscr{C}\right\}}_{(\checkmark)} + \mathcal{O}(\mu^4), \tag{3.82}$$

where we introduced three vectors:

$$\mathscr{A} \stackrel{\text{def}}{:=} -\nabla(\mathcal{D}\mathring{h}) - (\nabla \cdot \mathbf{u})\nabla \mathring{h},$$

$$\mathscr{B} \stackrel{\text{def}}{:=} -\nabla(\nabla \cdot \mathbf{u}),$$

$$\mathscr{C} \stackrel{\text{def}}{:=} \mathbf{V}_d\big|_{\mathring{r} = -\mathring{h}}.$$

In order to compute the term $\mathscr{C}$ in terms of other dynamic variables, we consider the asymptotic representation for $\mathscr{U}(\lambda, \theta, \mathring{r}, t)$ and evaluate (3.82) at $\mathring{r} = \mathring{r}_\sigma(\lambda, \theta, t)$. By definition (3.80) we must have $\mathscr{U}(\lambda, \theta, \mathring{r}_\sigma, t) \equiv \mathbf{u}$. Consequently, at $\mathring{r} = \mathring{r}_\sigma(\lambda, \theta, t)$ the expression in braces $(\checkmark) \equiv \mathbf{0}$ must vanish. Thus, we obtain

$$\mathscr{C} = -(\mathring{r}_\sigma + \mathring{h})\,\mathscr{A} - \frac{(\mathring{r}_\sigma + \mathring{h})^2}{2}\,\mathscr{B}.$$

Thus, the substitution of this expression for $\mathscr{C}$ into (3.82) yields

$$\mathscr{U}(\lambda, \theta, \mathring{r}, t) = \mathbf{u}(\lambda, \theta, t) + \mu^2 \mathbb{G}^{-1} \cdot \left\{ (\mathring{r} - \mathring{r}_\sigma)\,\mathscr{A} + \frac{(\mathring{r} + \mathring{h})^2 - (\mathring{r}_\sigma + \mathring{h})^2}{2}\,\mathscr{B} \right\} + \mathcal{O}(\mu^4).$$

In other words, the distribution of the 'horizontal' velocity is approximatively quadratic to the asymptotic order $\mathcal{O}(\mu^4)$. The last formula can be used to reconstruct approximatively the 3D velocity field by having in hands the solution of the dispersive system only. As another side result of the formula above we obtain easily the required expression for the dispersive correction $\mathbf{U}_d$:

$$\mathbf{U}_d = \mathbb{G}^{-1} \cdot \left\{ (\mathring{r} - \mathring{r}_\sigma)\,\mathscr{A} + \frac{(\mathring{r} + \mathring{h})^2 - (\mathring{r}_\sigma + \mathring{h})^2}{2}\,\mathscr{B} \right\} + \mathcal{O}(\mu^2).$$

By applying the depth-averaging operator we obtain also the required closure relation to close the base model:

$$\mathcal{U} = \mathbb{G}^{-1} \cdot \left\{ \left[\frac{\mathcal{H}}{2} - (\mathring{r}_\sigma + \mathring{h}) \right] \mathscr{A} + \left[\frac{\mathcal{H}^2}{6} - \frac{(\mathring{r}_\sigma + \mathring{h})^2}{2} \right] \mathscr{B} \right\} + \mathcal{O}(\mu^2).$$

The base model in physical variables has the same expressions (3.55), (3.56), but vector functions $\mathcal{U}$ and $\mathcal{V}$ have to be set accordingly to the closure presented in this section.

3.6.1 Further Simplifications

Some expressions and equations above can be further simplified by noticing that

$$\nabla(\mathcal{D}\,\mathring{h}) = \nabla(\mathcal{D}\,h) - \nabla(\mathcal{D}\,\eta_{00}) = \nabla(\mathcal{D}\,h) + \mathcal{O}\Big(\frac{\varsigma}{\mu}\Big),$$

$$\nabla\mathring{h} = \nabla h - \nabla\eta_{00} = \nabla h + \mathcal{O}\Big(\frac{\varsigma}{\mu}\Big).$$

Thus, $\nabla(\mathcal{D}\,\mathring{h})$ and $\nabla\mathring{h}$ can be asymptotically interchanged with $\nabla(\mathcal{D}\,h)$ and ∇h correspondingly since $\mathcal{U}$ and $\mathcal{V}$ always appear in equations with coefficient μ^2. Consequently, we have

$$\mu^2 \mathcal{U} = \mu^2 \mathbb{G}^{-1} \cdot \Big\{ \Big[\frac{\mathcal{H}}{2} - (r_\sigma + h) \Big] \mathscr{A}^\star + \Big[\frac{\mathcal{H}^2}{6} - \frac{(r_\sigma + h)^2}{2} \Big] \mathscr{B} \Big\} + \mathcal{O}(\mu^4 + \varsigma\mu + \varsigma^2),$$

where $r_\sigma \equiv \mathring{r}_\sigma - \eta_{00}$, with $-h \leqslant r_\sigma \leqslant \varepsilon\eta$ and

$$\mathscr{A}^\star \stackrel{\text{def}}{:=} -\nabla(\mathcal{D}h) - (\nabla \cdot \mathbf{u})\nabla h.$$

Let us summarize the developments made so far in this section. First, the 'horizontal' fluid velocity was defined in equation (3.80). Then, we made a simplifying assumption (3.81), which allowed us to derive the following closure relation:

$$\mathcal{U} = \mathbb{G}^{-1} \cdot \Big\{ \Big[\frac{\mathcal{H}}{2} - (r_\sigma + h) \Big] \mathscr{A}^\star + \Big[\frac{\mathcal{H}^2}{6} - \frac{(r_\sigma + h)^2}{2} \Big] \mathscr{B} \Big\}. \tag{3.83}$$

The base model with velocity choice (3.80) has the same form (3.55), (3.56) in dimensional variables. The invariance of equations with respect to the choice of 'horizontal' velocity is among the main advantages of our modelling approach.

3.6.2 Base Model in Terms of the Linear Velocity Components

Similarly as we did in Sect. 3.4.1, we can recast the base model (3.55), (3.56) in terms of the components of linear velocity u and v (and we refer to Sect. 3.4.1 for their definition):

$$(\mathcal{H} R \sin\theta)_t + \big[\mathcal{H} u\big]_\lambda + \big[\mathcal{H} v \sin\theta\big]_\theta = -\Big\{ (\mathcal{H} U)_\lambda + (\mathcal{H} V \sin\theta)_\theta \Big\},$$

$$\begin{aligned}(\mathcal{H} u R \sin\theta)_t &+ \Big[\mathcal{H} u^2 + g\frac{\mathcal{H}^2}{2}\Big]_\lambda + \Big[\mathcal{H} u v \sin\theta\Big]_\theta \\ &= g\mathcal{H} h_\lambda - \mathcal{H} u v \cos\theta - \digamma\mathcal{H} v R \sin\theta + \wp_\lambda - \check{\wp} h_\lambda \\ &- \Big\{ (\mathcal{H} U R \sin\theta)_t + (2\mathcal{H} U u)_\lambda + \big(\mathcal{H}(U v + V u)\sin\theta\big)_\theta \Big\},\end{aligned}$$

$$(\mathcal{H} v R \sin\theta)_t + \Big[\mathcal{H} u v\Big]_\lambda + \Big[\Big(\mathcal{H} v^2 + g\frac{\mathcal{H}^2}{2}\Big)\sin\theta\Big]_\theta =$$

$$= g\,\mathcal{H}\,h_\theta\,\sin\theta \,+\, g\,\frac{\mathcal{H}^2}{2}\,\cos\theta \,+\, \mathcal{H}\,u^2\,\cos\theta \,+$$
$$F\,\mathcal{H}\,u\,R\,\sin\theta \,+\, \bigl(\wp_\theta \,-\, \check{\wp}\,h_\theta\bigr)\,\sin\theta$$
$$-\Bigl\{\bigl(\mathcal{H}\,V\,R\,\sin\theta\bigr)_t \,+\, \bigl(\mathcal{H}\,U\,v \,+\, \mathcal{H}\,V\,u\bigr)_\lambda \,+\, \bigl(2\,\mathcal{H}\,V\,v\,\sin\theta\bigr)_\theta\Bigr\},$$

where $U \stackrel{\text{def}}{:=} R\,\mathcal{U}^1 \sin\theta,\ V \stackrel{\text{def}}{:=} R\,\mathcal{U}^2$ and $\mathcal{U}^{1,2}$ were defined in (3.83). The computation of non-hydrostatic pressure contributions are computed precisely as it is explained in Sect. 3.4.1. In the second equation above we omitted intentionally in the right-hand side three terms $\mathcal{H}\,(U\,v \,+\, V\,u)\,\cos\theta$ and $F\,\mathcal{H}\,V\,R\,\sin\theta$ since in dimensionless form they have the asymptotic order $\mathcal{O}\,(\varsigma\,\mu)$.

3.6.3 Weakly Nonlinear Model

In order to obtain a weakly nonlinear model with a velocity variable defined on an arbitrary surface, it is sufficient to simplify accordingly the system of equations given in the previous section. Namely, all nonlinearities in the dispersive terms are to be neglected. The first group of terms to be simplified consists in non-hydrostatic pressure corrections $\nabla\wp \,-\, \check{\wp}\,\nabla h$, which is present for any choice of the velocity variable $\mathbf{u}$ (or equivalently for any closure relation $\mathcal{U} \,=\, \mathcal{U}\,(\mathcal{H},\,\mathbf{u})$). This simplification was explained above in Sect. 3.4.2.

The second group of terms contains the vector $\mathcal{U}$. They are present in both continuity and momentum conservation equations. Vector $\mathcal{U}$ appears always with dimensionless coefficient μ^2. Taking into account the BOUSSINESQ (i.e. weakly nonlinear) regime and definition $\mathcal{H} \,=\, h \,+\, \varepsilon\,\eta$, we obtain that instead of closure relation (3.83), we have to use consistently

$$\mathcal{U} \,=\, \mathcal{U}_0 \,+\, \mathbb{G}^{-1}\cdot\Bigl[\Bigl(\frac{h}{2} \,+\, r_\sigma\Bigr)\Bigl(\nabla(\mathbf{u}\cdot\nabla h) \,+\, (\nabla\cdot\mathbf{u})\,\nabla h\Bigr) \,-\, \Bigl(\frac{h^2}{6} \,-\, \frac{(r_\sigma \,+\, h)^2}{2}\Bigr)\,\nabla(\nabla\cdot\mathbf{u})\Bigr], \tag{3.84}$$

where

$$\mathcal{U}_0 \stackrel{\text{def}}{:=} \Bigl(\frac{h}{2} \,+\, r_\sigma\Bigr)\,\mathbb{G}^{-1}\cdot\nabla h_t\,.$$

In other words, since coefficients $\mathscr{A}^\star$ and $\mathscr{B}$ are linear in velocities, then it was sufficient to replace $\mathcal{H}$ by h. The same operation has to be performed consistently in the right-hand sides as well:

$$(\mathcal{H} U)_\lambda + [\mathcal{H} V \sin\theta]_\theta \rightsquigarrow (h U)_\lambda + [h V \sin\theta]_\theta ,$$

$$(\mathcal{H} U R \sin\theta)_t \rightsquigarrow (h U R \sin\theta)_t , \qquad (\mathcal{H} V R \sin\theta)_t \rightsquigarrow (h V R \sin\theta)_t .$$

The components U and V are computed as above

$$U = R\,\mathcal{U}^1 \sin\theta , \qquad V = R\,\mathcal{U}^2 ,$$

with the only difference is that $\mathcal{U}^{1,2}$ are given by Eq. (3.84).

Similar transformations (in this case linearizations) $\mathcal{H} \rightsquigarrow h$, $U \rightsquigarrow U_0$ and $V \rightsquigarrow V_0$ have to be done in the remaining terms as well:

$$2[\mathcal{H} u U]_\lambda + [\mathcal{H}(U v + V u)\sin\theta]_\theta \rightsquigarrow$$
$$2[h u U_0]_\lambda + [h(U_0 v + V_0 u)\sin\theta]_\theta ,$$

$$[\mathcal{H}(U v + V u)]_\lambda + 2[\mathcal{H} v V \sin\theta]_\theta \rightsquigarrow$$
$$[h(U_0 v + V_0 u)]_\lambda + 2[h v V_0 \sin\theta]_\theta ,$$

where U_0 and V_0 are defined through components of the vector $\mathcal{U}_0$ as

$$U_0 \stackrel{\text{def}}{:=} R\,\mathcal{U}_0^1 \sin\theta , \qquad V_0 \stackrel{\text{def}}{:=} R\,\mathcal{U}_0^2 .$$

Obtained in this way weakly nonlinear model is a spherical analogue of well-known NWOGU system on the plane [254]. If we vanish the dispersive velocity correction $\mathcal{U} \equiv \mathbf{0}$ we shall obtain the spherical counterpart of the classical PEREGRINE system [264].

3.6.4 State-of-the-Art

Let us mention a few publications which report the derivation or use of dispersive wave models on a sphere with the velocity defined on an arbitrary surface lying in the fluid bulk. A detailed derivation of the fully nonlinear weakly dispersive wave model with this choice of the velocity is given in [197, 286]. However, the resulting equations turn out to be cumbersome and in numerical simulations the authors employ only the weakly nonlinear spherical BOUSSINESQ-type equations. For example, in [164] a numerical coupling between 3D NAVIER–STOKES (for the landslide area) and 2D spherical BOUSSINESQ (for the far field propagation) is reported.

To our knowledge, the fully nonlinear model derived in the present study and earlier in [286] has never been used for large scale numerical simulations. It can be partially explained by the complexity of equations and by the lack of robust and efficient numerical discretizations for dispersive PDEs on a sphere.

3.7 Discussion

After the developments presented hereinabove in a globally spherical geometry, we finish the present chapter by outlining the main conclusions and perspectives of our study.

3.7.1 *Conclusions*

In this work we derived a generic weakly dispersive but fully nonlinear model on a rotating, possibly deformed, sphere. This model contains a free contravariant $1-$tensor variable $\mathcal{U}$ (or its covariant equivalent $\mathcal{V}$). In order to close the system, one has to specify $\mathcal{U}$ as a function of two other model variables, i.e. $\mathcal{U} = \mathcal{U}(\mathcal{H}, \mathbf{u})$. This functional dependence is called the *closure* relation. By choosing various closures, we show how one can obtain from the base model by simple substitutions some well-known models (or, at least, their fully nonlinear counterparts). Other choices of closure $\mathcal{U} = \mathcal{U}(\mathcal{H}, \mathbf{u})$ lead to completely new models, whose properties are to be studied separately. Moreover, for any choice of the closure relation the base model has a nice conservative structure. Thus, this work can be considered as an effort towards further systematization of dispersive wave models on a spherical geometry. Moreover, the governing equations of the base model are given for the convenience in terms of the covariant/contravariant and linear velocity variables. For every model we give also its weakly nonlinear counterparts in case simpler models are needed. Of course, the classical nonlinear shallow water or SAINT-VENANT equations on a rotating sphere can be simply obtained from the base model by neglecting all non-hydrostatic terms.

Two popular closures were proposed in this chapter. In our work we always tried to use only the minimal assumptions about the three-dimensional flow. For instance, in contrast to [197, 286] we do not assume the flow to be irrotational. We note also the fact that the bottom was assumed to be non-stationary. It allows to model tsunami generation by seismic [98, 103, 106] and landslide [26, 109] mechanisms.

3.7.2 Perspectives

In the present chapter the base model derivation was presented in a spherical geometry. This choice was made by the authors due to the importance of applications in Meteorology, Climatology and Oceanography on global planetary scales. However, in this study we prepare a setting which could be fruitfully used in more general geometries. For instance, we believe that the techniques presented in this chapter could be used to derive long wave models for shallow flows on compact manifolds. Curvilinear coordinates are routinely used in Fluid Mechanics. However, we believe that the right setting to work with Fluid Mechanics equations in complex geometries is the RIEMANNIAN geometry. In future studies we plan to show successful applications of this technology on more general geometries.

In the following chapter we shall discuss the numerical discretization of nonlinear long wave models on globally spherical geometries. Namely, for the sake of simplicity, we shall take a particular *avatar* of the base model and we shall show how to discretize it using modern finite volumes schemes. After a direct generalization it can be easily extended to the base model as well, if it is needed, of course. The numerical solution of fully nonlinear dispersive wave equations with the velocity defined on an arbitrary level in the fluid bulk still constitutes a challenging problem which will be addressed in our future studies.

Chapter 4
Numerical Simulation on a Globally Spherical Geometry

In this chapter we focus on numerical aspects while the model derivation was described in Chap. 3. The algorithm we propose is based on the splitting approach. Namely, equations are decomposed on a uniform elliptic equation for the dispersive pressure component and a hyperbolic part of shallow water equations (on a sphere) with source terms. This algorithm is implemented as a two-step predictor–corrector scheme. On every step we solve separately elliptic and hyperbolic problems. Then, the performance of this algorithm is illustrated on model idealized situations with even bottom, where we estimate the influence of sphericity and rotation effects on dispersive wave propagation. The dispersive effects are quantified depending on the propagation distance over the sphere and on the linear extent of generation region. Finally, the numerical method is applied to a couple of real-world events. Namely, we undertake simulations of the BULGARIAN 2007 and CHILEAN 2010 tsunamis. Whenever the data is available, our computational results are confronted with real measurements.

Until recently, the modelling of long wave propagation on large scales has been performed in the framework of Nonlinear Shallow Water Equations (NSWE) implemented under various software packages [199]. This model is hydrostatic and non-dispersive [296]. Among popular packages we can mention, for example, the TUNAMI code [175] based on a conservative finite difference leap–frog scheme on real bathymetries. This code has been extensively used for tsunami wave modelling by various groups (see e.g. [336]). The code MOST uses the directional splitting approach [309, 310] and is also widely used for the simulation of tsunami wave propagation and run-up [301, 329]. The MGC package [289] is based on a modified MACCORMACK finite difference scheme [126], which discretizes NSWE in spherical coordinates. Obviously, the MGC code can also work in CARTESIAN coordinates as well. This package was used to simulate the wave run-up on a real-world beach [158] and tsunami wave generation by underwater landslides [26]. Recently the VOLNA code was developed using modern second order finite volume

G. Khakimzyanov et al., *Dispersive Shallow Water Waves*, Lecture Notes in Geosystems Mathematics and Computing, https://doi.org/10.1007/978-3-030-46267-3_4

schemes on unstructured grids [113]. Nowadays this code is essentially used for the quantification of uncertainties of the tsunami risk [23].

All numerical models described above assume the wave to be non-dispersive. However, in the presence of wave components with higher frequencies (or equivalently shorter wavelengths), the frequency dispersion effects may influence the wave propagation. Even in 1982 in [241] it was pointed out:

> [...] the considerations and estimates for actual tsunamis indicate that nonlinearity and dispersion can appreciably affect the tsunami wave propagation at large distances.

Later this conclusion was reasserted in [261] as well. The catastrophic SUMATRA event in 2004 [299] along with subsequent events brought a lot of new data all around the globe and also from satellites [204]. The detailed analysis of this data allowed to understand better which models and algorithms should be applied at various stages of a tsunami life cycle to achieve the desired accuracy [79, 250]. The main conclusion can be summarized as follows: for a complete and satisfactory description of a tsunami wave life cycle on global scales, one has to use a nonlinear dispersive wave model with moving (in the generation region [106]) realistic bathymetry. For trans-oceanic tsunami propagation one has to include also EARTH's sphericity and rotation effects. A whole class of suitable mathematical models combining all these features was presented in the previous chapter.

At the present time we have a rather limited amount of published research literature devoted to numerical issues of long wave propagation in a spherical ocean. In many works (see e.g. [141, 168]) EARTH's sphericity is not taken explicitly into account. Instead, the authors project EARTH's surface (or at least a sub-region) on a tangent plane to EARTH in some point and computations are then performed on a flat space using a BOUSSINESQ-type (Weakly Nonlinear and Weakly Dispersive—WNWD) model without taking into account the CORIOLIS force. We notice that some geometric defects are unavoidable in this approach. However, even in this simplified framework the importance of dispersive effects has been demonstrated by comparing the resulting wave field with hydrostatic (NSWE) computations.

In [223] the authors studied the trans-oceanic propagation of a hypothetical tsunami generated by an eventual giant landslide which may take place at LA PALMA island, which the most north-westerly island of the CANARY Islands, SPAIN. Similarly the authors employed a WNWD model, but this time written in spherical coordinates with EARTH's rotation effects. However, the employed model could handle only static bottoms. As a result, the initial fields were generated using a different hydrodynamic model and, then, transferred into the WNWD model as the initial condition to compute the wave long time evolution. However, when waves approach the shore, another limitation of weakly nonlinear models becomes apparent—in coastal regions nonlinear effects grow quickly and, thus, the computations should be stopped before the wave reaches the coast. Otherwise, the numerical results may lose their validity. In [223] it was also shown that the wave dispersion may play a significant rôle on the resulting wave field. Namely, NSWE predict the first significant wave hitting the shore, while WNWD equations predict rather an undular bore in which the first wave amplitude is not necessarily the highest [224]. Of course, these undular bores cannot be described in the framework of NSWE [153, 263].

An even more detailed study of tsunami dispersion was undertaken recently [142], where also a WNWD model was used, but the dispersion effect was estimated for several real-world events. The authors came to interesting 'uncertain' conclusions:

> [...] However, undular bores, which are not included in shallow-water theory, may evolve during shoaling. Even though such bores may double the wave height locally, their effect on inundation is more uncertain because the individual crests are short and may be strongly affected by dissipation due to wave breaking.

It was also noted that near coasts WNWD model provides unsatisfactory results, that is why fully nonlinear dispersive models should be employed to model all stages from tsunami generation to the inundation. The same year a fully nonlinear dispersive model on a sphere including CORIOLIS effect was derived in [197]. However, in contrast to another paper from the same group [152], the horizontal velocity variable is taken as a trace of 3D fluid velocity on a certain surface laying between the bottom and free surface. The proposed model may have some drawbacks. First of all, the well-posed character of the CAUCHY problem is not clear. A very similar (and much simpler) NWOGU's model [254] is known to possess instabilities for certain configurations of the bottom [222]. We underline also that the authors of [197] did not present so far any numerical simulations with their fully nonlinear model. A study of dispersive and CORIOLIS effects were performed in the WNWD counterpart of their fully nonlinear equations. In order to solve numerically their spherical BOUSSINESQ-type system a CARTESIAN TVD scheme previously described in [285] was generalized to spherical coordinates. This numerical model was implemented as a part of well-known FUNWAVE(-TVD) code [286].

A fully nonlinear weakly dispersive model on a sphere with the depth-averaged velocity variable was first derived in [129]. Later it was shown in [131] that the same model can be derived without using the potential flow assumption. Moreover, this model admits an elegant conservative structure with the mass, momentum and energy conservations. In particular, the energy conservation allows to control the amount of numerical viscosity in simulations. The same conservative structure can be preserved while deriving judiciously weakly nonlinear models as well. Only the expressions of the kinetic energy and various fluxes vary from one model to another. In this way one may obtain the whole hierarchy of simplified shallow water models on a sphere enjoying the same formal conservative structure [290]. In particular, it allows to develop a unique numerical algorithm, which can be applied to all models in this hierarchy by changing only the fluxes and source terms in the numerical code.

In this chapter we develop a numerical algorithm to model shallow water wave propagation on a rotating sphere in the framework of a fully nonlinear weakly dispersive model, which will be described in the following section. For numerical illustrations we consider first model problems on the perfect sphere (i.e. the bottom is even). In this way we assess the influence of dispersive, sphericity and rotation effects depending on the propagation distance and on the size of the wave generation region. These methods are implemented in NLDSW_SPHERE code which is used to produce numerical results reported below.

The present chapter is organized as follows. The governing equations that we tackle in our study are set in Sect. 4.1. The numerical algorithm is described in Sect. 4.3. Several numerical illustrations are described in Sect. 4.4. Namely, we start with tests over a perfect rotating sphere in Sect. 4.4.1. Then, as an illustration of medium scale wave propagation we study the BULGARIAN 2007 tsunami in Sect. 4.4.2. On large trans-oceanic scales we simulate the 2010 CHILEAN tsunami in Sect. 4.4.3. Finally in Sect. 4.5 we outline the main conclusions and perspectives of our study.

4.1 Problem Formulation

The detailed derivation of the fully nonlinear model considered in the present study can be found in the previous Chap. 3. Here we only repeat the governing equations:

$$(\mathcal{H} R \sin\theta)_t + \bigl[\mathcal{H} u\bigr]_\lambda + \bigl[\mathcal{H} v \sin\theta\bigr]_\theta = 0, \tag{4.1}$$

$$(\mathcal{H} u R \sin\theta)_t + \Bigl[\mathcal{H} u^2 + g\,\frac{\mathcal{H}^2}{2}\Bigr]_\lambda + \bigl[\mathcal{H} u v \sin\theta\bigr]_\theta$$
$$= g\,\mathcal{H} h_\lambda - \mathcal{H} u v \cos\theta - \mathcal{F}\,\mathcal{H} v R \sin\theta + \wp_\lambda - \check{\wp}\, h_\lambda, \tag{4.2}$$

$$(\mathcal{H} v R \sin\theta)_t + \bigl[\mathcal{H} u v\bigr]_\lambda + \Bigl[\Bigl(\mathcal{H} v^2 + g\,\frac{\mathcal{H}^2}{2}\Bigr)\sin\theta\Bigr]_\theta$$
$$= g\,\mathcal{H} h_\theta \sin\theta + g\,\frac{\mathcal{H}^2}{2}\cos\theta + \mathcal{H} u^2\cos\theta$$
$$+ \mathcal{F}\,\mathcal{H} u R \sin\theta + \bigl(\wp_\theta - \check{\wp}\, h_\theta\bigr)\sin\theta, \tag{4.3}$$

where R is the radius of a virtual sphere rotating with a constant angular velocity Ω around the axis $O\,z$ of a fixed CARTESIAN coordinate system $O\,x\,y\,z$. The origin O of this coordinate system is chosen so that the plane $O\,x\,y$ coincides with sphere's equatorial plane.

In order to describe conveniently the fluid flow we choose also a spherical coordinate system $O\,\lambda\,\theta\,r$ whose origin is located at sphere's center and it rotates with the sphere. Here λ is the longitude increasing in the rotation direction starting from a certain meridian ($0 \leqslant \lambda < 2\pi$). The other angle $\theta \stackrel{\text{def}}{:=} \dfrac{\pi}{2} - \phi$ is the *complementary latitude* $(-\frac{\pi}{2} < \phi < \frac{\pi}{2})$. Finally, r is the radial distance from sphere's center. The NEWTONIAN gravity force[1] acts on fluid particles and its vector $\mathbf{g}$ is directed towards virtual sphere's center. The total water depth $\mathcal{H} \stackrel{\text{def}}{:=} \eta + h > 0$ is supposed to be small comparing to sphere's radius, i.e. $\mathcal{H} \ll R$. That is why we can suppose that the gravity acceleration $g \stackrel{\text{def}}{:=} |\mathbf{g}|$ and fluid density ρ are constants throughout the fluid layer. The functions $h(\lambda, \theta, t)$ (the bottom profile) and $\eta(\lambda, \theta, t)$ (the free surface excursion) are given as deviations from the still water level $\eta_{00}(\theta)$, which is not spherical due to the rotation effect (see Chap. 3).

By u and v we denote the linear components of the velocity vector:

$$u \stackrel{\text{def}}{:=} R\,u^1 \sin\theta, \qquad v \stackrel{\text{def}}{:=} R\,u^2,$$

where $u^1 = \dot{\lambda}$ and $u^2 = \dot{\theta}$. The CORIOLIS parameter $\mathcal{F} \stackrel{\text{def}}{:=} 2\Omega\cos\theta$ is expressed through the complementary latitude θ and additionally we can assume that

$$\theta_0 \leqslant \theta \leqslant \pi - \theta_0, \tag{4.4}$$

where $\theta_0 = \text{const} \ll 1$ is a small angle. In other words, the poles are excluded from our computations. In practice, it is not a serious limitations since on the EARTH poles are covered with ice and no free surface flow takes place there. The quantities $\wp$ and $\check{\wp}$ are dispersive components of the depth-integrated pressure $\mathscr{P}$ and fluid pressure at the bottom $\check{p}$, respectively:

$$\mathscr{P} = \frac{g\,\mathcal{H}^2}{2} - \wp, \qquad \check{p} = g\,\mathcal{H} - \check{\wp}.$$

These dispersive components $\wp$ and $\check{\wp}$ can be computed according to the following formulas [188]:

$$\wp = \frac{\mathcal{H}^3}{3}\,\mathscr{R}_1 + \frac{\mathcal{H}^2}{2}\,\mathscr{R}_2, \qquad \check{\wp} = \frac{\mathcal{H}^2}{2}\,\mathscr{R}_1 + \mathcal{H}\,\mathscr{R}_2, \tag{4.5}$$

where

$$\mathscr{R}_1 \stackrel{\text{def}}{:=} \mathcal{D}(\nabla\cdot\mathbf{u}) - (\nabla\cdot\mathbf{u})^2, \qquad \mathscr{R}_2 \stackrel{\text{def}}{:=} \mathcal{D}^2 h, \qquad \mathbf{u} \stackrel{\text{def}}{:=} \left(u^1, u^2\right).$$

To complete model presentation we remind also the definitions of various operators in spherical coordinates that we use:

[1] Here we understand the force per unit mass, i.e. the acceleration.

$$\mathfrak{D} \stackrel{\text{def}}{:=} \partial_t + \mathbf{u} \cdot \boldsymbol{\nabla}, \qquad \boldsymbol{\nabla} \stackrel{\text{def}}{:=} (\partial_\lambda, \partial_\theta), \qquad \mathbf{u} \cdot \boldsymbol{\nabla} \equiv u^1 \partial_\lambda + u^2 \partial_\theta,$$

$$\boldsymbol{\nabla} \cdot \mathbf{u} \equiv u^1_\lambda + \frac{1}{\mathfrak{J}} (\mathfrak{J} u^2)_\theta, \qquad \mathfrak{J} \stackrel{\text{def}}{:=} -R^2 \sin\theta.$$

In the most detailed form, functions $\mathscr{R}_{1,2}$ can be equivalently rewritten as

$$\mathscr{R}_1 \equiv (\boldsymbol{\nabla} \cdot \mathbf{u})_t + \frac{1}{R \sin\theta} \Big[u (\boldsymbol{\nabla} \cdot \mathbf{u})_\lambda + v (\boldsymbol{\nabla} \cdot \mathbf{u})_\theta \sin\theta \Big] - (\boldsymbol{\nabla} \cdot \mathbf{u})^2,$$

$$\mathscr{R}_2 \equiv (\mathfrak{D} h)_t + \frac{1}{R \sin\theta} \Big[u (\mathfrak{D} h)_\lambda + v (\mathfrak{D} h)_\theta \sin\theta \Big],$$

with

$$\boldsymbol{\nabla} \cdot \mathbf{u} \equiv \frac{1}{R \sin\theta} \Big[u_\lambda + (v \sin\theta)_\theta \Big],$$

$$\mathfrak{D} h \equiv h_t + \frac{1}{R \sin\theta} \Big[u h_\lambda + v h_\theta \sin\theta \Big].$$

The model (4.1)–(4.3) is referred to as 'fully nonlinear' one since it was derived without any simplifying assumptions on the wave amplitude [188]. In other words, all nonlinear (but weakly dispersive) terms are kept in this model. So far we shall refer to this model as Fully Nonlinear Weakly Dispersive (FNWD) model. The FNWD model should be employed to simulate water wave propagation in coastal and even in slightly deeper regions over uneven bottoms. Weak dispersive effects in the FNWD model ensure that we shall obtain more accurate results than with simple NSWE. Moreover, the FNWD model contains the terms coming from moving bottom effects [183]. Consequently, we can model also the wave generation process by fast or slow bottom motions [106], thus, allowing to model tsunami waves from their generation until the coasts [113]. In this way we extend the validity region of existing WNWD models [142, 223, 224] by including the wave generation regions along with the coasts where the nonlinearity becomes critical.

Concerning the *linear* dispersive properties, it is generally believed that models with the depth-averaged velocity can be further improved in the sense of Bona–Smith [35] and Nwogu [254]. However, nonlinear dispersive wave models tweaked in this way (see e.g. [226]) have a clear advantage only in the linear one-dimensional (1D) situations. For nonlinear 3D computations (especially involving the moving bottom [103]) the advantage of 'improved' models becomes more obscure comparing to dispersive wave models with the depth-averaged velocity adopted in our study [70, 288]. Moreover, the mathematical model after such transformations (or 'improvements' as they are called in the literature) often loses the energy conservation and GALILEAN's invariance properties. A successful attempt in this direction was achieved only recently [74].

Equations (4.1)–(4.3) admit also a non-conservative form:

$$\mathcal{H}_t + \frac{1}{R \sin\theta}\Big[(\mathcal{H}u)_\lambda + (\mathcal{H}v\sin\theta)_\theta\Big] = 0, \tag{4.6}$$

$$u_t + \frac{1}{R\sin\theta}uu_\lambda + \frac{1}{R}vu_\theta + \frac{g}{R\sin\theta}\eta_\lambda = \frac{1}{R\sin\theta}\frac{\wp_\lambda - \check{\wp}h_\lambda}{\mathcal{H}} - \frac{uv}{R}\cot\theta - \digamma v, \tag{4.7}$$

$$v_t + \frac{1}{R\sin\theta}uv_\lambda + \frac{1}{R}vv_\theta + \frac{g}{R}\eta_\theta = \frac{\wp_\theta - \check{\wp}h_\theta}{R\mathcal{H}} + \frac{u^2}{R}\cot\theta + \digamma u. \tag{4.8}$$

In the numerical algorithm presented below we use both conservative and non-conservative forms for our convenience.

If in conservative (4.1)–(4.3) or non-conservative (4.6)–(4.8) governing equations we neglect dispersive contributions, i.e. $\wp \rightsquigarrow 0$, $\check{\wp} \rightsquigarrow 0$, then we recover NSWE on a rotating attracting sphere [62]:

$$\mathcal{H}_t + \frac{1}{R \sin\theta}\Big[(\mathcal{H}u)_\lambda + (\mathcal{H}v\sin\theta)_\theta\Big] = 0,$$

$$u_t + \frac{1}{R\sin\theta}uu_\lambda + \frac{1}{R}vu_\theta + \frac{g}{R\sin\theta}\eta_\lambda = -\frac{uv}{R}\cot\theta - \digamma v,$$

$$v_t + \frac{1}{R\sin\theta}uv_\lambda + \frac{1}{R}vv_\theta + \frac{g}{R}\eta_\theta = \frac{u^2}{R}\cot\theta + \digamma u.$$

In order to obtain a well-posed problem, we have to prescribe initially the free surface deviation from its equilibrium position along with the initial velocity field. Moreover, the appropriate boundary conditions have to be prescribed as well. The curvilinear coast line is approximated by a family of closed polygons and on edges e we prescribe the wall boundary condition:

$$\mathbf{u}\cdot\mathbf{n}\Big|_e = 0, \tag{4.9}$$

where $\mathbf{n}$ is an exterior normal to edge e. In this way, we replace the wave run-up problem by wave/wall interaction, where a solid wall is located along the shoreline on a certain prescribed depth h_{w}.

4.2 Derivation of the Equation for Pressure

In this section we provide the complete derivation of the elliptic equation (2.9) for the dispersive component $\wp$ of the depth-integrated pressure p. It is convenient to start with the base model (4.1)–(4.3) written in the following equivalent form (see [188] for more details):

$$(\mathcal{J}\mathcal{H})_t + \left[\mathcal{J}\mathcal{H}u^1\right]_\lambda + \left[\mathcal{J}\mathcal{H}u^2\right]_\theta = 0, \tag{4.10}$$

$$(\mathcal{J}\mathcal{H}\mathbf{v})_t + \left[\mathcal{J}\mathcal{H}\mathbf{v}u^1\right]_\lambda + \left[\mathcal{J}\mathcal{H}\mathbf{v}u^2\right]_\theta + g\,\nabla\Big(\mathcal{J}\,\frac{\mathcal{H}^2}{2}\Big)$$
$$= g\,\frac{\mathcal{H}^2}{2}\,\nabla\mathcal{J} + \Big[g\,\mathcal{H}\,\nabla h + \mathcal{H}\,\boldsymbol{S} + \nabla\wp - \check{\wp}\,\nabla h\Big]\mathcal{J}, \tag{4.11}$$

where $\mathbf{v} = \big(v_1,\, v_2\big)$ is the covariant velocity vector:

$$v_1 = (\Omega + u^1)\,R^2\sin^2\theta\,, \qquad v_2 = R^2u^2\,. \tag{4.12}$$

Finally, the vector $\boldsymbol{S} = \big(0,\, s_2\big)$ with

$$s_2 = \left[2\,\Omega\,u^1 + (u^1)^2\right]R^2\sin\theta\,\cos\theta\,.$$

Equation (4.11) can be rewritten in a non-conservative form:

$$\mathcal{D}\mathbf{v} = -g\,\nabla\eta + \frac{\nabla\wp - \check{\wp}\,\nabla h}{\mathcal{H}} + \boldsymbol{S}, \tag{4.13}$$

where $\mathcal{D}\,\mathbf{v} \equiv \mathbf{v}_t + (\mathbf{u}\cdot\nabla)\,\mathbf{v}\,.$

As in the globally flat case, we shall express first the dispersive pressure component on the bottom $\check{\wp}$ as a function of $\wp$ and other variables. From definitions (4.5) we have

$$\check{\wp} = \frac{3\,\wp}{2\,\mathcal{H}} + \frac{\mathcal{H}}{4}\,\mathcal{R}_2, \tag{4.14}$$

where the term $\mathcal{R}_2$ can be fully expanded:

$$\mathcal{R}_2 \stackrel{\text{def}}{:=} \mathcal{D}^2 h \equiv \mathcal{D}\,(\mathcal{D}\,h) = (\mathcal{D}\,h)_t + \mathbf{u}\cdot\nabla(\mathcal{D}\,h)$$
$$= \left[h_t + \mathbf{u}\cdot\nabla h\right]_t + \mathbf{u}\cdot\nabla\left[h_t + \mathbf{u}\cdot\nabla h\right] = \underbrace{h_{tt} + 2\,\mathbf{u}\cdot\nabla h_t}_{\stackrel{\text{def}}{=:}\,\mathcal{B}} + \mathbf{u}_t\cdot\nabla h + \mathbf{u}\cdot\nabla(\mathbf{u}\cdot\nabla h)\,.$$

The term $\mathcal{B}$ contains all the terms involving the bottom motion. The last term in $\mathcal{R}_2$ can be further transformed[2] equivalently as

[2]Indeed,

$$\mathbf{u}\cdot\nabla(\mathbf{u}\cdot\nabla h) = u^1\left[u^1h_\lambda + u^2h_\theta\right]_\lambda + u^2\left[u^1h_\lambda + u^2h_\theta\right]_\theta =$$

$$\mathbf{u}\cdot\nabla(\mathbf{u}\cdot\nabla h) = \big[(\mathbf{u}\cdot\nabla)\,\mathbf{u}\big]\cdot\nabla h + \mathbf{u}\cdot\big[(\mathbf{u}\cdot\nabla)\,\nabla h\big].$$

Consequently,

$$\begin{aligned}\mathscr{R}_2 = \mathscr{B} + \mathbf{u}_t\cdot\nabla h + \big[(\mathbf{u}\cdot\nabla)\,\mathbf{u}\big]\cdot\nabla h + \mathbf{u}\cdot\big[(\mathbf{u}\cdot\nabla)\,\nabla h\big] \\ \equiv \mathscr{B} + (\mathcal{D}\mathbf{u})\cdot\nabla h + \mathbf{u}\cdot\big[(\mathbf{u}\cdot\nabla)\,\nabla h\big],\end{aligned} \tag{4.15}$$

where in component-wise form we have

$$\begin{aligned}\mathcal{D}u^1 &= u^1_t + u^1 u^1_\lambda + u^2 u^1_\theta\,,\\ \mathcal{D}u^2 &= u^2_t + u^1 u^2_\lambda + u^2 u^2_\theta\,,\\ (\mathcal{D}\mathbf{u})\cdot\nabla h &= \big(u^1_t + u^1 u^1_\lambda + u^2 u^1_\theta\big)\,h_\lambda + \big(u^2_t + u^1 u^2_\lambda + u^2 u^2_\theta\big)\,h_\theta\,,\\ \mathbf{u}\cdot\big[(\mathbf{u}\cdot\nabla)\,\nabla h\big] &= (u^1)^2 h_{\lambda\lambda} + 2u^1u^2 h_{\lambda\theta} + (u^2)^2 h_{\theta\theta}\,.\end{aligned}$$

By substituting the last expression in (4.15) into Eq. (4.14) we obtain

$$\check{\wp} = \frac{3\wp}{2\mathcal{H}} + \frac{\mathcal{H}}{4}\Big\{\mathscr{B} + (\mathcal{D}\mathbf{u})\cdot\nabla h + \mathbf{u}\cdot\big[(\mathbf{u}\cdot\nabla)\,\nabla h\big]\Big\}. \tag{4.16}$$

Now let us express the convective term $\mathcal{D}\,\mathbf{u}$ in terms of $\mathcal{D}\,\mathbf{v}$. According to formulas (4.12), the covariant vector of the velocity $\mathbf{v}$ can be rewritten as

$$\mathbf{v} = \Omega\,\mathscr{G} + \mathbb{G}\cdot\mathbf{u}\,,$$

where

$$\mathscr{G} \overset{\text{def}}{:=} \begin{pmatrix} g_{11} \\ 0 \end{pmatrix}, \qquad \mathbb{G} \overset{\text{def}}{:=} \begin{pmatrix} g_{11} & 0 \\ 0 & g_{22} \end{pmatrix},$$

with g_{11}, g_{22} being covariant components of the metric tensor on a sphere [188], i.e.

$$g_{11} = R^2\sin^2\theta\,, \qquad g_{22} = R^2\,.$$

Then, we have

$u^1u^1_\lambda h_\lambda + (u^1)^2 h_{\lambda\lambda} + u^1u^2_\lambda h_\theta + u^1u^2 h_{\lambda\theta} + u^2u^1_\theta h_\lambda + u^2u^1 h_{\lambda\theta} + u^2u^2_\theta h_\theta + (u^2)^2 h_{\theta\theta} =$

$\big(u^1u^1_\lambda + u^2u^1_\theta\big)h_\lambda + \big(u^1u^2_\lambda + u^2u^2_\theta\big)h_\theta + u^1\big(u^1(h_\lambda)_\lambda + u^2(h_\lambda)_\theta\big) + u^2\big(u^1(h_\theta)_\lambda + u^2(h_\theta)_\theta\big) =$

$\big[(\mathbf{u}\cdot\nabla)\,\mathbf{u}\big]\cdot\nabla h + \mathbf{u}\cdot\big[(\mathbf{u}\cdot\nabla)\,\nabla h\big].$

□

$$\mathcal{D}\mathbf{v} = \Omega\,\mathcal{D}\mathcal{G} + \mathcal{D}\,(\mathbb{G}\cdot\mathbf{u}) = u^2\,\Omega\,\mathcal{G}_\theta + \mathbb{G}\cdot\mathcal{D}\mathbf{u} + u^1\,u^2\,\mathcal{G}_\theta =$$
$$\mathbb{G}\cdot\mathcal{D}\mathbf{u} + u^2\,(\Omega + u^1)\,\mathcal{G}_\theta\,, \tag{4.17}$$

where

$$\mathcal{G}_\theta = \begin{pmatrix} 2\,R^2\,\sin\theta\,\cos\theta \\ 0 \end{pmatrix}.$$

By inverting equation (4.17), we obtain:

$$\mathcal{D}\mathbf{u} = \mathbb{G}^{-1}\cdot\mathcal{D}\mathbf{v} - u^2\,(\Omega + u^1)\,\mathbb{G}^{-1}\cdot\mathcal{G}_\theta\,,$$

and using the non-conservative equation (4.13) we arrive to the required representation:

$$\mathcal{D}\mathbf{u} = \mathbb{G}^{-1}\cdot\Big\{ - g\,\nabla\eta + \frac{\nabla\wp - \check{\wp}\,\nabla h}{\mathcal{H}}\Big\} + \mathbb{G}^{-1}\cdot\boldsymbol{\Lambda}\,, \tag{4.18}$$

with

$$\mathbb{G}^{-1} = \begin{pmatrix} g^{11} & 0 \\ 0 & g^{22} \end{pmatrix}, \qquad g^{11} \equiv \frac{1}{g_{11}} = \frac{1}{R^2\,\sin^2\theta}\,, \qquad g^{22} \equiv \frac{1}{g_{22}} = \frac{1}{R^2}\,,$$

$$\boldsymbol{\Lambda} = \begin{pmatrix} \Lambda_1 \\ \Lambda_2 \end{pmatrix} = \boldsymbol{S} - u^2\,(\Omega + u^1)\,\mathcal{G}_\theta = R^2\,\sin\theta\,\cos\theta\begin{pmatrix} -2\,u^2\,(\Omega + u^1) \\ u^1\,(2\,\Omega + u^1) \end{pmatrix}.$$

Finally, by substituting expression (4.18) into Eq. (4.16) we obtain:

$$\check{\wp} = \frac{3\,\wp}{2\,\mathcal{H}} + \frac{\mathcal{H}}{4}\Big\{\frac{\nabla\wp\cdot\nabla h - \check{\wp}\,|\nabla h|^2}{\mathcal{H}} + \mathcal{Q}\Big\},$$

where

$$\nabla\wp\cdot\nabla h \equiv g^{11}\,\wp_\lambda\,h_\lambda + g^{22}\,\wp_\theta\,h_\theta = \frac{1}{R^2}\Big[\frac{\wp_\lambda\,h_\lambda}{\sin^2\theta} + \wp_\theta\,h_\theta\Big],$$

$$|\nabla h|^2 \equiv g^{11}\,h_\lambda^2 + g^{22}\,h_\theta^2 = \frac{1}{R^2}\Big[\frac{h_\lambda^2}{\sin^2\theta} + h_\theta^2\Big],$$

$$\mathcal{Q} \stackrel{\text{def}}{:=} \big[\boldsymbol{\Lambda} - g\,\nabla\eta\big]\cdot\nabla h + \mathcal{B} + \mathbf{u}\cdot\big((\mathbf{u}\cdot\nabla)\,\nabla h\big)\,,$$

$$\nabla\eta\cdot\nabla h \equiv g^{11}\,\eta_\lambda\,h_\lambda + g^{22}\,\eta_\theta\,h_\theta = \frac{1}{R^2}\Big[\frac{\eta_\lambda\,h_\lambda}{\sin^2\theta} + \eta_\theta\,h_\theta\Big],$$

$$\boldsymbol{\Lambda}\cdot\nabla h \equiv g^{11}\,\Lambda_1\,h_\lambda + g^{22}\,\Lambda_2\,h_\theta$$

$$= \cot\theta\Big[-2u^2(\Omega + u^1)h_\lambda + u^1(2\Omega + u^1)h_\theta \sin^2\theta\Big].$$

Consequently, the dispersive part of the fluid pressure at the bottom can be expressed in terms of other variables as

$$\check{\wp} = \frac{1}{\Upsilon}\left\{\frac{6\wp}{\mathcal{H}} + \mathcal{H}\mathcal{Q} + \nabla\wp\cdot\nabla h\right\}, \tag{4.19}$$

where we introduced a new dependent variable

$$\Upsilon \overset{\text{def}}{:=} 4 + |\nabla h|^2.$$

Now we can proceed to the derivation of an equation for $\wp$. From definitions (4.5) it follows that

$$\wp = \frac{\mathcal{H}^3}{12}\mathcal{R}_1 + \frac{\mathcal{H}}{2}\check{\wp}, \tag{4.20}$$

where $\mathcal{R}_1$ was defined as

$$\mathcal{R}_1 \overset{\text{def}}{:=} \mathcal{D}(\nabla\cdot\mathbf{u}) - (\nabla\cdot\mathbf{u})^2.$$

Let us transform the last expression for $\mathcal{R}_1$ using the definition of the divergence $\nabla\cdot(\cdot)$ and material derivative $\mathcal{D}$ operators

$$\mathcal{D}(\nabla\cdot\mathbf{u}) = (\nabla\cdot\mathbf{u})_t + u^1(\nabla\cdot\mathbf{u})_\lambda + u^2(\nabla\cdot\mathbf{u})_\theta$$

$$= \nabla\cdot\mathbf{u}_t + u^1\left\{u^1_\lambda + \frac{(\mathcal{J}u^2)_\theta}{\mathcal{J}}\right\}_\lambda + u^2\left\{u^1_\lambda + \frac{(\mathcal{J}u^2)_\theta}{\mathcal{J}}\right\}_\theta$$

$$= \nabla\cdot\mathbf{u}_t + (u^1u^1_\lambda)_\lambda - (u^1_\lambda)^2 + (u^2u^1_\theta)_\lambda - u^2_\lambda u^1_\theta + u^1\frac{(\mathcal{J}u^2)_{\lambda\theta}}{\mathcal{J}} + u^2\left[\frac{(\mathcal{J}u^2)_\theta}{\mathcal{J}}\right]_\theta$$

$$= \underbrace{(u^1_t)_\lambda}_{\star_1} + \frac{(\mathcal{J}u^2_t)_\theta}{\mathcal{J}} + \underbrace{\left[u^1u^1_\lambda + u^2u^1_\theta\right]_\lambda}_{\star_1} + \frac{[u^1(\mathcal{J}u^2)_\lambda]_\theta}{\mathcal{J}} - \frac{u^1_\theta(\mathcal{J}u^2)_\lambda}{\mathcal{J}}$$

$$+ \left[u^2\frac{(\mathcal{J}u^2)_\theta}{\mathcal{J}}\right]_\theta - u^2_\theta\frac{(\mathcal{J}u^2)_\theta}{\mathcal{J}} - (u^1_\lambda)^2 - u^2_\lambda u^1_\theta.$$

It is not difficult to see that terms marked with ($\star_1$) can be aggregated into $(\mathcal{D}u^1)_\lambda$. Consequently, we have

$$\mathcal{D}\,(\nabla\cdot\mathbf{u}) \;=\; (\mathcal{D}\,u^1)_\lambda \;+\; \underbrace{\frac{(\mathcal{J}\,u_t^2)_\theta \;+\; (\mathcal{J}\,u^1\,u_\lambda^2)_\theta}{\mathcal{J}}}_{\star_2} \;-\; u_\theta^1\,u_\lambda^2 \;+\; \underbrace{\frac{[\mathcal{J}\,u^2\,u_\theta^2]_\theta}{\mathcal{J}}}_{\star_2}$$
$$+\;\frac{[\mathcal{J}_\theta\,(u^2)^2]_\theta}{\mathcal{J}} \;-\; \underbrace{\left[\frac{u^2\,(\mathcal{J}\,u^2)_\theta}{\mathcal{J}^2}\,\mathcal{J}_\theta \;+\; \frac{u_\theta^2\,(\mathcal{J}\,u^2)_\theta}{\mathcal{J}} \;+\; (u_\lambda^1)^2\right]}_{\bigstar} \;-\; u_\lambda^2\,u_\theta^1\,.$$

The terms marked above with ($\star_2$) give another interesting combination:

$$\frac{[\mathcal{J}\,(u_t^2 \;+\; u^1\,u_\lambda^2 \;+\; u^2\,u_\theta^2)]_\theta}{\mathcal{J}} \;\equiv\; \frac{(\mathcal{J}\,\mathcal{D}\,u^2)_\theta}{\mathcal{J}}\,.$$

According to the definition of the divergence $\nabla\cdot\mathbf{u}$, the term ($\bigstar$) can be transformed as

$$(u_\lambda^1)^2 \;+\; \left[\frac{(\mathcal{J}\,u^2)_\theta}{\mathcal{J}}\right]^2 \;\equiv\; (\nabla\cdot\mathbf{u})^2 \;-\; 2\,u_\lambda^1\,\frac{(\mathcal{J}\,u^2)_\theta}{\mathcal{J}}\,.$$

Consequently, we can rewrite $\mathcal{D}\,(\nabla\cdot\mathbf{u})$ using mentioned above simplifications:

$$\mathcal{D}\,(\nabla\cdot\mathbf{u}) \;=\; \left(\mathcal{D}\,u^1\right)_\lambda \;+\; \frac{(\mathcal{J}\,\mathcal{D}\,u^2)_\theta}{\mathcal{J}} \;-\; (\nabla\cdot\mathbf{u})^2$$
$$+\;2\,u_\lambda^1\,u_\theta^2 \;+\; 2\,u_\lambda^1\,u^2\,\frac{\mathcal{J}_\theta}{\mathcal{J}} \;+\; 2\,u^2\,u_\theta^2\,\frac{\mathcal{J}_\theta}{\mathcal{J}} \;+\; (u^2)^2\,\frac{\mathcal{J}_{\theta\theta}}{\mathcal{J}} \;-\; 2\,u_\theta^1\,u_\lambda^2\,.$$

Finally, taking into account the fact that $\mathcal{J}_{\theta\theta} \;\equiv\; -\,\mathcal{J}$, we obtain:

$$\mathcal{D}\,(\nabla\cdot\mathbf{u}) = \nabla\cdot(\mathcal{D}\,\mathbf{u}) - (\nabla\cdot\mathbf{u})^2 + 2\left(u_\lambda^1\,u_\theta^2 - u_\theta^1\,u_\lambda^2\right) + 2\,u^2\left(u_\lambda^1 + u_\theta^2\right)\,\cot\theta - (u^2)^2.$$

By assembling all the ingredients together, we obtain the following expression for $\mathscr{R}_1$:

$$\mathscr{R}_1 \;=\; \nabla\cdot(\mathcal{D}\,\mathbf{u}) - 2\,(\nabla\cdot\mathbf{u})^2 + 2\left(u_\lambda^1\,u_\theta^2 - u_\theta^1\,u_\lambda^2\right) + 2\,u^2\left(u_\lambda^1 + u_\theta^2\right)\,\cot\,\theta - (u^2)^2\,.$$

By substituting the last expression for $\mathscr{R}_1$ into Eq. (4.20) and using formulas (4.18), (4.19) for $\mathcal{D}\,\mathbf{u}$ and $\check{\wp}$ correspondingly, we obtain the first version of the required equation for $\wp$:

$$\wp = \frac{\mathcal{H}^3}{12}\left[\nabla\cdot\left\{\mathbb{G}^{-1}\cdot\left(\boldsymbol{\Lambda} - g\,\nabla\eta + \frac{\nabla\wp}{\mathcal{H}} - \frac{6\,\nabla h}{\mathcal{H}^2\,\Upsilon} - \nabla h\,\frac{\mathcal{Q}}{\Upsilon} - \frac{(\nabla\wp\cdot\nabla h)\,\nabla h}{\mathcal{H}\,\Upsilon}\right)\right\}\right.$$

$$- 2(\nabla \cdot \mathbf{u})^2 + 2\left(u_\lambda^1 u_\theta^2 - u_\theta^1 u_\lambda^2\right) + 2u^2\left(u_\lambda^1 + u_\theta^2\right)\cot\theta - (u^2)^2\Big]$$

$$+ \frac{\mathcal{H}}{2\Upsilon}\left\{\frac{6\wp}{\mathcal{H}} + \mathcal{H}\mathcal{Q} + \nabla\wp \cdot \nabla h\right\}. \tag{4.21}$$

We can notice that the multiplication of covariant vectors $\boldsymbol{\Lambda}$, $\nabla\eta$, $\nabla\wp$ and ∇h by matrix $\mathbb{G}^{-1}$ transforms them into contravariant vectors which enter into the definition of the divergence operator, for example,

$$\nabla \cdot \nabla \eta \overset{\text{def}}{:=} \frac{\left(\mathcal{J} g^{11}\eta_\lambda\right)_\lambda + \left(\mathcal{J} g^{22}\eta_\theta\right)_\theta}{\mathcal{J}} = \frac{1}{R^2 \sin\theta}\left[\frac{\eta_{\lambda\lambda}}{\sin\theta} + \left(\eta_\theta \sin\theta\right)_\theta\right] \equiv \boldsymbol{\Delta}\eta,$$

$$\nabla \cdot \boldsymbol{\Lambda} \overset{\text{def}}{:=} \frac{\left(\mathcal{J} g^{11}\Lambda_1\right)_\lambda + \left(\mathcal{J} g^{22}\Lambda_2\right)_\theta}{\mathcal{J}} = \frac{1}{R^2 \sin\theta}\left[\frac{\Lambda_{1,\lambda}}{\sin\theta} + \left(\Lambda_2 \sin\theta\right)_\theta\right].$$

We can notice also that the following identity holds:

$$\nabla \cdot \left[\frac{6\nabla h}{\mathcal{H}^2\Upsilon}\wp\right] = \frac{6}{\mathcal{H}^2\Upsilon}\nabla\wp \cdot \nabla h + 6\wp\nabla \cdot \left[\frac{\nabla h}{\mathcal{H}^2\Upsilon}\right].$$

Taking into account the last two remarks, we can rewrite equation (4.21) in a more compact form:

$$\mathcal{L}\wp = \mathcal{F}^\star, \tag{4.22}$$

where the linear operator $\mathcal{L}$ and the right-hand side $\mathcal{F}^\star$ are defined as

$$\mathcal{L}\wp \overset{\text{def}}{:=} \nabla \cdot \left\{\frac{\nabla\wp}{\mathcal{H}} - \frac{(\nabla\wp \cdot \nabla h)\nabla h}{\mathcal{H}\Upsilon}\right\} - 6\wp\left\{\frac{2}{\mathcal{H}^3}\frac{\Upsilon - 3}{\Upsilon} + \nabla \cdot \left[\frac{\nabla h}{\mathcal{H}^2\Upsilon}\right]\right\},$$

$$\mathcal{F}^\star \overset{\text{def}}{:=} g\,\boldsymbol{\Delta}\eta + \nabla \cdot \left\{\frac{\mathcal{Q}\nabla h}{\Upsilon} - \boldsymbol{\Lambda}\right\} - \frac{6\mathcal{Q}}{\mathcal{H}\Upsilon}$$

$$+ 2(\nabla \cdot \mathbf{u})^2 - 2\left(u_\lambda^1 u_\theta^2 - u_\theta^1 u_\lambda^2\right) - 2u^2\left(u_\lambda^1 + u_\theta^2\right)\cot\theta + (u^2)^2.$$

By expanding divergences of covariant vectors, we can express the operator $\mathcal{L}(\cdot)$ through partial derivatives:

$$\mathcal{L}\wp \equiv \frac{1}{R^2\sin\theta}\left\{\frac{1}{\sin\theta}\left[\frac{\wp_\lambda}{\mathcal{H}} - \frac{\nabla\wp \cdot \nabla h}{\mathcal{H}\Upsilon}h_\lambda\right]_\lambda + \left[\left(\frac{\wp_\theta}{\mathcal{H}} - \frac{\nabla\wp \cdot \nabla h}{\mathcal{H}\Upsilon}h_\theta\right)\sin\theta\right]_\theta\right\}$$

$$-6\wp\left\{\frac{2}{\mathcal{H}^3}\frac{\Upsilon - 3}{\Upsilon} + \frac{1}{R^2\sin\theta}\left[\frac{1}{\sin\theta}\left(\frac{h_\lambda}{\mathcal{H}^2\Upsilon}\right)_\lambda + \left(\frac{h_\theta}{\mathcal{H}^2\Upsilon}\sin\theta\right)_\theta\right]\right\}.$$

After multiplying both sides of Eq. (4.22) by $R^2 \sin\theta$ and switching to linear components of the velocity u and v, we obtain the following equation for $\wp$:

$$\left[\frac{1}{\sin\theta}\left\{\frac{\wp_\lambda}{\mathcal{H}}-\frac{\nabla\wp\cdot\nabla h}{\mathcal{H}\Upsilon}\,h_\lambda\right\}\right]_\lambda+\left[\left\{\frac{\wp_\theta}{\mathcal{H}}-\frac{\nabla\wp\cdot\nabla h}{\mathcal{H}\Upsilon}\,h_\theta\right\}\sin\theta\right]_\theta$$
$$-6\wp\left[\frac{2}{\mathcal{H}^3}\,\frac{(\Upsilon-3)R^2\sin\theta}{\Upsilon}+\left\{\left(\frac{h_\lambda}{\mathcal{H}^2\Upsilon\sin\theta}\right)_\lambda+\left(\frac{h_\theta}{\mathcal{H}^2\Upsilon}\sin\theta\right)_\theta\right\}\right]=\mathcal{F}, \tag{4.23}$$

where the right-hand side $\mathcal{F}$ is defined as

$$\mathcal{F}\stackrel{\text{def}}{:=}\left[\frac{1}{\sin\theta}\left\{g\,\eta_\lambda+\frac{\mathcal{Q}}{\Upsilon}\,h_\lambda-\Lambda_1\right\}\right]_\lambda+\left[\left\{g\,\eta_\theta+\frac{\mathcal{Q}}{\Upsilon}\,h_\theta-\Lambda_2\right\}\sin\theta\right]_\theta$$
$$-\;R^2\,\frac{6\,\mathcal{Q}}{\mathcal{H}\,\Upsilon}\,\sin\theta+\frac{2}{\sin\theta}\left\{u_\lambda+(v\,\sin\theta)_\theta\right\}^2$$
$$-\;2\left(u_\lambda\,v_\theta-v_\lambda\,u_\theta\right)-2\,(u\,v)_\lambda\cot\theta-\left(v^2\cos\theta\right)_\theta.$$

The quantities $\mathcal{Q}$, $\mathcal{B}$ and $\boldsymbol{\Lambda}$ are expressed through linear velocity components as

$$\mathcal{Q}\equiv\left(\boldsymbol{\Lambda}-g\,\nabla\eta\right)\cdot\nabla h+\frac{1}{R^2\sin\theta}\left\{\frac{u^2}{\sin\theta}\,h_{\lambda\lambda}+2\,u\,v\,h_{\lambda\theta}+v^2\,h_{\theta\theta}\,\sin\theta\right\}+\mathcal{B},$$

$$\mathcal{B}\equiv h_{tt}+2\left\{\frac{u}{R\,\sin\theta}\,h_{\lambda t}+\frac{v}{R}\,h_{\theta t}\right\},$$

$$\Lambda_1\stackrel{\text{def}}{:=}-\left(2\,u\,v\,\cot\theta+\digamma\,v\,R\right)\sin\theta, \qquad \Lambda_2\stackrel{\text{def}}{:=}u^2\cot\theta+\digamma\,u\,R,$$

$$\boldsymbol{\Lambda}\cdot\nabla h\equiv\frac{\Lambda_1\,h_\lambda}{R^2\,\sin^2\theta}+\frac{\Lambda_2\,h_\theta}{R^2}.$$

This concludes naturally the derivation of the elliptic equation (4.23) for the dispersive part of the depth-integrated pressure $\wp$.

Remark 4.1 Let us assume that the angular velocity vanishes, i.e. $\Omega\equiv 0$. In a vicinity of a fixed admissible point $(\lambda^\star,\,\theta^\star)$ we introduce a non-degenerate transformation of coordinates:

$$x^\checkmark\stackrel{\text{def}}{:=}R\,(\lambda-\lambda^\star)\,\sin\theta^\star, \qquad y^\checkmark\stackrel{\text{def}}{:=}-R\,(\theta-\theta^\star),$$

along with corresponding velocity components:

$$u^\checkmark\equiv\dot{x}^\checkmark=R\,\dot\lambda\,\sin\theta\equiv\varsigma\,u, \qquad v^\checkmark\equiv\dot{y}^\checkmark=-R\,\dot\theta=-v,$$

where $\varsigma \stackrel{\text{def}}{:=} \dfrac{\sin\,\theta^{\star}}{\sin\,\theta}$. By assuming that the considered neighbourhood is small in the latitude, i.e. the quantity $\varepsilon \stackrel{\text{def}}{:=} |\,\theta - \theta^{\star}\,| \ll 1$ is small. After transforming equation (4.23) to new independent $(x^{\checkmark},\, y^{\checkmark})$ and new dependent $(u^{\checkmark},\, v^{\checkmark})$ variables and neglecting the terms of the order $\mathcal{O}(\varepsilon)$ and $\mathcal{O}(R^{-1})$ one can obtain the familiar to us elliptic equation for $\wp$ in the globally plane case. This fact is a further confirmation of the correctness of our derivation procedure described above.

4.3 Numerical Algorithm

The system of Eqs. (4.1)–(4.3) is not of CAUCHY–KOVALEVSKAYA's type, since momentum balance equations (4.2), (4.3) involve mixed derivatives with respect to time and space of the velocity components u, v. We already encountered this difficulty in the globally flat case (see Chap. 2). A direct (e.g. finite difference) approximation of governing equations would lead to a complex system of fully coupled nonlinear algebraic equations in a very high dimensional space. In the globally flat case it was found out that it is more judicious to perform a preliminary decoupling[3] of the system into a scalar elliptic equation and a system of hyperbolic equations with source terms (see Chap. 2). In the present chapter we realize the same idea for FNWD equations (4.1)–(4.3) on a rotating attracting sphere. Similarly we shall rewrite the governing equations as a scalar elliptic equation to determine the dispersive component of the depth-integrated pressure $\wp$ and a hyperbolic system of shallow water type equations with some additional source terms. With this splitting we can apply the most appropriate numerical methods for elliptic and hyperbolic problems correspondingly.

The derivation of the elliptic equation for non-hydrostatic pressure component $\wp$ can be found in Sect. 4.2. Here we provide only the final result:

$$\left[\frac{1}{\sin\theta}\left\{\frac{\wp_{\lambda}}{\mathcal{H}} - \frac{\nabla\wp\cdot\nabla h}{\mathcal{H}\Upsilon}\,h_{\lambda}\right\}\right]_{\lambda} + \left[\left\{\frac{\wp_{\theta}}{\mathcal{H}} - \frac{\nabla\wp\cdot\nabla h}{\mathcal{H}\Upsilon}\,h_{\theta}\right\}\sin\,\theta\right]_{\theta} - \mathring{\mathcal{K}}\,\wp = \mathcal{F}, \tag{4.24}$$

where

$$\mathring{\mathcal{K}} \stackrel{\text{def}}{:=} \mathcal{K}_{00} + \frac{\partial\mathcal{K}_{01}}{\partial\lambda} + \frac{\partial\mathcal{K}_{02}}{\partial\theta},$$

with

[3] We underline that this decoupling procedure involves no approximation and the resulting system of equations is completely equivalent to the base model (4.1)–(4.3).

$$\mathcal{K}_{00} \overset{\text{def}}{:=} R^2 \frac{12(\Upsilon - 3)}{\mathcal{H}^3 \Upsilon} \sin\theta, \quad \mathcal{K}_{01} \overset{\text{def}}{:=} \frac{6 h_\lambda}{\mathcal{H}^2 \Upsilon \sin\theta}, \quad \mathcal{K}_{02} \overset{\text{def}}{:=} \frac{6 h_\theta}{\mathcal{H}^2 \Upsilon} \sin\theta.$$

We used the variable Υ defined here as

$$\Upsilon \overset{\text{def}}{:=} 4 + \nabla h \cdot \nabla h \equiv 4 + |\nabla h|^2.$$

The scalar product $\nabla \wp \cdot \nabla h$ can be easily expressed as

$$\nabla \wp \cdot \nabla h \equiv \frac{1}{R^2} \left\{ \frac{\wp_\lambda h_\lambda}{\sin^2\theta} + \wp_\theta h_\theta \right\}.$$

The right-hand side $\mathcal{F}$ is defined as

$$\mathcal{F} \overset{\text{def}}{:=} \Bigg[\underbrace{\frac{1}{\sin\theta} \left\{ g\,\eta_\lambda + \frac{\mathcal{Q}}{\Upsilon} h_\lambda - \Lambda_1 \right\}}_{\overset{\text{def}}{=:}\, \mathring{\mathcal{F}}_1} \Bigg]_\lambda + \Bigg[\underbrace{\left\{ g\,\eta_\theta + \frac{\mathcal{Q}}{\Upsilon} h_\theta - \Lambda_2 \right\} \sin\theta}_{\overset{\text{def}}{=:}\, \mathring{\mathcal{F}}_2} \Bigg]_\theta$$
$$- R^2 \frac{6\mathcal{Q}}{\mathcal{H}\Upsilon} \sin\theta + \frac{2}{\sin\theta} \left\{ u_\lambda + (v \sin\theta)_\theta \right\}^2$$
$$- 2\big(u_\lambda v_\theta - v_\lambda u_\theta\big) - 2(u\,v)_\lambda \cot\theta - \big(v^2 \cos\theta\big)_\theta,$$

where

$$\mathcal{Q} \overset{\text{def}}{:=} \big(\boldsymbol{\Lambda} - g\nabla\eta\big) \cdot \nabla h +$$
$$\frac{1}{R^2 \sin\theta} \left\{ \frac{u^2}{\sin\theta} h_{\lambda\lambda} + 2uv\,h_{\lambda\theta} + v^2 h_{\theta\theta} \sin\theta \right\}$$
$$+ \underbrace{h_{tt} + 2\left\{ \frac{u}{R\sin\theta} h_{\lambda t} + \frac{v}{R} h_{\theta t} \right\}}_{\overset{\text{def}}{=:}\, \mathcal{B}},$$

with vector $\boldsymbol{\Lambda} \overset{\text{def}}{:=} (\Lambda_1, \Lambda_2)^\top$ whose components are

$$\Lambda_1 \overset{\text{def}}{:=} -\big(2uv\cot\theta + \digamma v R\big) \sin\theta, \qquad \Lambda_2 \overset{\text{def}}{:=} u^2 \cot\theta + \digamma u R.$$

The term $\mathcal{B}$ contains all the terms coming from bottom motion effects. If the bottom is stationary, then $\mathcal{B} \equiv 0$.

The particularity of Eq. (4.24) is that it does not contain time derivatives of dynamic variables $\mathcal{H}$, u and v. This equation is very similar to the elliptic equation derived in the globally flat case [191]. The differences consist only in terms coming from EARTH's sphericity and rotation effects. It is not difficult to show that under total water depth positivity assumption $\mathcal{H} > 0$ and condition (4.4), Eq. (4.24) is uniformly elliptic. In order to have the uniqueness result of the DIRICHLET problem for (4.24), the coefficient $\mathring{\mathcal{K}}$ has to be positive defined [206], i.e. $\mathring{\mathcal{K}} > 0$. In the case of an even bottom (i.e. $h \equiv h_0 = \text{const}$) the positivity condition takes the following form:

$$\mathring{\mathcal{K}} = \frac{12\,R^2 \sin\theta}{\mathcal{H}^3\,\mathring{\Upsilon}} \Big[\mathring{\Upsilon} - 3 - \frac{\Omega^2}{2g}\Big\{\mathcal{H}\sin^2\theta - \eta_\theta \sin(2\theta) + \frac{8\,\mathcal{H}}{\mathring{\Upsilon}}\cos(2\theta)\Big\}\Big] > 0, \qquad (4.25)$$

where this time the variable $\mathring{\Upsilon}$ is defined simply as

$$\mathring{\Upsilon} \overset{\text{def}}{:=} 4 + \frac{\Omega^4 R^2}{4g^2}\sin^2(2\theta).$$

Using the values of parameters for our planet, i.e.

$$R = 6.38 \times 10^6\,\mathrm{m}, \qquad \Omega = 7.29 \times 10^{-5}\,\mathrm{s}^{-1}, \qquad g = 9.81\,\frac{\mathrm{m}}{\mathrm{s}^2},$$

and assuming (4.4) along with the fact that the gradient of the free surface elevation η is bounded, one can show that $\mathring{\mathcal{K}} \gg 1$. Thus, the condition (4.25) is trivially verified. In this case one can construct a finite difference operator with positive definite (grid-)operator. Theoretically, it is not excluded that for large bottom variations locally the coefficient $\mathring{\mathcal{K}}$ might become negative. In such cases, the conditioning of the discrete system is worsened and more iterations are needed to achieve the desired accuracy in solving equation (4.24). In the globally flat case the analysis of these cases was performed in Chap. 2. In practice, sometimes it is possible to avoid such complications by applying a prior smoothing operator to the bathymetry function $h(\lambda, \theta, t)$.

4.3.1 Numerical Scheme Construction

The finite difference counterpart of Eq. (4.24) can be obtained using the so-called integro-interpolation method [185]. In this section we construct a second order approximation to be able to work on coarser grids for a fixed desired accuracy.

In spherical coordinates we consider a uniform grid with spacings Δ_λ and Δ_θ along the axes $O\lambda$ and $O\theta$ correspondingly. In order to derive a numerical scheme, first we integrate equation (4.24) over a rectangle $A\,B\,C\,D$ depicted in Fig. 4.1a, with vertices A, B, C and D being the geometrical centers of adjacent cells. After

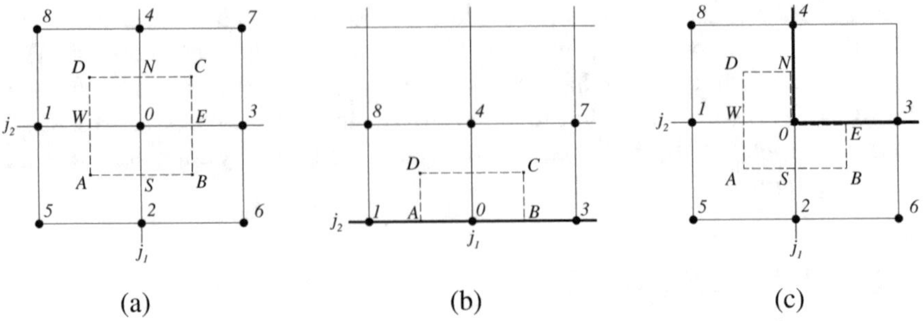

Fig. 4.1 Integration contours and finite difference stencils for Eq. (4.24) in an internal (**a**), boundary (**b**) and corner (**c**) nodes

applying GREEN's formula to the resulting integral, we obtain:

$$\int_{BC} \mathcal{F}^1 \, \mathrm{d}\theta - \int_{AD} \mathcal{F}^1 \, \mathrm{d}\theta + \int_{DC} \mathcal{F}^2 \, \mathrm{d}\lambda - \int_{AB} \mathcal{F}^2 \, \mathrm{d}\lambda \\ - \iint_{ABCD} \mathring{\mathcal{K}} \wp \, \mathrm{d}\lambda \, \mathrm{d}\theta = \iint_{ABCD} \mathcal{F} \, \mathrm{d}\lambda \, \mathrm{d}\theta \, . \tag{4.26}$$

From the double integral on the right-hand side we can extract a contour part as well:

$$\iint_{ABCD} \bar{\nabla} \cdot \mathring{\mathcal{F}} \, \mathrm{d}\lambda \, \mathrm{d}\theta = \int_{BC} \mathring{\mathcal{F}}_1 \, \mathrm{d}\theta - \int_{AD} \mathring{\mathcal{F}}_1 \, \mathrm{d}\theta + \int_{DC} \mathring{\mathcal{F}}_2 \, \mathrm{d}\lambda - \int_{AB} \mathring{\mathcal{F}}_2 \, \mathrm{d}\lambda \, ,$$

where $\bar{\nabla} \cdot (\cdot) \stackrel{\text{def}}{:=} (\cdot_1)_\lambda + (\cdot_2)_\theta$ is the 'flat' divergence operator, vectors $\mathring{\mathcal{F}} \stackrel{\text{def}}{:=} (\mathring{\mathcal{F}}_1, \mathring{\mathcal{F}}_2), (\mathcal{F}^1, \mathcal{F}^2)$ with components defined as

$$\mathcal{F}^1 = \frac{1}{\sin\theta} \left\{ \frac{\wp_\lambda}{\mathcal{H}} - \frac{\nabla\wp \cdot \nabla h}{\mathcal{H}\Upsilon} h_\lambda \right\}, \quad \mathcal{F}^2 = \left\{ \frac{\wp_\theta}{\mathcal{H}} - \frac{\nabla\wp \cdot \nabla h}{\mathcal{H}\Upsilon} h_\theta \right\} \sin\theta \, ,$$

$$\mathring{\mathcal{F}}_1 = \frac{1}{\sin\theta} \left\{ g\,\eta_\lambda + \frac{\mathcal{Q}}{\Upsilon} h_\lambda - \Lambda_1 \right\}, \quad \mathring{\mathcal{F}}_2 = \left\{ g\,\eta_\theta + \frac{\mathcal{Q}}{\Upsilon} h_\theta - \Lambda_2 \right\} \sin\theta \, .$$

In order to compute approximatively the integrals, we use the trapezoidal numerical quadrature rule along with a second order finite difference approximation of the derivatives in cell centers [154]. As a result one can obtain a finite difference approximation for Eq. (4.24) with nine points stencil in every internal node of the grid.

In an analogous way one can construct finite difference approximations in boundary nodes adjacent to a non-permeable wall. Let us assume that the node

$\mathbf{x}_{j_1, j_2} = (\lambda_{j_1}, \theta_{j_2})$ belongs to the boundary Γ lying along a parallel represented with a bold solid line in Fig. 4.1b. The interior of the computational domain lies in North-ward direction to this parallel. If nodes $\mathbf{x}_{j_1 - 1, j_2}$ and $\mathbf{x}_{j_1 + 1, j_2}$ also belong to Γ, then, the integration contour $A\,B\,C\,D$ is chosen such that vertices C and D coincide with adjacent cell centers (sharing the common node $\mathbf{x}_{j_1, j_2}$) and vertices A, B coincide with centers of edges belonging to Γ (see again Fig. 4.1b). In this case the rectangle side $A\,B$ belongs completely to the boundary Γ. That is why boundary conditions for the function $\wp$ have to be used while computing the integrals in the integro-differential equation (4.26).

Boundary Conditions Treatment

In the present study we consider only the case of wall boundary conditions. In the situation depicted in Fig. 4.1b the boundary condition (4.9) becomes simply

$$v\,\big|_{\Gamma} \equiv 0.$$

Thanks to Eq. (4.8), we obtain the following boundary condition for the variable $\wp$:

$$\frac{1}{R}\left\{\frac{\wp_{\theta} - \check{\wp}\,h_{\theta}}{\mathcal{H}} - g\,\eta_{\theta}\right\} + \frac{u^2}{R}\cot\theta + \mathcal{F}\,u = 0, \qquad \mathbf{x} \in \Gamma. \tag{4.27}$$

In Sect. 4.2 it is shown that the dispersive component of the fluid pressure at the bottom $\check{\wp}$ is related to other quantities $\mathcal{H}$, u, v and $\wp$ as

$$\check{\wp} = \frac{1}{\Upsilon}\left\{\frac{6\,\wp}{\mathcal{H}} + \mathcal{H}\,\mathcal{Q} + \nabla\wp \cdot \nabla h\right\}. \tag{4.28}$$

After substituting the last formula into Eq. (4.27), the required boundary condition takes the form:

$$\left\{\frac{1}{R}\left(\frac{\wp_{\theta}}{\mathcal{H}} - \frac{\nabla\wp \cdot \nabla h}{\mathcal{H}\,\Upsilon}\,h_{\theta}\right) - \frac{6\,h_{\theta}}{R\,\mathcal{H}^2\,\Upsilon}\,\wp\right\}\bigg|_{\Gamma} =$$
$$\left\{\frac{1}{R}\left(g\,\eta_{\theta} + \frac{\mathcal{Q}}{\Upsilon}\,h_{\theta}\right) - \frac{u^2}{R}\cot\theta - \mathcal{F}\,u\right\}\bigg|_{\Gamma},$$

or by using some notations introduced above we can simply write a more compact form:

$$\left\{\mathcal{F}^2 - \mathcal{K}_{02}\,\wp - \mathring{\mathcal{F}}_2\right\}\big|_{A\,B} = 0. \tag{4.29}$$

The last boundary condition is used while approximating the integrals over the rectangle side $A\,B$. Consider the double integral in the left-hand side of the integro-differential equation (4.26). It can be approximated as

$$\iint_{ABCD} \mathring{\mathcal{K}}\,\wp \,\mathrm{d}\lambda\,\mathrm{d}\theta \approx \iint_{ABCD} \mathcal{K}_{00}\,\wp\,\mathrm{d}\lambda\,\mathrm{d}\theta +$$

$$\left\{ \int_{BC} \mathcal{K}_{01}\,\mathrm{d}\theta - \int_{AD} \mathcal{K}_{01}\,\mathrm{d}\theta + \int_{DC} \mathcal{K}_{02}\,\mathrm{d}\lambda - \int_{AB} \mathcal{K}_{02}\,\mathrm{d}\lambda \right\} \wp\,(\mathbf{0}),$$

and after introducing the following approximations:

$$\int_{AB} \mathcal{F}^2\,\mathrm{d}\lambda \approx \mathcal{F}^2\,(\mathbf{0})\,\Delta_\lambda, \quad \int_{AB} \mathcal{K}_{02}\,\mathrm{d}\lambda \approx \mathcal{K}_{02}\,(\mathbf{0})\,\Delta_\lambda, \quad \int_{AB} \mathring{\mathcal{F}}_2\,\mathrm{d}\lambda \approx \mathring{\mathcal{F}}_2\,(\mathbf{0})\,\Delta_\lambda,$$

we come to an important conclusion: *all the terms coming from integrals* $\int_{AB}$ *vanish thanks to the boundary condition* (4.29). Consequently, the implementation of boundary conditions turns out to be trivial in the integro-interpolation method employed in our study. For boundary cells of other types, we use the same method of integro-interpolating approximations to obtain difference equations. As an illustration, in Fig. 4.1c we depict another one of the eight possible configurations of angular nodes. Since we write a difference equation for every grid node (interior and boundary), the total number of equations coincides with the total number of discretization points.

Computational Miscellanea

In the most general case a realistic computational domain for the wave propagation has a complex shape. Generally it is not convex and due to the existence of islands it might be multiply connected. It has some implications for linear solvers that we can use to solve the difference equations described above. First of all, due to the large scale nature of problems considered in this study, we privilege iterative schemes for the sake of computational efficiency. Then, due to geometrical and topological reasons described above, we adopt a simple but efficient method of Successive Over–Relaxation (SOR) [334]. This method contains a free parameter ϖ, which can be used to accelerate the convergence. The optimal value of $\varpi^\star$, which ensures the fastest convergence, is in general unknown. This question was studied theoretically for the POISSON equation with DIRICHLET boundary conditions in a rectangle [272]. So, in this case the optimal value of $\varpi^\star$ was shown to belong to the interval $(1,\,2)$. For example, if the mesh is taken uniform in each side ℓ of a square with the spacing $h = \frac{\ell}{N}$, then

$$\varpi^\star = \frac{2}{1 + \sin\frac{\pi h}{\ell}}.$$

If we take $\ell = 1$ and $N = 30$ we obtain the optimal value $\varpi^\star \approx 1.81$. Another observation is that in the limit $N \to +\infty$ the optimal value $\varpi^\star \to 2$. In our numerical experiments we observed the same tendency: with the mesh refinement the optimal relaxation parameter $\varpi^\star$ for the discretized equation (4.24) approaches 2 as well. In practice, we took the values $\varpi^\star \in [1.85, 1.95]$ depending on the degree of refinement.

For given functions $\wp$ and $\check{\wp}$ the system of conservative equations (4.1)–(4.3) is of hyperbolic type under the conditions (4.4) and water depth positivity $\mathcal{H}(\mathbf{x}, t) > 0, \forall t > 0$. Thanks to this property we have in our disposal the whole arsenal of numerical tools that have been developed for hyperbolic systems of equations [144, 145]. In the one-dimensional case we opted for predictor–corrector schemes [187, 191] with a free scheme parameter $\theta^n_{j_1, j_2}$. A judicious choice of this parameter ensures the TVD property and the monotonicity of solutions at least for scalar equations [192]. Some aspects of predictor–corrector schemes in two spatial dimensions are described in [291]. In the present study we employ the predictor–corrector scheme with $\theta^n_{j_1, j_2} \equiv 0$, which minimizes the numerical dissipation (and makes the scheme somehow more fragile). This choice is not probably the best for hyperbolic NSWE, but for non-hydrostatic FNWD models it works very well due to the inherent dispersive regularization property of solutions [73]. Moreover, on the predictor stage we compute directly the quantities $\mathcal{H}$, u and v (instead of computing the fluxes) since they are needed to compute the coefficients along with the right-hand side $\mathcal{F}$ of Eq. (4.24).

Let us describe briefly the numerical algorithm we use to solve the extended system of Eqs. (4.1)–(4.3), (4.24) (for more details see also [154, 191]). At the initial moment of time we are given by the free surface elevation $\eta(\mathbf{x}, 0)$ and velocity vector $\mathbf{u}(\mathbf{x}, 0)$. Moreover, if the bottom is not static, additionally we have to know also the quantities $h_t(\mathbf{x}, 0)$ and $h_{tt}(\mathbf{x}, 0)$. These data suffice to determine the initial distribution of the depth-integrated pressure $\wp$ by solving numerically the elliptic equation (4.24). Finally, the dispersive pressure component on the bottom $\check{\wp}$ is computed by a finite difference analogue of Eq. (4.28). By recurrence, let us assume that we know the same data on the nth time layer $t = t^n$: $\mathcal{H}^n$, $\mathbf{u}^n$, $\wp^n$ and $\check{\wp}^n$. Then, we employ the predictor–corrector scheme, each time step of this scheme consists of two stages. On the predictor stage we compute the quantities $\mathcal{H}^{n+\frac{1}{2}}$, $\mathbf{u}^{n+\frac{1}{2}}$ in cell centers as a solution of explicit discrete counterparts of Eqs. (4.6)–(4.8) (with right-hand sides taken from the time layer t^n). Then, one solves a difference equation to determine $\wp^{n+\frac{1}{2}}$. The coefficients and the right-hand side are evaluated using new values $\mathcal{H}^{n+\frac{1}{2}}$ and $\mathbf{u}^{n+\frac{1}{2}}$. From formula (4.28) one infers the value of $\check{\wp}^{n+\frac{1}{2}}$. All the values computed at the predictor stage $\mathcal{H}^{n+\frac{1}{2}}$, $\mathbf{u}^{n+\frac{1}{2}}$, $\wp^{n+\frac{1}{2}}$ and $\check{\wp}^{n+\frac{1}{2}}$ are then used at the corrector stage to determine the new values $\mathcal{H}^{n+1}$ and $\mathbf{u}^{n+1}$. At the corrector stage we employ the conservative form of Eqs. (4.1)–(4.3). In the very last step we compute also the values of $\wp^{n+1}$ and $\check{\wp}^{n+1}$. The algorithm described becomes a spherical analogue of the well-known LAX–WENDROFF scheme if one neglects the dispersive terms.

An important property of the proposed numerical algorithm is its well-balanced character [146] if the bottom is steady (i.e. $\mathscr{B} \equiv 0$) and the sphere is not too deformed (i.e. $\mathring{\mathcal{K}} > 0$). In other words, it preserves exactly the so-called lake-at-rest states where the fluid is at rest $\mathbf{u}^n \equiv \mathbf{0}$ and the free surface is unperturbed $\eta^n \equiv 0$. Then, it can be rigorously shown that this particular state will be preserved in the following layer $t = t^{n+1}$ as well. It is achieved by balanced discretizations of left and right-hand sides in the momentum equations (4.2), (4.3). This task is not a priori trivial since the equilibrium free surface shape is not spherical due to EARTH's rotation effects [188].

4.4 Numerical Illustrations

Currently, there is a well-established set of test problems [300] which are routinely used to validate numerical codes for tsunami propagation and run-up. These tests can be used also for inter-comparison of various algorithms in 1D and 2D [170]. However, currently, there do not exist such (generally admitted) tests for nonlinear dispersive wave models on a rotating sphere. The material presented below can be considered as a further effort to constitute such a database.

4.4.1 Wave Propagation Over a Flat Rotating Sphere

Consider a simple bounded spherical domain which occupies the region from 100 to 300° from the WEST to the EAST and from −60° to 65° from the SOUTH to the NORTH. From now on we use for simplicity the geographical latitude $\phi \equiv \frac{\pi}{2} - \theta$ instead of the variable θ. The considered domain is depicted in Fig. 4.2 and contains a large portion of the PACIFIC OCEAN, excluding, of course, the poles (see restriction (4.4)). The idealization consists in the fact that we assume the (undisturbed) water depth is constant, i.e. $h \equiv 4\,\mathrm{km}$. The initial condition consists of a GAUSSIAN-shaped bump put on the free surface:

$$\eta(\lambda, \phi, 0) = \alpha_0 \mathrm{e}^{-\varpi \rho^2(\lambda, \phi)}, \tag{4.30}$$

with zero velocity field in the fluid bulk. Function $\rho(\lambda, \phi)$ is a great-circle distance between the points (λ, ϕ) and (λ_0, ϕ_0), i.e.

$$\rho(\lambda, \phi) \stackrel{\text{def}}{:=} R \cdot \arccos\left\{ \cos\phi \cos\phi_0 \cos(\lambda - \lambda_0) + \sin\phi \sin\phi_0 \right\}. \tag{4.31}$$

In our numerical simulations we take the initial amplitude $\alpha_0 = 5\,\mathrm{m}$, the GAUSSIAN center is located at $(\lambda_0, \phi_0) = (280°, -40°)$. The parameter ϖ is chosen from three values 8×10^{-10}, 8×10^{-11} and $8 \times 10^{-12}\,\mathrm{m}^{-2}$. It corresponds

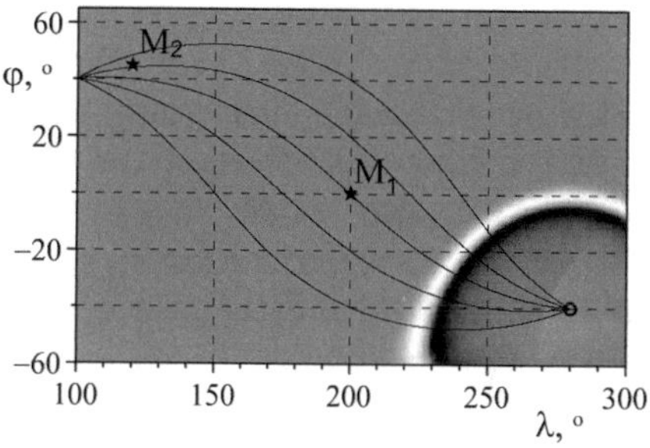

(a)

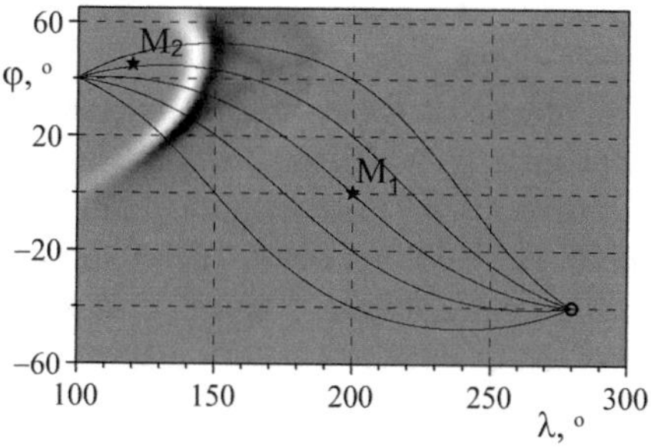

(b)

Fig. 4.2 The computational domain and free surface elevations computed with the FNWD model for the source $\mathscr{W}_3$ at $t = 6$ h (**a**) and $t = 23$ h (**b**). The locations of two synthetic wave gauges are depicted with the symbol filled star. The center of the source region is shown with symbol open circle

to effective linear source sizes equal approximatively to $\mathscr{W}_1 \approx 107.3\,\mathrm{km}$, $\mathscr{W}_2 \approx 339\,\mathrm{km}$ and $\mathscr{W}_3 \approx 1073\,\mathrm{km}$, respectively. The effective source size is defined as the diameter of the circle $\mathcal{S}_{10}$ serving as the level-set $\dfrac{\alpha_0}{10}$ of the initial free surface elevation $\eta(\lambda, \phi, 0)$, i.e.

$$\mathcal{S}_{10} \stackrel{\text{def}}{:=} \left\{ (\lambda, \phi) \mid \eta(\lambda, \phi, 0) = \frac{\alpha_0}{10} \right\}.$$

On the boundary of the computational domain we prescribed SOMMERFELD-type non-radiation boundary conditions [154]. In our opinion, this initial condition has the advantage of being symmetric, comparing to the asymmetric source proposed in [197]. Indeed, if one neglects the EARTH rotation effect (i.e. $\Omega \equiv 0$), our initial condition will generate symmetric solutions in the form of concentric circles drawn on sphere's surface. If one does not observe them numerically, it should be the first red flag. In the presence of EARTH's rotation (i.e. $\Omega > 0$), the deviation of wave fronts from concentric circles for $t > 0$ characterizes CORIOLIS's force effects (see Sect. 4.4.1).

Oscillations of the free surface were recorded in our simulations by two synthetic wave gauges located in points $\mathcal{M}_1 = (200°, 0°)$ and $\mathcal{M}_2 = (120°, 45°)$ (see Fig. 4.2). All simulations were run with the resolution 6001×3751 of nodes (unless explicitly stated to the contrary). The physical simulation time was set to $T = 30\,\mathrm{h}$. In our numerical experiments we observed that the CPU time for NSWE runs is about five times less than FNWD computations. Sequential FNWD runs took about 6 days. This gives the first idea of the 'price' we pay to have non-hydrostatic effects on planetary scales.

Sphericity Effects

The effects of EARTH's sphericity are studied by performing direct comparisons between our FNWD spherical model (4.1)–(4.3) and the same FNWD on the plane (see Chap. 1 for the derivation and Chap. 2 for the numerics). In the plane case the initial condition was constructed in order to have the same linear sizes $\mathscr{W}_{1,2,3}$ as in the spherical case. Namely, it is given by formula (4.30) with function $\rho(x, y)$ replaced by the EUCLIDEAN distance to the center (x_0, y_0):

$$\rho(x, y) = \sqrt{(x - x_0)^2 + (y - y_0)^2}.$$

The computational domain was a plane rectangle with sides lengths approximatively equal to those of the spherical rectangle depicted in Fig. 4.2. The source was located in the point (x_0, y_0) such that the distance to the SOUTH–WEST corner is the same as in the spherical configuration. Synthetic wave gauges $\mathcal{M}_{1,2}$ were located on the plane in order to preserve the distance from the source. Moreover, in order to isolate sphericity effects, we turn off EARTH's rotation in computations presented in this section, i.e. $\Omega \equiv 0$.

In Fig. 4.3 one can see the synthetic gauge records computed with the spherical and 'flat' FNWD models. It can be clearly seen that sphericity effects become more and more important when the source size $\mathscr{W}$ increases. We notice also that in all cases the wave amplitudes predicted with the spherical FNWD model are higher than in 'flat' FNWD computations. In general, the differences recorded at wave gauge $\mathcal{M}_1$ are rather moderate. However, when we look farther from the source, e.g. at location $\mathcal{M}_2$ the discrepancies become flagrant—the difference in wave amplitudes may reach easily several times (as always the 'flat' model underestimates the wave). The explanation of this phenomenon is purely *geometrical*. In the plane case the geometrical spreading along the rays starting at the origin is monotonically decreasing (the wave amplitude decreases like $\mathcal{O}(r^{-\frac{1}{2}})$, where r is the distance from the source). On the other hand, in the spherical geometry the rays starting from the origin (λ_0, ϕ_0) go along initially divergent great circles that intersect in a diametrically opposite point to (λ_0, ϕ_0) (this point has coordinates $(100°, 40°)$ in our case). As a result, the amplitude first decreases in geometrically divergent areas, but then there is a wave focusing phenomenon when the rays convergent in one point. A few such (divergent/convergent) rays are depicted in Fig. 4.2 with solid lines emanating from the point (λ_0, ϕ_0). We intentionally put a wave gauge into the point $\mathcal{M}_2$ close to the focusing area in order to illustrate this phenomenon. Otherwise, the amplification could be made even larger.

Such amplification phenomena should take place, in principle, on the whole sphere. It did not happen in our numerical simulation since we employ SOMMERFELD-type no-radiation boundary conditions. It creates sub-regions which are not attainable by the waves. It happens since the rays emanating from the origin (λ_0, ϕ_0) cross the boundary of the computational domain (and, thus, the waves propagating along these rays are lost). To give an example of such 'shaded' regions

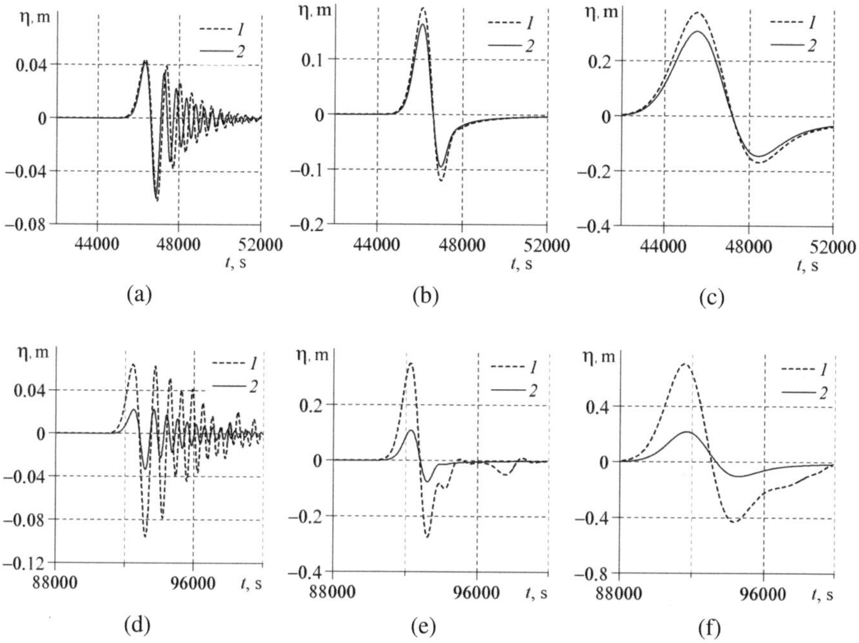

Fig. 4.3 Synthetic wave gauge records registered in points $\mathcal{M}_1$ (**a–c**) and $\mathcal{M}_2$ (**d–f**). FNWD predictions on a sphere are given with dashed lines (1) and the flat model is depicted with the solid line (2). The effective linear source sizes are $\mathscr{W}_1$ (**a, d**), $\mathscr{W}_2$ (**b, e**) and $\mathscr{W}_3$ (**c, f**)

we can mention a neighbourhood of the point $(100°, \ -60°)$. That is why the wave field loses its symmetry as it can be seen in Fig. 4.2b: the edges of the wave front get smeared due to the escape of information through the boundaries. On the other hand, in Fig. 4.2a we show the free surface profile at earlier times where boundary effects did not affect yet the wave front which conserved a circle-like shape.

We should mention an earlier work [80] where the importance of sphericity effects was also underlined. In that work the authors employed a WNWD model to simulate the Indian Ocean tsunami of the 25th December 2004 [299]. The computational domain was significantly smaller than in our computations presented above (no more than 40° in each direction). That is why the sphericity effects were less pronounced than in Fig. 4.3. More precisely, *only* 30% discrepancy was reported in [80] comparing to the 'flat' computation.

Coriolis Effects

Let us estimate now the effect of Earth's rotation on the wave propagation process. The main effect comes from the Coriolis force induced by the rotation of our planet. This force appears in the right-hand sides of Eqs. (4.2), (4.3) and (4.24).

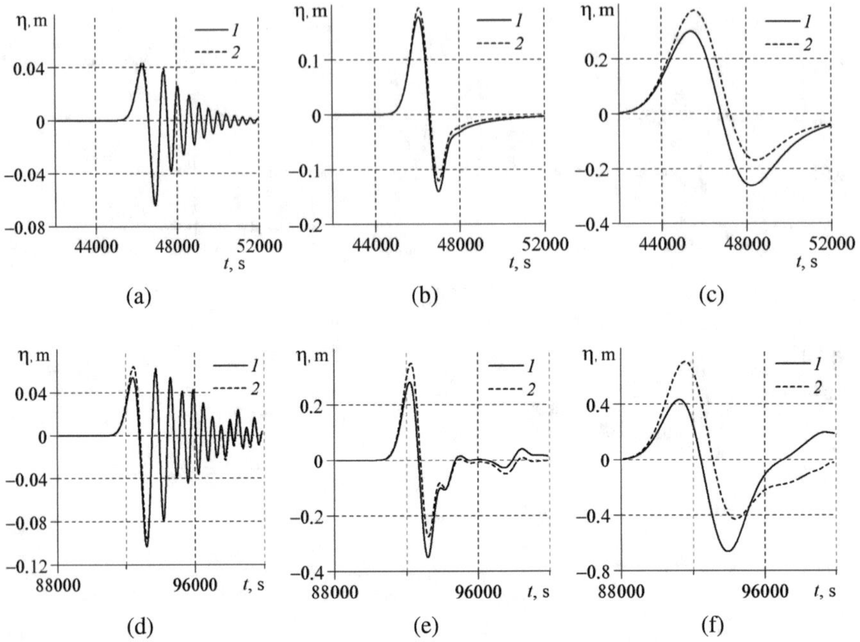

Fig. 4.4 Synthetic wave gauge records registered in points $\mathcal{M}_1$ (**a**–**c**) and $\mathcal{M}_2$ (**d**–**f**). FNWD predictions on a sphere with CORIOLIS's force are shown with solid lines (1) and without CORIOLIS is depicted with dashed lines (2). The effective linear source sizes are $\mathscr{W}_1$ (**a**, **d**), $\mathscr{W}_2$ (**b**, **e**) and $\mathscr{W}_3$ (**c**, **f**)

In Fig. 4.4 we show synthetic wave gauge records (for precisely the same test case and recorded in the same locations $\mathcal{M}_{1,2}$ as described in previous Sect. 4.4.1) with ($\Omega = 7.29 \times 10^{-5}\,\mathrm{s}^{-1}$) and without ($\Omega = 0$) EARTH's rotation. We also considered initial conditions of different spatial extents $\mathscr{W}_{1,2,3}$. In particular, we can see that CORIOLIS's force effect also increases with the source region size. However, EARTH's rotation seems to reduce somehow wave amplitudes (i.e. elevation waves are decreased and depression waves are increased in absolute value). We conclude also that CORIOLIS's force can alter significantly only the waves travelling at large distances.

Numerical simulations on a sphere, which rotates faster than EARTH, show that CORIOLIS and centrifugal forces produce also much more visible effects on the wave propagation. In particular, important residual vortices remain in the generation region and the wave propagation speed is also reduced.

Concerning earlier investigations conducted in the framework of WNWD models, it was reported in [80] that CORIOLIS's force may change the maximal wave amplitude up to 15%, in [223]—1.5–2.5% and in [197]—up to 5%. Our results generally agree with these findings for corresponding source sizes. In the latter reference it is mentioned also that CORIOLIS's force influence increases with source extent and it retains a portion of the initial perturbation in the source region, thus contributing to the formation of the residual wave field [253].

Dispersive Effects

In order to estimate the contribution of the frequency dispersion effects on wave propagation, we are going to compare numerical predictions obtained with the FNWD and (hydrostatic non-dispersive) NSWE models on a *rotating* sphere.[4] Both codes are fed with the same initial condition (4.31) as described above. In these simulations we use a fine grid with the angular resolution of $40''$ in order to resolve numerically shorter wave components. As it is shown in Fig. 4.5a, FNWD model generates a dispersive tail behind the main wave front. The tail consists of shorter waves with smaller (decreasing) amplitudes. Obviously, the NSWE model does not reproduce this effect (see Fig. 4.5b).

Computations based on the FNWD model show that dispersive effects are more pronounced for more compact initial perturbations. Signals in wave gauges (see Fig. 4.4a, d) show that the initial condition of effective width $\mathscr{W}_1$ generates a dispersive tail quasi-absent in $\mathscr{W}_2$ and inexistent in $\mathscr{W}_3$. Another interesting particularity of dispersive wave propagation is the fact that during long distances the maximal amplitude may move into the dispersive tail. In other words, the amplitude of the first wave may become smaller than amplitude of waves from the dispersive tail. Moreover, the number of the highest wave may increase with the propagation

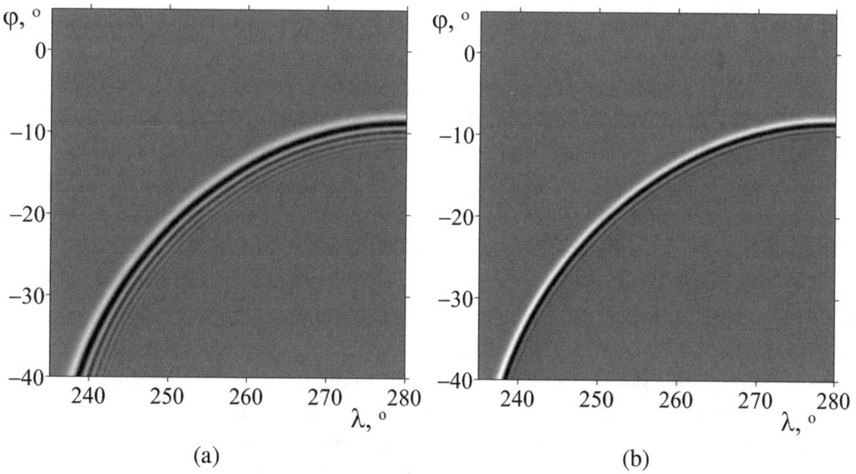

Fig. 4.5 Zoom on the wave field after $t = 5$ h of free propagation over a sphere computed with FNWD (**a**) and NSWE (**b**) for the initial condition $\mathscr{W}_1$

[4]We use the same angular velocity $\Omega = 7.29 \times 10^{-5}\,\mathrm{s}^{-1}$.

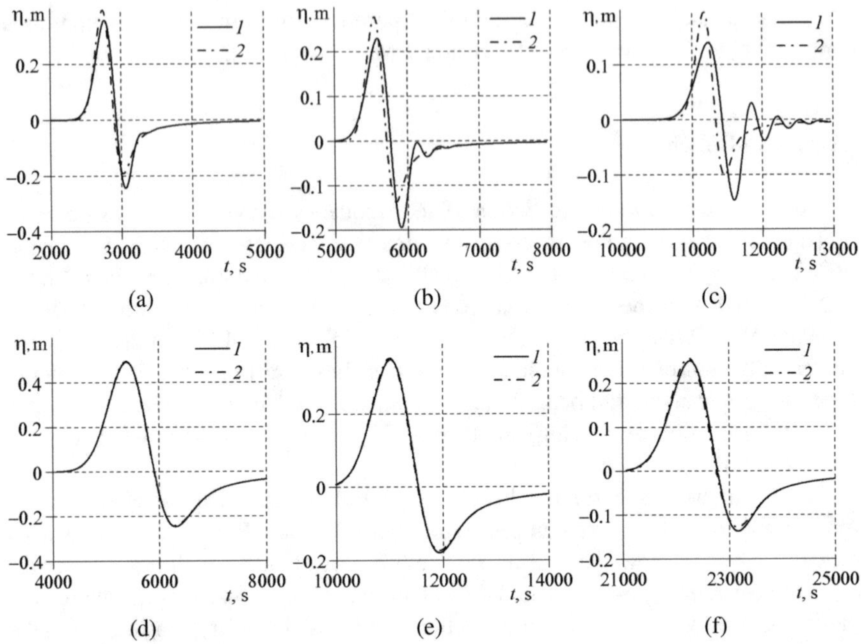

Fig. 4.6 Synthetic wave gauge records predicted by FNWD (1) and NSWE (2) spherical models for initial sources of sizes $\mathscr{W}_1$ (**a–c**) and $\mathscr{W}_2$ (**d–f**). Wave gauges are located in $\mathcal{M}_3$ (**a**), $\mathcal{M}_4$ (**b**, **d**), $\mathcal{M}_5$ (**c**, **e**) and $\mathcal{M}_6$ (**f**)

distance [261]. A similar effect was observed in the framework of WNWD models in [142, 223, 224].

Moreover, the influence of dispersion is known to grow with the traveled distance [10, 281]. It can be also observed in Fig. 4.4. For example, for the initial condition of the size $\mathscr{W}_2$, the frequency dispersion effect is not apparent yet on wave gauge $\mathcal{M}_1$ (see Fig. 4.4b, line 1), but it starts to emerge in gauge $\mathcal{M}_2$ (see Fig. 4.4e, line 1). Concerning the source of size $\mathscr{W}_1$, here the dispersion becomes to play its rôle even before the point $\mathcal{M}_1$. In order to estimate better the distance on which dispersive effects may become apparent, we create additional synthetic wave gauges in points $\{\mathcal{M}_i\}_{i=3}^{6}$ with coordinates $\{(\lambda_i, \phi_i)\}_{i=3}^{6}$. The longitude is taken to be that of the source center, i.e. $\lambda_i \equiv \lambda_0 = 280°, \forall i \in 3, \ldots, 6$, while the latitude takes the following values: $\phi_3 = -35°$, $\phi_4 = -30°$, $\phi_5 = -20°$, $\phi_6 = 0°$. Thus, these wave gauges are located much closer to the source (λ_0, ϕ_0) than $\mathcal{M}_1$. The recorded data is represented in Fig. 4.6. In panels Fig. 4.6a–c one can see that for smallest source $\mathscr{W}_1$ the dispersion is already fully developed in gauge $\mathcal{M}_5$ (and becomes apparent in gauge $\mathcal{M}_4$), i.e. at the distance of ≈2200 km from the source. Indeed, in panel Fig. 4.6c the second wave of depression has the amplitude larger than the first wave. One can notice in general that the dispersion always yields a slight reduction of the leading wave (compare with NSWE curves in dash dotted lines (2)). To our knowledge this fact was first reported in numerical simulations in [68]. Concerning the initial condition of size $\mathscr{W}_2$, the dispersion does not seem to appear even at the point $\mathcal{M}_6$, located $\approx$4400 km from the source (see Fig. 4.6d–f). As a result, we conclude that, at least for tsunami applications, the frequency dispersion effect is mainly determined by the size of the generation region and thus, by the wavelength of initially excited waves. The dispersion effect is decreasing when the source extent increases. A common point between FNWD and NSWE models is that on relatively short distances the period of the main wave increases and the wave amplitude decreases.

Some Rationale on the Dispersion

In order to estimate approximatively the distance ℓ_d on which dispersive effects become significant for a given wavelength λ we consider a model situation. For simplicity we assume the bottom to be even with constant depth d. Let us assume also that wave dynamics is described by the so-called linearized BENJAMIN–BONA–MAHONY (BBM) equation [29, 263]:

$$\eta_t + \upsilon_0 \eta_x = \nu \eta_{xxt},$$

which can be obtained from linearized FNWD plane model reported in [191]. Here $\upsilon_0 \overset{\text{def}}{:=} \sqrt{g d}$ is the linear long wave speed and $\nu \overset{\text{def}}{:=} \frac{d^2}{6}$. For this equation, the dispersion relation $\omega(k)$, the phase speeds v and v_k, corresponding to the given wavelength λ and wavenumber k, respectively, are given by the following formulas:

$$\omega(k) = \frac{\upsilon_0 k}{1 + \nu k^2}, \qquad \text{v} = \frac{\upsilon_0}{1 + \nu \left(\frac{2\pi}{\lambda}\right)^2}, \qquad \text{v}_k = \frac{\upsilon_0}{1 + \nu \left(\frac{2\pi}{\lambda_k}\right)^2},$$

where $\lambda_k \overset{\text{def}}{:=} \dfrac{2\pi}{k}$.

During the time t the generated wave travels the distance $\ell_d = \text{v} \cdot t$, while a shorter wave of length $\lambda_k < \lambda$ will travel a shorter distance

$$\ell_k = \text{v}_k \cdot t = \ell_d \frac{\text{v}_k}{\text{v}} < \ell_d.$$

The difference in traveled distances is a manifestation of the frequency dispersion. Now, we assume that the wavelength of a separated wave is related to λ as

$$\lambda_k = \delta\lambda, \qquad \delta \in (0, 1),$$

and we assume that by time t the dispersion had enough time to act, i.e.

$$\ell_k = \ell_d - (1+\delta)\frac{\lambda}{2}.$$

From this equality we can determine ℓ_d:

$$\ell_d\left(1 - \frac{\nu_k}{\nu}\right) = \frac{1+\delta}{2}\lambda,$$

and if we substitute the expressions of phase speeds ν and ν_k we obtain:

$$\ell_d \left\{ 1 - \frac{\dfrac{\upsilon_0}{1+\nu\left(\frac{2\pi}{\lambda_k}\right)^2}}{\dfrac{\upsilon_0}{1+\nu\left(\frac{2\pi}{\lambda}\right)^2}} \right\} = \frac{1+\delta}{2}\lambda.$$

By using the fact that $\lambda_k = \delta\lambda$ we obtain the final expression for the *dispersion distance*:

$$\ell_d = \frac{1}{2(1-\delta)}\left[\lambda + \frac{3}{2\pi^2}\delta^2\frac{\lambda^3}{d^2}\right] \approx \frac{1}{2(1-\delta)}\left[\lambda + 0.152\,\delta^2\frac{\lambda^3}{d^2}\right]. \tag{4.32}$$

The dependence of the dispersive distance ℓ_d on the wavelength λ is depicted in Fig. 4.7.

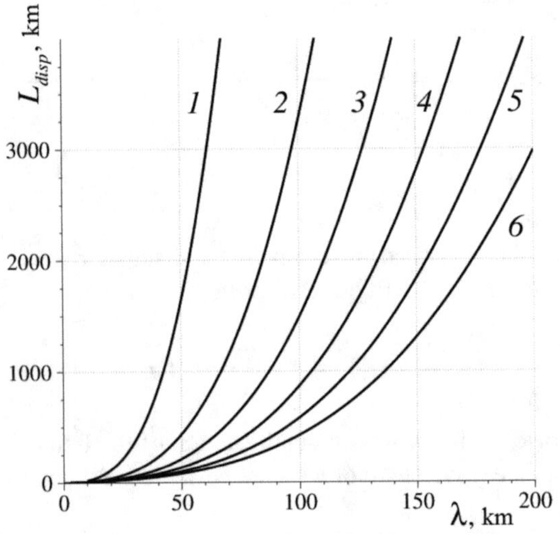

Fig. 4.7 Dependence of the dispersion distance ℓ_d on the wavelength for parameter $\delta = 0.33$ and for water depth $d = 1$ km (1), $d = 2$ km (2), $d = 3$ km (3), $d = 4$ km (4), $d = 5$ km (5) and $d = 6$ km (6)

Let us mention that an analogue of formula (4.32) was obtained earlier in [241, 261] based on the (linearized) KORTEWEG–DE VRIES equation [43, 200].

Numerical Application

In order to apply formula (4.32) (and check its validity) to our results, we have to estimate first the generated wavelength λ and parameter δ. We saw above that for the initial condition of extent $\mathscr{W}_1$ the wave gauge $\mathcal{M}_5$ registered a well-separated secondary wave with the period about three times smaller. Thus, we can set $\delta = 0.33$. Concerning the wavelength of the generated wave, in such estimations it is roughly identified[5] with the size of the generated region $\mathscr{W}_1$ [197]. Under these assumptions, Eq. (4.32) gives us the dispersive distance $\ell_d = 1058\,\text{km}$. Thus, we can conclude that dispersive effects should be apparent in wave gauge $\mathcal{M}_5$, which is in the perfect agreement with our numerical simulations. Let us take a source region with a larger horizontal extent $\mathscr{W}_2$. If we take $\delta = 0.33$ (as above) and $\lambda = 339\,\text{km}$, formula (4.32) gives us $\ell_d = 30{,}139\,\text{km}$. Thus, we conclude that dispersive effects will not be seen even in the farthest wave gauge $\mathcal{M}_2$.

Alternative Approaches

There exist other criteria of the importance of frequency dispersion effects. One popular criterium is the so-called KAJIURA number $\mathcal{K}a$ employed, for example, in [197]:

$$\mathcal{K}a \stackrel{\text{def}}{:=} \left(\frac{6\,d}{\ell}\right)^{\frac{1}{3}} \cdot \frac{\mathscr{W}}{d},$$

where $\mathscr{W}$ is the source region width and ℓ is the distance traveled by the wave. It is believed that dispersive effects manifest if $\mathcal{K}a < 4$. Let us check it against our numerical simulations. In the case depicted in Fig. 4.6c we have $\mathscr{W} = 107.3\,\text{km}$ and $\ell = 2200\,\text{km}$. Thus, $\mathcal{K}a \approx 6 > 4$. According, to KAJIURA's criterium, the dispersion should not appear yet at the point $\mathcal{M}_5$. However, it contradicts clearly our direct numerical simulations shown in Fig. 4.6c. It was already noticed in [142] that criterium $\mathcal{K}a < 4$ is too stringent. It was proposed instead to use a more sophisticated criterium based on the *normalized dispersion time*:

$$\vartheta \stackrel{\text{def}}{:=} 6\sqrt{g\,d}\,\frac{d^2}{\lambda^3}\,t\,.$$

[5]However, in the general case this approximation might be rather poor [261].

If $\vartheta > 0.1$ the wave propagation is considered to be dispersive. In our simulations $t = 12{,}000\,\mathrm{s}$. It is the time needed for the wave to travel to the location $\mathcal{M}_5$. Thus, in our case $\vartheta \approx 0.18 > 0.1$ and the dispersion should be already visible in location $\mathcal{M}_5$ for the source size $\mathscr{W}_1$, which is in perfect agreement with our synthetic wave record (see again Fig. 4.6c).

In our opinion, the use of the criterium (4.32) based on the dispersion distance ℓ_d is preferable, since it gives directly an estimation of the distance, where the dispersive effects will become non-negligible. Outside the circle of radius ℓ_d and centered at the source the use of hydrostatic non-dispersive models is not advised.

Let us mention also a few earlier works where the importance of dispersive effects was studied for real-world events using WNWD models. In [151, 168] the SUMATRA 2004 INDIAN ocean tsunami was studied and it was shown that in the deep WEST part of the INDIAN ocean the discrepancy in wave amplitudes reaches 20% (between WNWD and NSWE models). For the same tsunami event the discrepancy of 60% was reported in [80]. Numerical simulations of the TOHOKU 2011 tsunami event reported in [197] confirmed again the difference of 60% between hydrostatic and non-hydrostatic model predictions.

4.4.2 Bulgarian 2007 Tsunami

The basin of the BLACK SEA is subject to a relatively important seismic activity and there exists a potential hazard of tsunami wave generation, which may be caused not only by earthquakes, but also by underwater landslides, which can be triggered even by weak seismic events. Geophysical surveys show that large portions of the BLACK SEA continental shelf contain unstable masses [124], which have to be taken into account while assessing tsunami hazard for the population and/or underwater infrastructure. Moreover, some past events are well documented [180].

Some anomalous oscillations of the sea level were registered on the 7th of May 2007 at BULGARIAN coasts. In [267] it was conjectured that a landslide may have provoked these oscillations and the authors considered four possible locations of this hypothetical landslide. In all these cases the landslide started at the water depth about 100 m and at the distances of 30–50 km from the coast. The suggested volume of the landslide is between 30 and 60 millions of m^3. Landslide thickness is about 20–40 m. In [267] the authors represented the landslide as a system of interconnected solid blocks which can move along the slope under the force of gravity whose action is compensated by frictional effects. The hydrodynamics was described by NSWE incorporating some viscous effects solved with the Finite Element Method (FEM). For all four considered cases simulations show that the landslide achieves the speed of about 20 m/s in 200–300 s after the beginning of the motion. Landslide's motion stops at the depth around 1000 m after running out about 20 km from its initial position.

The hypothesis of landslide mechanism is studied by confronting numerical predictions with eyewitness reports and coastal wave gauge data. In particular, we know the values of lowest (i.e. negative) and highest amplitudes for seven locations along the coast, respectively:

1. Shabla: $-1.5\,\mathrm{m}\,,\,0.9\,\mathrm{m}$
2. Bolata: $-1.3\,\mathrm{m}\,,\,0.9\,\mathrm{m}$
3. Dalboka: $-2.0\,\mathrm{m}\,,\,1.2\,\mathrm{m}$
4. Kavarna: $-1.8\,\mathrm{m}\,,\,0.9\,\mathrm{m}$
5. Balchik: $-1.5\,\mathrm{m}\,,\,1.2\,\mathrm{m}$
6. Varna: $-0.7\,\mathrm{m}\,,\,0.4\,\mathrm{m}$
7. Galata: $-0.2\,\mathrm{m}\,,\,0.1\,\mathrm{m}\,.$

The comparisons from [267] show that one can seemingly adopt the landslide mechanism hypothesis. However, even the most plausible landslide scenario (among four considered) does not give a satisfactory agreement with *all* available field data. Moreover, the maximal synthetic amplitudes are shifted to the SOUTH (towards EMINE), which was not observed during the real event.

In a companion study [318] the authors investigated also the hypothesis of a meteo-tsunami responsible of anomalous waves recorded on the 7th of May 2007 at BULGARIAN coasts. Their analysis showed that the weather conditions could provoke anomalous waves near BULGARIAN coasts. The numerical simulations show again a good agreement of maximal amplitudes at some locations, even if we cannot speak yet about a good general agreement. In particular, numerical predictions seriously underestimate wave amplitudes in NORTHERN parts of the coast such as SHABLA and overestimate them in SOUTHERN ones (e.g. large waves were not observed in BURGAS Bay, but they existed in numerical predictions).

In the present study we continue to develop the landslide-generated hypothesis of anomalous waves. In contrast to the previous study [267], we employ the FNWD model presented above and the landslide will be modelled using the quasi-deformable body paradigm [26]. The driving force was taken as the sum of the gravity, buoyancy, friction and water drag forces acting on elementary volumes. Quasi-deformability property means that the landslide can deform in order to follow complex bathymetry profiles by preserving the general shape (see [26, 109, 118] for more details). However, the deformation process is such that the horizontal components of the velocity vector are the same throughout the sliding body (as in the absolutely rigid case). This model has been validated against experimental data [154] and direct numerical simulations of the free surface hydrodynamics [156]. Moreover, it was already successfully applied to study numerically a real-world tsunami which occurred in PAPUA NEW GUINEA on the 17th of July 1998 [186]. Important parameters, which enter in our landslide model and, thus, that have to be prescribed are:

V: landslide volume
C_w: added mass coefficient
C_d: drag coefficient
$C_f \equiv \tan\theta^*$: friction coefficient and θ^* is the friction angle
$\gamma \stackrel{\text{def}}{:=} \frac{\rho_s}{\rho_w} > 1$: ratio between water ρ_w and sliding mass ρ_s densities.

The initial shape of the landslide is given by the following formula:

$$h_s^0(x, y) = \begin{cases} \frac{T}{4}\left[1+\cos\left(\frac{2\pi(x-x_c^0)}{B_x}\right)\right]\cdot\left[1+\cos\left(\frac{2\pi(y-y_c^0)}{B_y}\right)\right], & (x, y) \in \mathscr{D}_0, \\ 0, & (x, y) \notin \mathscr{D}_0, \end{cases}$$

where $\mathscr{D}_0 = \left[x_c^0 - \frac{B_x}{2}, x_c^0 + \frac{B_x}{2}\right] \times \left[y_c^0 - \frac{B_y}{2}, y_c^0 + \frac{B_y}{2}\right]$ is the domain occupied by the sliding mass, $B_{x,y}$ are horizontal extensions of the landslide along the axes $O\,x$ and $O\,y$, respectively, (x_c^0, y_c^0) is the position of its barycenter and T is its thickness.

Remark 4.2 For the sake of notation compactness, in the landslide description above, we used CARTESIAN coordinates. This approximation is valid since the landslide size is small enough not to 'feel' substantially EARTH's sphericity. We place the origin in the left side center of the spherical rectangular computational domain, i.e. in the point $(27°, 43°)$. Then, the local CARTESIAN coordinates are introduced in the following way:

$$x = R\,\frac{\pi}{180}\,(\lambda - 27)\cos\left(\frac{\pi}{180}\,43\right), \qquad y = R\,\frac{\pi}{180}\,(\phi - 43).$$

If (λ_c^0, ϕ_c^0) are spherical coordinates of the landslide barycenter, then its local CARTESIAN coordinates (x_c^0, y_c^0) are computed accordingly:

$$x_c^0 = R\,\frac{\pi}{180}\,(\lambda_c^0 - 27)\cos\left(\frac{\pi}{180}\,43\right), \qquad y_c^0 = R\,\frac{\pi}{180}\,(\phi_c^0 - 43).$$

During the modelling of landslide events in the BLACK SEA we used the parameters of some historical events [180]. We noticed that the most sensitive parameter is the initial location of the landslide. That is why in the present study we focus specifically on this aspect in order to shed some light on this unknown parameter. In this perspective we chose 40 different initial locations along the BULGARIAN coastline which were located mainly at the depth of $200\,\mathrm{m}$, $1000\,\mathrm{m}$ and $1500\,\mathrm{m}$. These locations are depicted with black rectangles in Fig. 4.8. Other parameters are given in Table 4.1. More information regarding the modelling of this event can be found in [157]. The volume V of the landslide in our simulations is equal to $62.5 \times 10^6\ \mathrm{m}^3$, which is close to the value used in [267]. We notice that there are two competing effects in our problem. If we increase the initial landslide depth, the amplitude of generated waves will be seriously reduced. However, this effect can be compensated by increasing the landslide thickness T. In general, the amplitude of waves is proportional to T.

In these numerical simulations we use the finest angular resolution (in this section) of $3.75''$, since the domain is relatively compact. It corresponds to the

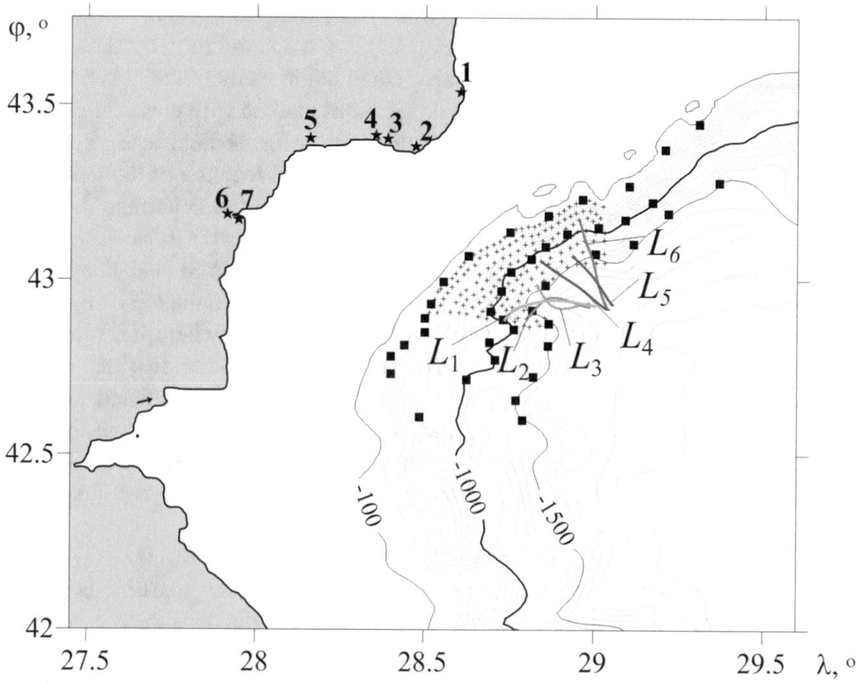

Fig. 4.8 Computational domain with isolines of the bathymetry function. With star symbols (⋆) we denote coastal locations (towns) where the maximal and minimal wave heights are known. Small black squares (■) and pluses (+) denote various initial positions of the landslide considered in our study. The coloured lines show the most probable landslide trajectories. Finally, the little black arrow (on the left) shows the starting point and direction along the coastline where we record maximal and minimal wave heights

Table 4.1 Physical parameters used to simulate the hypothetical landslide motion during Bulgarian tsunami of the 7th of May 2007

Parameter	Value
Added mass coefficient, C_w	1.0
Drag coefficient, C_d	1.0
Densities ratio, γ	2.0
Friction angle, θ^*	1°
Landslide thickness, T	10 m
Landslide length, B_x	5000 m
Landslide width, B_y	5000 m
Landslide volume, V	$62.5 \times 10^6\,\mathrm{m}^3$

grid of 2881 × 1921 nodes. The bathymetry data was obtained by applying bilinear interpolation to data retrieved from 'GEBCO One Minute Grid—2008'. The computational (CPU) time of each run was about 35 h for this resolution. Some information on the wave propagation can be obtained using the so-called radiation diagrams, which represents the spatial distribution of maximal and minimal wave amplitudes[6] during the whole simulation time. After computing the first 40 scenarii (marked with little black rectangles in Fig. 4.8), we could delimit[7] the area where the hypothetical landslide could take place. Then, in the second time, this area was refined with additional 171 initial landslide locations marked with pluses in Fig. 4.8. In this way, by comparing simulation results with available field data, we could choose the most probable scenarii L_1—L_6 of the initial location of the landslide. Finally, we performed the third optimisation cycle in order to determine the most likely landslide thickness T. The optimal parameters are given in Table 4.2.

The radiation diagrams for this tsunami generated by inferred landslides L_1—L_6 are shown in Fig. 4.9. One can see, in particular, that the radiation of maximal (positive) amplitudes is directed towards the coastal towns where the most significant oscillations of the sea level were observed. We notice also that in our simulations L_1—L_6 we do not obtain a two-tongue structure predicted in [267, Fig. 2]. However, such radiation diagrams are very sensitive to the initial location of the landslide. For some starting points we observed (as in an earlier work [186]), for example, an abrupt termination of landslide motion, which affected quite a lot the radiation diagram.

Minimal and maximal wave amplitudes recorded along the BULGARIAN coast are shown in Fig. 4.10. The starting point of the path along the coastline is shown with a little black arrow in Fig. 4.8. We mention also that the distance is computed on our grid (thus, the coastline is approximated *in fine* by a polygon). Hence, there might be little discrepancies (in the sense of overestimations) with 'real-world' distances. Figure 4.10 shows an overall good agreement of the waves generated by the landslides L_1—L_6 with the field data. Moreover, in contrast to the previous study [267], our scenarii does not trigger large wave amplitudes in Southern parts of the BULGARIAN coast. It is remarkable also that all these landslides (L_1—L_6) stopped in the same region. Consequently, we can only suggest to study this area of

Table 4.2 Geographical coordinates of the initial positions, thicknesses and initial sea depths of landslides L_1—L_6, which could be involved in the genesis of the Bulgarian tsunami of the 7th of May 2007

Landslide number	Longitude (°)	Latitude (°)	Thickness (m)	Initial depth (m)
L_1	28.7342	42.8871	110	1024
L_2	28.8003	42.8898	200	1323
L_3	28.8306	42.9835	320	1429
L_4	28.8416	43.0551	280	1178
L_5	28.9353	43.0689	300	1385
L_6	28.9517	43.1735	110	811

[6]Here we mean maximal positive and minimal negative waves with respect to the still water level.

[7]Here we stress again that the computational domain was the same, but we delimit the search area for the landslide initial position.

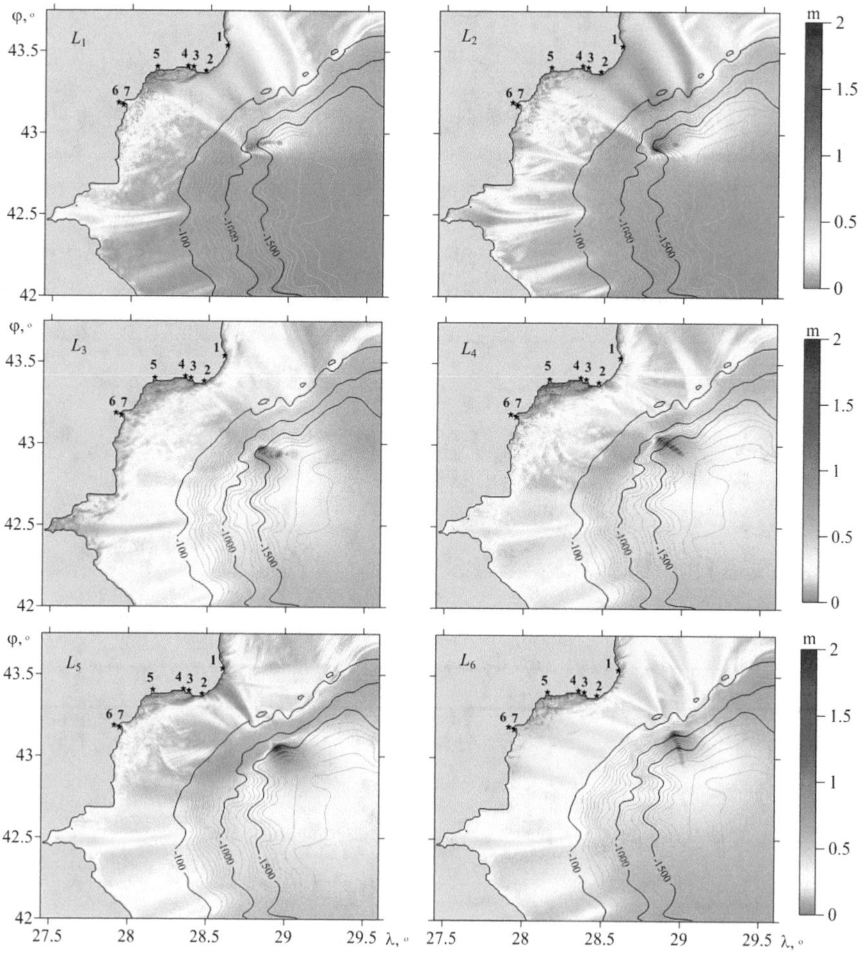

Fig. 4.9 Radiation diagrams obtained with FNWD computations for six most probable landslides L_1—L_6

the BLACK SEA in the perspective to discover eventually the deposits of this past landslide event.

We underline also that the geometrical extensions of the 'optimal' landslides L_1—L_6 can be changed without losing sensibly the good agreement with observations demonstrated in Fig. 4.10. For example, we made separate runs of the FNWD code with $B_x = B_y = 2500$ m (i.e. two times shorter landslides in horizontal dimensions) with similar amplitudes recorded in observation points. However, with $B_x = B_y = 2500\,\text{m}$, the optimal value of the landslide thickness becomes much less realistic (about 3–4 times bigger than the optimal thickness in numerical simulations with $B_x = B_y = 5000\,\text{m}$). In general, such modifications may alter

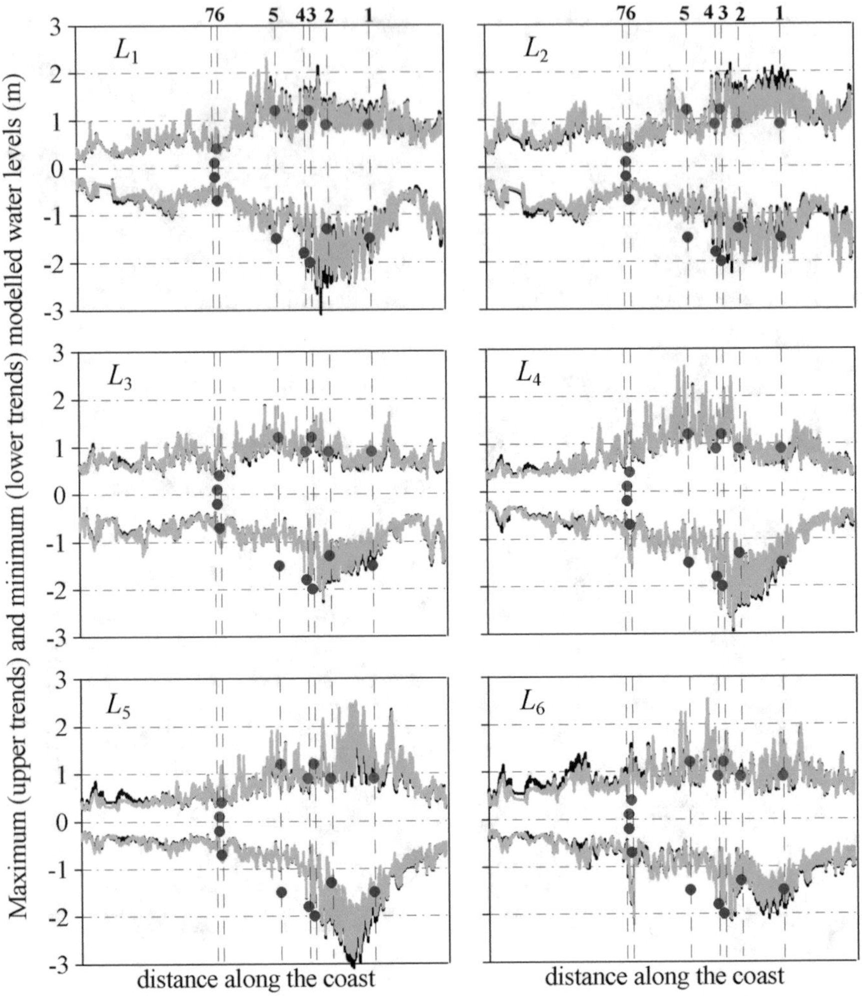

Fig. 4.10 Minimal and maximal wave amplitudes along the BULGARIAN coast recorded during the whole numerical simulation of landslide scenarii L_1—L_6. Grey and black solid lines are the numerical predictions of the FNWD and NSWE models correspondingly. Red and blue circles represent the available field observations

seriously the landslide trajectory and velocity. However, in these particular cases the new trajectories followed closely the corresponding coloured solid lines depicted in Fig. 4.8.

The synthetic marigrams for the seven coastal observation points of the FNWD simulation of the landslide scenario L_1 are presented in Fig. 4.11. It can be easily seen on this figure that the main oscillation period was between 3 and 12 min, which is also in good agreement with the observations. Notice also that at most observation

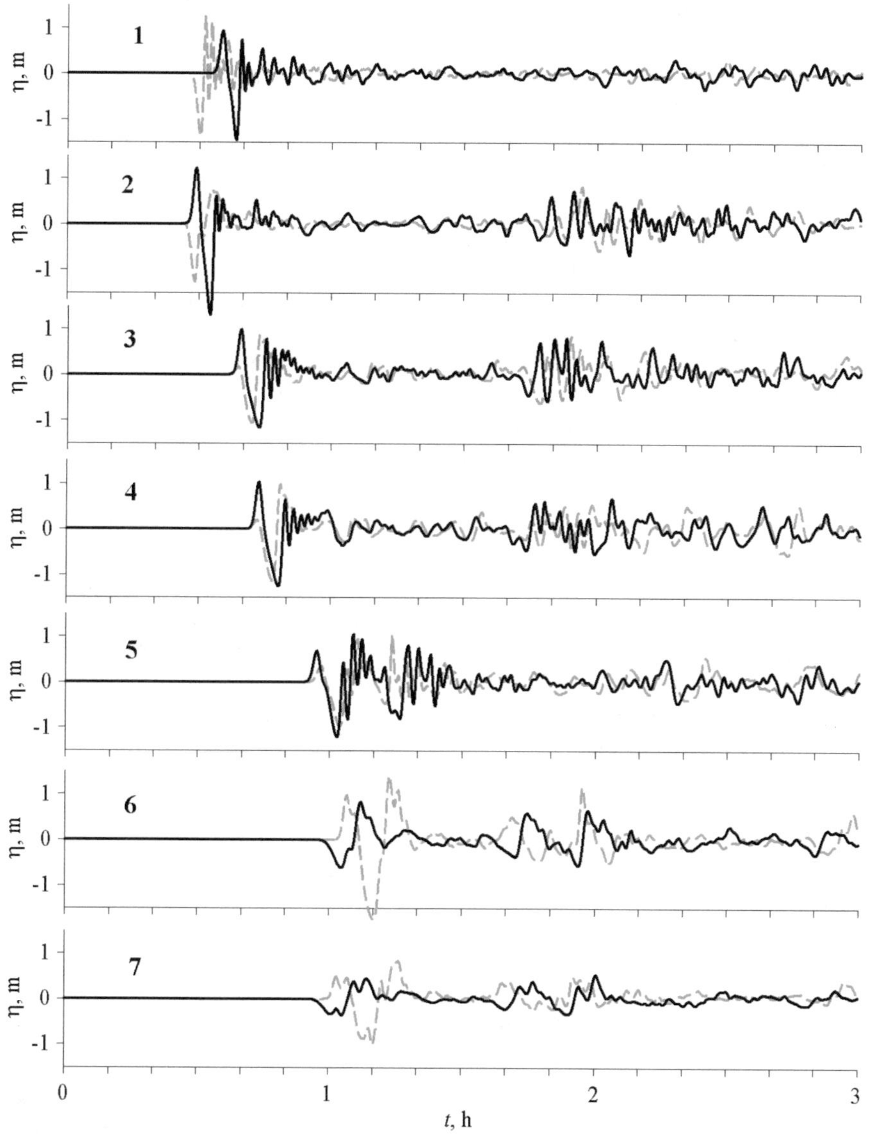

Fig. 4.11 Synthetic marigrams obtained in FNWD simulations for landslide scenarii L_1 (black lines) and L_6 (grey lines)

points the first wave was positive. For the sake of comparison, we also present the synthetic marigrams for the landslide scenario L_6, where the first wave was negative at most locations.

Dispersive Effects

The influence of the frequency dispersion in this particular tsunami event will be estimated by computing the relative[8] differences between radiation diagrams (i.e. maximal amplitudes) computed with FNWD and NSWE models. In this way, in Fig. 4.12 we show how the incorporation of non-hydrostatic effects modifies extreme (positive) wave amplitudes generated by landslide scenarii L_1—L_6. In particular, one can see that relative differences can reach up to 60% in deep parts of the BLACK SEA. The results presented in Fig. 4.12 show that the non-dispersive NSWE model gives a reliable prediction of the maximal wave heights in significant portions of the coastal area (see light blue/light green areas in Fig. 4.12). In particular, the observation points are located in this area. We are tempted to draw the conclusion that dispersive effects are not crucial in this particular problem. We are convinced that the main reason for this weak rôle of the dispersion here is the focus on relatively shallow areas. In the previous study [186] it was observed that in numerical simulations with wider submarine landslides (e.g. $B_x = 10{,}000$ m and $B_y = 20{,}000$ m) the dispersive effects became clearly visible only after propagation through deeper areas. Also, we should stress out that the wave propagation becomes sensibly more dispersive for narrower landslides ($B_x = B_y = 2500$ m). As a general rule, we can mention also that an abrupt termination of the landslide motion[9] further contributes into the importance of non-hydrostatic effects.

The main goal of this section is to demonstrate that FNWD models can be successfully applied to study *in silico* real-world events on all scales from regional (as in this section) to global ones (as demonstrated in the following Sect. 4.4.3). We can even perform extensive parametric studies with spherical FNWD models. To give an example, the determination of six *optimal* scenarii L_1—L_6 provided in Table 4.2 required in total the simulation of more than 210 hypothetical landslide events.

4.4.3 *Chilean 2010 Tsunami*

In order to illustrate the application of our spherical FNWD model to a real-world large scale seismically generated tsunami event we consider the CHILEAN tsunami which took place on the 27th of February 2010. Earthquake epicenter was located under the Ocean 117 km to the NORTH from CONCEPCIÓN at the depth of about 35 km below the bottom. This event was estimated to have the seismic moment

[8]In the relative difference we divide by the magnitude of the NSWE prediction.

[9]We remind here that the landslide motion is computed in our model according to the balance of forces acting on the sliding mass and the second law of NEWTON. In more detail the quasi-deformable model is described in [26, 109].

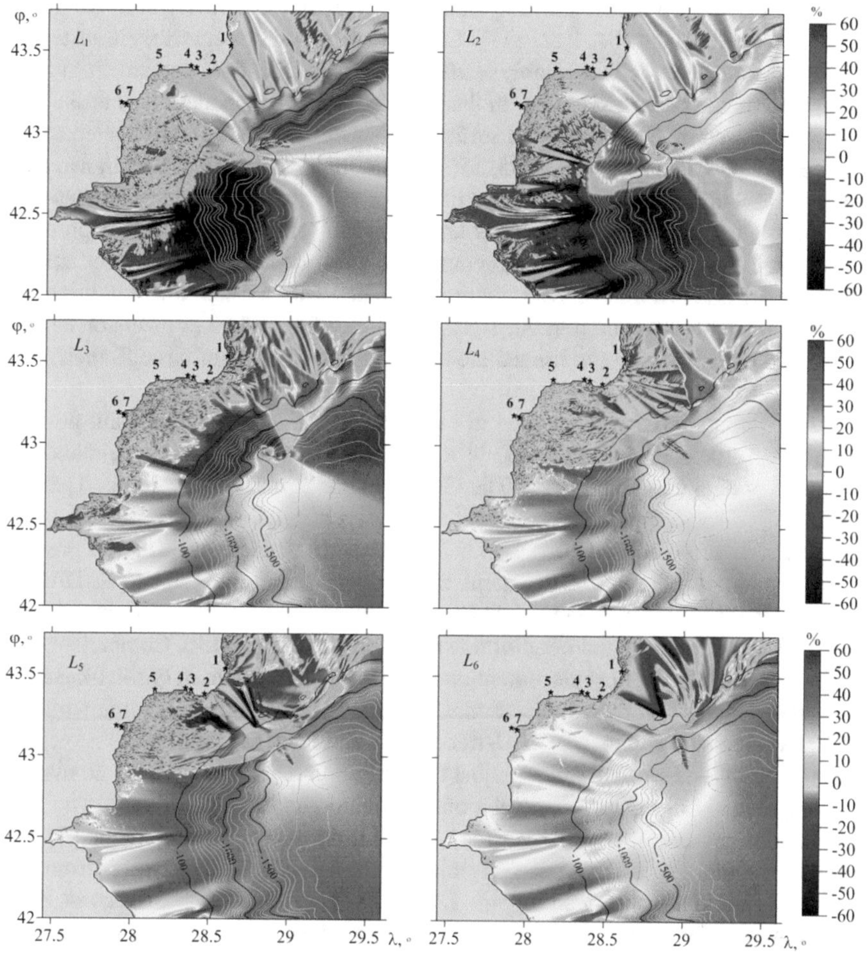

Fig. 4.12 The spatial distribution of relative differences in the maximal positive wave amplitudes computed according to FNWD and NSWE models for the optimal landslide scenarii L_1—L_6

magnitude $M_W = 8.8$. This earthquake generated a tsunami wave, which was observed in the whole PACIFIC OCEAN. The most important aspect for us is that this wave was registered at DART buoys. Many scientific works are devoted to the investigation of this particular event. Here we mention a few numerical studies which go along the lines of our own work [2, 258, 330]. Contrary to the catastrophic TOHOKU 2011 event, where researchers had to introduce local landslide hypothesis in order to explain some extreme run-up values [304], CHILEAN 2010 event seems to be purely seismic since the available data on this tsunami can be reproduced fairly well starting from the initial water column disturbances caused by the earthquake solely.

In order to reconstruct the displacement field of the EARTH surface, some authors use GPS data [88, 247, 317]. Later, these seismic scenarii were tested in [2] to confront them with available tsunami field data. The final agreement quality depended on the chosen scenario. In the present work we consider the fourth alternative proposed by USGS. EARTH surface displacements were reconstructed using the celebrated Okada solution [255, 256]. Then, this displacement was transferred to the free surface as the initial condition for our hydrodynamic computations. This tsunami generation procedure is known as the ‘passive generation’ approach [110, 183]. We would like to mention that there are noticeable discrepancies among all these seismic inversions. Thus, there is an uncertainty in the initial condition for tsunami wave propagation [89, 90, 103]. Our choice for the USGS inversion can be explained essentially by the immediate availability of their data through their web site.

The computational domain used in our simulations covers a significant portion of the PACIFIC ocean—$\left[\,199^\circ,\ 300^\circ\,\right] \times \left[\,-\,60^\circ,\ 5^\circ\,\right]$. We used a $1'$ grid and the bathymetry data was taken from ‘The GEBCO One Minute Grid—2008’. The use of finer grids or computation in significantly larger domains does not seem to be feasible with serial codes (see Sect. 4.5.2). In order to validate our code we present the comparisons of predicted tsunami waves against three DART buoys: DART–32411, DART–32412 and DART–51406. The locations of these buoys are shown in Fig. 4.13a and the initial condition is represented in Fig. 4.13b. Comparisons of FNWD predictions against aforementioned DART data is shown in Fig. 4.14. Buoys data was downloaded from the National Oceanic and Atmospheric Administration (NOAA) web site. In the case of DART–32411 and DART–51406 buoys a vertical translation of data was needed to adjust the still water level. One can see an overall good agreement in Fig. 4.14 between our simulation and real-world data. The first oscillations present in DART data for $t < 1\,\mathrm{h}$ do not seem to be related to the studied tsunami event, since the wave did not have enough time to travel from the source region to the observation point. One can see also that synthetic records have somehow smaller amplitudes. It can be related to our choice of the initial condition (USGS) which did not take into account tsunami-related constraints during the inversion process [2].

Dispersive Effects

In order to estimate the influence of dispersive effects in this particular tsunami event, we perform two simulations—with FNWD and NSWE models. Moreover, we consider one case with the real bathymetry data and another one with an even bottom of constant $d = 4\,\mathrm{km}$ depth. Radiation diagrams for all these four cases are shown in Fig. 4.15. One can see that FNWD and NSWE predict significantly

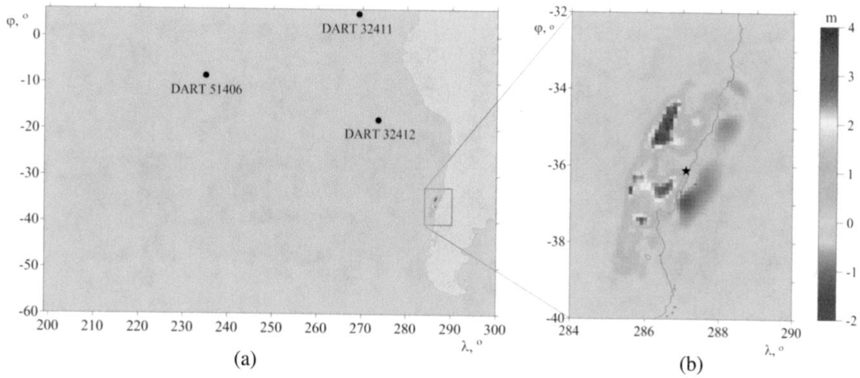

Fig. 4.13 The computational domain (**a**) and the initial wave elevation (**b**) computed according to USGS inversion of the Chilean 2010 earthquake. Symbol filled star on the right panel shows the earthquake epicenter

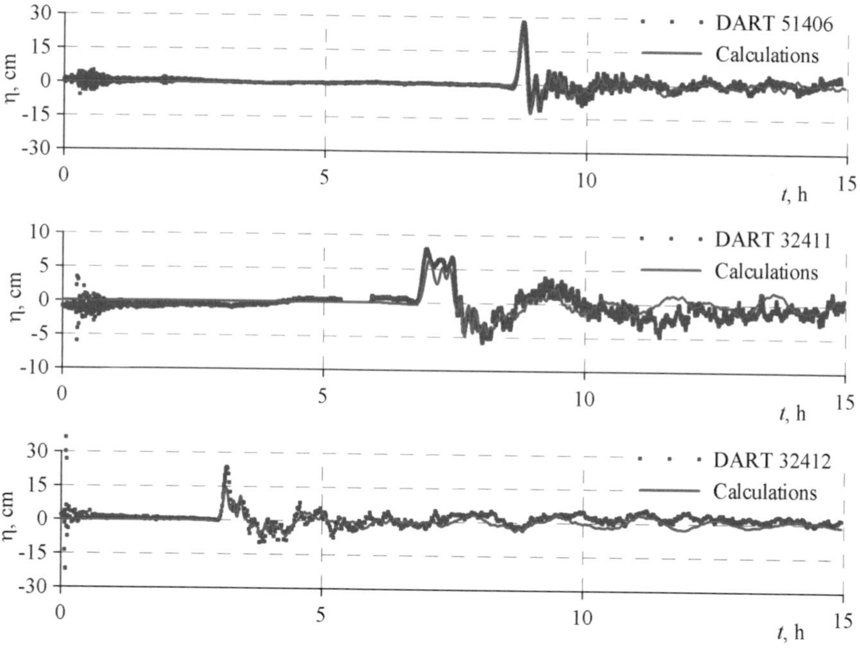

Fig. 4.14 Comparison of our numerical predictions with the spherical FNWD model with DART data

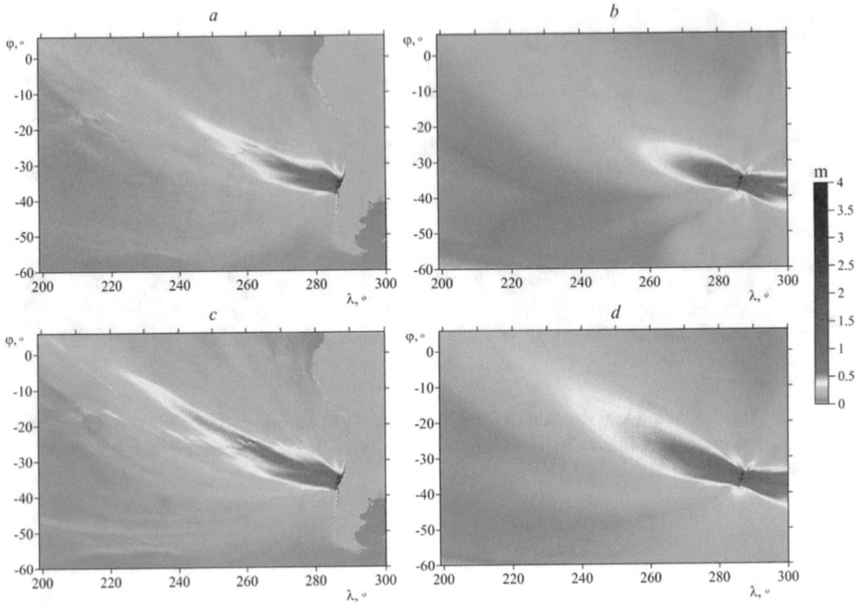

Fig. 4.15 The distribution of maximal positive wave amplitudes predicted with FNWD model (**a**, **b**) and NSWE (**c**, **d**). The real bathymetry data is used in computations (**a**, **c**), while the even bottom of constant depth $d = 4$ km is used in (**b**, **d**)

different radiation diagrams.[10] The difference becomes even more flagrant in the idealized constant depth case.[11] Bottom irregularities contribute equally to radiation diagrams even if they fail to alter the maximal radiation direction (at least in this particular tsunami event). It seems that here the initial condition has the dominant rôle in shaping the triggered tsunami wave.

In order to highlight the differences between NSWE and FNWD models predictions, we present in Fig. 4.16 the absolute and relative[12] differences among the corresponding radiation diagrams. The biggest absolute differences are concentrated along the main radiation direction and NSWE model seems to overestimate substantially the wave amplitude. The picture of relative differences has a much more complex structure even in the idealized case. The largest relative differences attain easily 60% not only along the main radiation direction, but also to the SOUTH from the epicenter. We can only conclude that the frequency dispersion has to be taken into account in this event. However, the dispersion effect may vary with

[10]Compare the top panels Fig. 4.15a, b, which show FNWD result with lower panels Fig. 4.15c, d representing NSWE predictions.

[11]Compare panels Fig. 4.15b and d.

[12]While computing the relative difference, we divide by the magnitude of the NSWE prediction as we did it above in Sect. 4.4.2.

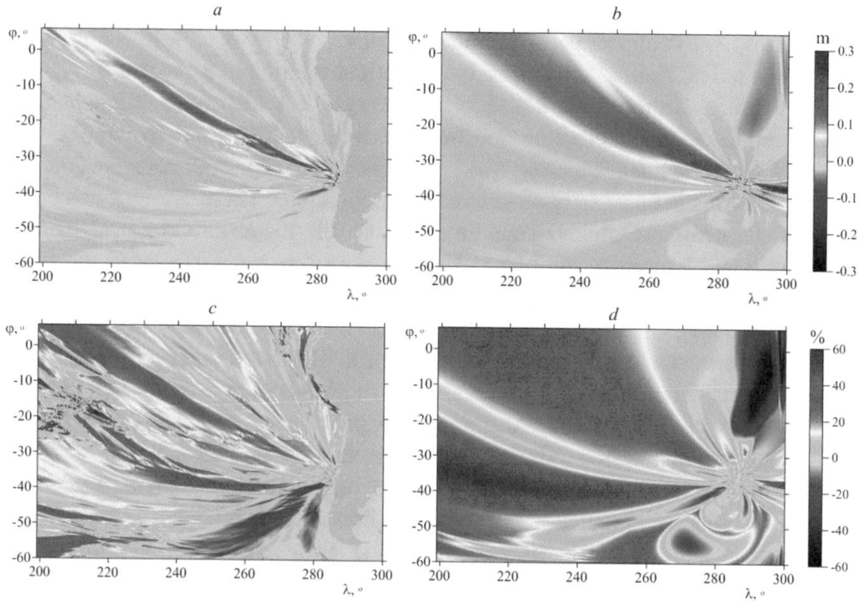

Fig. 4.16 The absolute (**a**, **b**) and relative (**c**, **d**) differences among radiation diagrams computed according to FNWD and NSWE models on a real bathymetry (**a**, **c**) and on an even constant bottom (**b**, **d**)

the tsunami initial condition [155]. Consequently, it is not excluded that for other seismic inversions the initial free surface shape may change such that the dispersion will play a more modest rôle.

4.5 Discussion

After the numerical developments and illustrations presented above, we finish this chapter by outlining the main conclusions and perspectives of all four chapters in general.

4.5.1 Conclusions

In this work we presented a numerical algorithm to simulate the generation and propagation of long surface waves in the framework of a Fully Nonlinear Weakly Dispersive (FNWD) model on a globally spherical domain. Our model includes the rotation effects (CORIOLIS and centrifugal forces) of the EARTH. Using a judicious choice of the still water level, we could 'hide' the terms corresponding to the centrifugal force. However, if the still water level is assumed to be spherical, then these terms have to appear in governing equations.

Our numerical method is based on the numerical solution of the extended system, which consists of an elliptic equation to determine the dispersive component $\wp$ of the depth-integrated pressure and of a quasi-linear first order hyperbolic part common with Nonlinear Shallow Water Equations (NSWE). Both parts are coupled via source terms in the right hand of hyperbolic equations. These source terms come from non-hydrostatic effects. Thus, one can use the favourite numerical method for elliptic and hyperbolic problems. For the elliptic part we employed the finite differences, while hyperbolic equations were discretized with a two-step predictor–corrector scheme. On every stage of this scheme we solve both sub-problems.

The performance of the proposed algorithm is illustrated on several test cases. First, we consider an idealized situation of wave propagation over an even bottom. However, in this ideal setting we study the importance of sphericity, rotation, Coriolis and dispersion effects. For instance, we showed that the force of Coriolis becomes important only on large propagation distances (unless the angular velocity Ω is increased) and the rotation reduces somehow the maximal positive wave amplitudes (but only positive). The frequency dispersion also appears at rather large distances, but it depends greatly on the size of the initial free surface elevation, i.e. more compact sources generate more dispersive waves. Contrary to the dispersion, the Coriolis effect becomes more important when we increase the generation area.

The question of dispersive effects importance is even more complex, since it can vary with the source size, but also with the source shape (i.e. the initial condition). Consequently, for real-world problems a special investigation is needed in each particular case. It does not seem realistic that using simple criteria today we can select the most pertinent hydrodynamic model. In practical problems the bathymetry profile may have an important effect as well.

In the present study we did not rise the question of the numerical dispersion at all. Here we can mention that for grid resolutions of $40''$ and smaller, the numerical dispersion seems to be completely negligible comparing to the physical one (for the initial condition of the horizontal extension $\mathscr{W}_1$, for shorter waves the grid spacing has to be further reduced, of course). However, we are not sure that a weakly dispersive model is able to reproduce such waves with acceptable accuracy, without even speaking of the inherent computational complexity on such fine grids.

General Conclusions

In these chapters our main goal was to present a unified framework to modelling and numerical simulations of nonlinear dispersive waves. Indeed, in Chap. 1 we presented a generalized derivation of dispersive wave models on a plane and the base model contained a free modelling variable—the dispersive component of the horizontal velocity vector. By making special choices of this free variable, we could recover some known and some new models. Then, in Chap. 3 the same approach was presented for globally spherical geometries with a similar degree of freedom at our disposal. The derivations on a sphere are more technical but in Chap. 3 we really follow the main lines of Chap. 1. Finally, there is a similar interplay between numerical Chaps. 2 and 4 on the globally flat and globally spherical geometries, respectively. The key ingredient in both cases consists in deriving an elliptic equation to determine the dispersive component $\wp$ of the depth-integrated pressure. Then, the governing equations are decoupled into an elliptic and hyperbolic parts. For each part we apply the most suitable numerical method. Even if the details are quite different in flat and spherical cases, the philosophy remains invariant. As we hope, the numerical tests presented in chapters with even numbers are convincing enough to show the operational qualities of the proposed algorithm.

4.5.2 Perspectives

One of the main drawbacks of non-hydrostatic FNWD models is the inherent computational complexity. At every time step we have to solve an elliptic equation at least once. Most of runs presented in this study took about a week of CPU time to be completed in serial implementations. Consequently, in the future, we plan to develop a parallel version of the NLDSW_SPHERE code in order to be able to handle much faster even higher grid resolutions. After this step we could try to implement fully nonlinear models with improved linear dispersion relation properties [74] derived in Chap. 3. Another improvement would consist in a better representation of the shoreline. In the current implementation the shoreline is approximated by a polygon with sides parallel to (spherical) coordinate axes. Perhaps, a local mesh refinement could improve this point but it would create other numerical difficulties. Finally, the main point which remains to be solved is the development of a genuine run-up algorithm in the spirit of local analytical in time solutions [187], which takes into account the non-hydrostatic nature of FNWD governing equations.

Appendix A
Long Wave Models Based on the Potential Flow Assumption

In Chap. 1 we presented a certain number of fully nonlinear weakly dispersive (FNWD) wave models. In this class we can mention the celebrated SGN model (cf. Sect. 1.1.5), ALESHKOV model [4] (cf. Sect. 1.1.6), LYNETT–LIU model [226] (cf. Sect. 1.1.6) and several others. All these models differ in the way to introduce the approximate velocity variable $\bar{\mathbf{u}}(\mathbf{x}, t)$ and the derivation procedure employs different assumptions on the structure of the horizontal velocity $\mathbf{u}(\mathbf{x}, y, t)$ of the 3D flow in question. In this appendix we shall present a different take on this problem. For instance, in PELINOVSKY–ZHELEZNYAK model [344], the variable $\bar{\mathbf{u}}$ is chosen to be the depth-averaged velocity of the 3D flow $\mathbf{u}(\mathbf{x}, y, t)$:

$$\bar{\mathbf{u}} \stackrel{\text{def}}{:=} \frac{1}{\mathcal{H}} \int_{-h}^{\eta} \mathbf{u}\, \mathrm{d}y\,.$$

Moreover, PELINOVSKY and ZHELEZNYAK make also the assumption that the 3D flow is irrotational. We would like to remind that the derivation of Green–Naghdi equations is based on completely different set of assumptions[1] (cf. Sect. 1.3 for a more detailed discussion). The original PELINOVSKY–ZHELEZNYAK model was derived in the case of the steady bottom. Unfortunately, this model cannot be employed to simulate tsunami generation processes by an underwater earthquake [90, 98, 105, 106, 110, 113, 114, 119, 121, 183] or landslide [26, 109, 118] to give two most important practical situations. In the appendix we generalize their derivation to the case of the moving bottom. Moreover, this derivation has another methodological interest: below we shall assume from the beginning that the underlying 3D flow is irrotational. This makes the derivation below interesting even

[1] To give briefly the main differences we remind that the GREEN–NAGHDI model is derived assuming that the vertical velocity v is a linear function of the vertical coordinate y and the components of the horizontal velocity $\mathbf{u}$ do not depend on y at all (to the considered approximation, of course).

G. Khakimzyanov et al., *Dispersive Shallow Water Waves*, Lecture Notes in Geosystems Mathematics and Computing, https://doi.org/10.1007/978-3-030-46267-3

without eventual applications to tsunami generation problems. To the best of our knowledge, the derivation procedure presented below is new and has never been presented before in the scientific literature.

A.1 Pelinovsky–Zheleznyak Model with Moving Bottom

A.1.1 The Mass Conservation Equation

Let us begin by the mass conservation. In the PELINOVSKY–ZHELEZNYAK model the continuity equation takes precisely the same form (1.76) as in the Green–Naghdi equations due to the choice of the velocity variable $\bar{\mathbf{u}}$. The derivation can be found above.

A.1.2 The Momentum Balance Equation

As it was clearly stated in the introduction to this appendix, it is supposed that the 3D flow is irrotational or potential, in other words. Mathematically, it means that there exists a velocity potential function $\phi : (\mathbf{x}, y, t) \longmapsto \phi(\mathbf{x}, y, t)$ such that equalities (1.13) are satisfied identically. The velocity potential ϕ verifies the LAPLACE equation (1.14) inside the fluid domain due to the incompressibility assumption [296]. Let us introduce also dimensionless variables as we did during the derivation of the base model (see Sect. 1.1.3). The velocity potential will be scaled as

$$\phi^* = \frac{\phi}{\ell\sqrt{g\,d}}\,.$$

After this scaling procedure, the LAPLACE equation (1.14) will read (we drop out from now on the asterisk signs for the sake of convenience):

$$\nabla^2\phi + \frac{1}{\mu^2}\,\phi_{yy} = 0, \tag{A.1}$$

while the bottom kinematic condition becomes:

$$h_t + \nabla\phi\cdot\nabla h + \frac{1}{\mu^2}\,\phi_y = 0, \qquad y = -h(\mathbf{x}, t)\,. \tag{A.2}$$

In order to simplify the governing equations, we shall assume the following ansatz for the velocity potential ϕ [331]:

$$\phi(\mathbf{x},\, y,\, t) \;=\; \check{\phi}(\mathbf{x},\, t) \;+\; \bigl[y \;+\; h(\mathbf{x},\, t)\bigr]\phi_1(\mathbf{x},\, t) \;+ \\ \bigl[y \;+\; h(\mathbf{x},\, t)\bigr]^2\phi_2(\mathbf{x},\, t) \;+\; \ldots \qquad \text{(A.3)}$$

It is not difficult to see the physical sense of the variable $\check{\phi}$:

$$\check{\phi}(\mathbf{x},\, t) \;\equiv\; \lim_{y \to -h} \phi(\mathbf{x},\, y,\, t)\,.$$

In order to determine the coefficient ϕ_1, we shall substitute expansion (A.3) into the scaled bottom boundary condition (A.2) to deduce that

$$\phi_1 \;=\; -\mu^2\,\frac{h_t \;+\; \nabla\check{\phi}\cdot\nabla h}{1 \;+\; \mu^2\,|\nabla h|^2} \;= \\ -\,\mu^2\bigl(h_t \;+\; \nabla\check{\phi}\cdot\nabla h\bigr) \;+\; \mu^4\,|\nabla h|^2\bigl(h_t \;+\; \nabla\check{\phi}\cdot\nabla h\bigr) \;+\; \mathcal{O}(\mu^6)\,.$$

The second asymptotic equality is valid only if the bottom does not vary too abruptly, i.e. $|\nabla h| \;<\; \dfrac{1}{\mu} \;=\; \dfrac{\ell}{d}$.

All other functional coefficients $\{\phi_j\}_{j \geqslant 2}$ from the expansion (A.3) can be determined by substituting this expansion into LAPLACE equation (A.1) and by equating the coefficients in front of equal powers of $(y \;+\; h)^j$, $j \;\geqslant\; 2$. Here we provide the expressions of three subsequent coefficients $\phi_{2,\,3,\,4}$ up to the asymptotic order $\mathcal{O}(\mu^6)$:

$$\phi_2 \;=\; -\frac{\mu^2}{2}\Bigl[(1 \;-\; \mu^2\,|\nabla h|^2)\Delta\check{\phi} \;-\; \mu^2\,(\Delta h)\cdot(h_t \;+\; \nabla\check{\phi}\cdot\nabla h) \\ -\; 2\,\mu^2\,\nabla h\cdot\nabla(h_t \;+\; \nabla\check{\phi}\cdot\nabla h)\Bigr] \;+\; \mathcal{O}(\mu^6)\,,$$

$$\phi_3 \;=\; \frac{\mu^4}{6}\Bigl[(\Delta h)\cdot(\Delta\check{\phi}) \;+\; 2\,(\nabla h)\cdot\nabla(\Delta\check{\phi}) \;+\; \Delta\bigl(h_t \;+\; \nabla\check{\phi}\cdot\nabla h\bigr)\Bigr] \;+\; \mathcal{O}(\mu^6)\,,$$

$$\phi_4 \;=\; \frac{\mu^4}{24}\,\Delta(\Delta\check{\phi}) \;+\; \mathcal{O}(\mu^6)\,,$$

where we defined a short-hand notation $\Delta \stackrel{\text{def}}{:=} \nabla^2 \stackrel{\text{def}}{:=} \partial^2_{x_1 x_1} \;+\; \partial^2_{x_2 x_2}$. We would like to stress out that this approach allows to obtain these formulas to any asymptotic order in the small parameter μ and not only to the order $\mathcal{O}(\mu^6)$. Of course, these computations can be recursively continued also for $j \;\geqslant\; 5$. However, we shall not need these expressions in this derivation since we are going to derive a specific FNWD model to the second order approximation. Nevertheless, the same approach can be used to derive formally FNWD models to any asymptotic order (see, e.g., [235] for a Hamiltonian derivation procedure).

From now on, we shall neglect the asymptotic terms of the order $\mathcal{O}(\mu^4)$ and higher. Henceforth, the velocity potential expansion (A.3) to this order becomes:

$$\phi = \check{\phi} - \mu^2 (y+h)\cdot(h_t + \nabla\check{\phi}\cdot\nabla h) - \mu^2 \frac{(y+h)^2}{2}\Delta\check{\phi} + \mathcal{O}(\mu^4). \qquad \text{(A.4)}$$

By applying the adapted depth-averaging operator to the last asymptotic expansion, we obtain:

$$\bar{\mathbf{u}} = \nabla\check{\phi} - \mu^2\Big[(h_t + \nabla\check{\phi}\cdot\nabla h)\cdot(\nabla h) + \frac{\mathcal{H}}{2}\nabla(h_t + \nabla\check{\phi}\cdot\nabla h) + \frac{\mathcal{H}}{2}(\Delta\check{\phi})\cdot(\nabla h) + \frac{\mathcal{H}^2}{6}\nabla(\Delta\check{\phi})\Big] + \mathcal{O}(\mu^4),$$

where $\bar{\mathbf{u}}$ is defined in this section as follows:

$$\bar{\mathbf{u}} \stackrel{\text{def}}{:=} \frac{1}{h+\varepsilon\eta}\int_{-h}^{\varepsilon\eta}\nabla\phi\,\mathrm{d}y.$$

By using the asymptotic identity $\nabla\check{\phi} = \bar{\mathbf{u}} + \mathcal{O}(\mu^2)$ in Eq. (A.4), we obtain:

$$\phi = \check{\phi} - \mu^2(y+h)\cdot(h_t + \bar{\mathbf{u}}\cdot\nabla h) - \mu^2\frac{(y+h)^2}{2}\nabla\cdot\bar{\mathbf{u}} + \mathcal{O}(\mu^4).$$

Consequently, we can compute the following important quantity:

$$|\nabla\phi|^2 = |\bar{\mathbf{u}}|^2 + \mu^2(\mathcal{H} - 2(y+h))\cdot\Big(\bar{\mathbf{u}}\cdot\nabla(h_t + \bar{\mathbf{u}}\cdot\nabla h) + (\nabla\cdot\bar{\mathbf{u}})\cdot(\bar{\mathbf{u}}\cdot\nabla h)\Big) + \mu^2\Big[\frac{\mathcal{H}^2}{3} - (y+h)^2\Big]\bar{\mathbf{u}}\cdot\nabla(\nabla\cdot\bar{\mathbf{u}}) + \mathcal{O}(\mu^4). \qquad \text{(A.5)}$$

After dropping out the terms of the asymptotic order $\mathcal{O}(\mu^4)$ and coming back to dimensional variables, we arrive to the following formulas for the velocity potential ϕ and the velocity components $\mathbf{u}$, v:

$$\phi = \check{\phi} - (y+h)\mathscr{D}h - \frac{(y+h)^2}{2}\nabla\cdot\bar{\mathbf{u}},$$

$$\mathbf{u} = \nabla\phi = \bar{\mathbf{u}} + \Big[\frac{\mathcal{H}}{2} - y - h\Big]\cdot\Big(\nabla\mathscr{D}h + (\nabla\cdot\bar{\mathbf{u}})\cdot\nabla h\Big) +$$

$$\left[\frac{\mathcal{H}^2}{6} - \frac{(y + h)^2}{2}\right] \nabla(\nabla \cdot \bar{\mathbf{u}}), \qquad \text{(A.6)}$$

$$v = \phi_y = -\mathscr{D}h - (y + h)\nabla \cdot \bar{\mathbf{u}}. \qquad \text{(A.7)}$$

The last formulas will be used to derive the equation of motion (i.e. momentum equation) of the PELINOVSKY–ZHELEZNYAK model. Moreover, we shall use the observation that the quantity $|\mathbf{u}|^2$ can be also computed (to the same asymptotic order of accuracy), thanks to Eq. (A.5):

$$|\mathbf{u}|^2 = |\bar{\mathbf{u}}|^2 + (\mathcal{H} - 2(y + h)) \cdot \Big[\bar{\mathbf{u}} \cdot \nabla \mathscr{D}h + (\nabla \cdot \bar{\mathbf{u}}) \cdot (\bar{\mathbf{u}} \cdot \nabla h)\Big] + \left[\frac{\mathcal{H}^2}{3} - (y + h)^2\right] \bar{\mathbf{u}} \cdot \nabla(\nabla \cdot \bar{\mathbf{u}}). \qquad \text{(A.8)}$$

By direct comparison of Eq. (A.7) with Eq. (1.86), we can see that the asymptotic representation of the vertical component v of the velocity coincides in the PELINOVSKY–ZHELEZNYAK model and SGN equations. We shall demonstrate that the asymptotic representations of the pressure variable p coincide as well (at least to the considered order $\mathcal{O}(\mu^4)$ and it is given by the same Eq. (1.80). Indeed, we shall use one consequence of the flow irrotationality:

$$\mathbf{u}_y = \nabla v. \qquad \text{(A.9)}$$

Then, the integral Equation (1.9) can be equivalently rewritten[2] in the following form:

$$\int_{\zeta}^{\eta} \Big(v_t + \tfrac{1}{2}(|\mathbf{u}|^2)_y + v\, v_y + p_y\Big)\, \mathrm{d}y = -g(\eta - \zeta). \qquad \text{(A.10)}$$

By using asymptotic representations (A.7) and (A.8), we obtain (to the given asymptotic order):

$$\frac{1}{2}(|\mathbf{u}|^2)_y = \bar{\mathbf{u}} \cdot \nabla v.$$

Consequently, we may write:

$$v_t + \tfrac{1}{2}(|\mathbf{u}|^2)_y = \mathscr{D}v.$$

Henceforth, the expression (1.80) follows from Eq. (A.10) if we take into account asymptotic formulas (1.78), (1.79).

[2]Under the partial irrotationality assumption (A.9), of course.

Let us demonstrate that asymptotic identities (1.81) also hold in the PELINOVSKY–ZHELEZNYAK model. Taking into account the identity (A.9) together with $\mathbf{u} = \nabla\phi$, we obtain:

$$\int_{-h}^{\eta} \bigl(\mathbf{u}_t + (\mathbf{u}\cdot\nabla)\mathbf{u} + v\,\mathbf{u}_y\bigr)\,\mathrm{d}y = \int_{-h}^{\eta} \Bigl(\mathbf{u}_t + \tfrac{1}{2}\nabla(|\mathbf{u}|^2) + v\,\nabla v\Bigr)\,\mathrm{d}y =$$

$$\underbrace{\int_{-h}^{\eta} \mathbf{u}_t\,\mathrm{d}y}_{(\Upsilon_1)} + \underbrace{\int_{-h}^{\eta} \tfrac{1}{2}\nabla(|\mathbf{u}|^2)\,\mathrm{d}y}_{(\Upsilon_2)} + \underbrace{\int_{-h}^{\eta} v\,\nabla v\,\mathrm{d}y}_{(\Upsilon_3)}.$$

Thanks to the previously established identity (A.8), we have the following intermediate result:

$$\int_{-h}^{\eta} \tfrac{1}{2}\nabla(|\mathbf{u}|^2)\,\mathrm{d}y = \frac{\mathcal{H}}{2}\nabla(|\bar{\mathbf{u}}|^2) + \frac{\mathcal{H}}{2}\bigl(\nabla\mathcal{H} - 2\nabla h\bigr)\cdot\bigl(\bar{\mathbf{u}}\cdot\nabla\mathscr{D}h + (\nabla\cdot\bar{\mathbf{u}})\cdot(\bar{\mathbf{u}}\cdot\nabla h)\bigr) + \left[\frac{\mathcal{H}^2}{3}\nabla\mathcal{H} - \frac{\mathcal{H}^2}{2}\nabla h\right]\bar{\mathbf{u}}\cdot\nabla(\nabla\cdot\bar{\mathbf{u}}). \qquad (A.11)$$

Later, we shall use the following relation:

$$\tfrac{1}{2}\nabla(|\bar{\mathbf{u}}|^2) = (\bar{\mathbf{u}}\cdot\nabla)\bar{\mathbf{u}} + \mathbf{b}, \qquad (A.12)$$

where $\mathbf{b} \stackrel{\text{def}}{:=} \bigl(\omega\bar{u}_2, -\omega\bar{u}_1\bigr)^{\top}$, with $(\bar{u}_1, \bar{u}_2)$ being the components of the depth-averaged horizontal velocity vector $\bar{\mathbf{u}}$ and $\omega \stackrel{\text{def}}{:=} \bar{u}_{2,x_1} - \bar{u}_{1,x_2}$ is related to the vertical component of the vorticity vector (see also Sect. 1.1.6). According to the asymptotic representation (A.6) of the horizontal velocity vector $\mathbf{u}$, we may obtain a better approximation of the quantity ω:

$$\omega \cong u_{2,x_1} - u_{1,x_2} \stackrel{(A.6)}{=} -\Bigl(\tfrac{1}{2}\mathcal{H}_{x_1} - h_{x_1}\Bigr)\cdot\bigl((\mathscr{D}h)_{x_2} + h_{x_2}\nabla\cdot\bar{\mathbf{u}}\bigr)$$
$$-\frac{\mathcal{H}}{2}h_{x_2}(\nabla\cdot\bar{\mathbf{u}})_{x_1} - \frac{\mathcal{H}}{3}\mathcal{H}_{x_1}(\nabla\cdot\bar{\mathbf{u}})_{x_2} + \Bigl(\tfrac{1}{2}\mathcal{H}_{x_2} - h_{x_2}\Bigr)\cdot\bigl((\mathscr{D}h)_{x_1} + h_{x_1}\nabla\cdot\bar{\mathbf{u}}\bigr) + \frac{\mathcal{H}}{2}h_{x_1}(\nabla\cdot\bar{\mathbf{u}})_{x_2} + \frac{\mathcal{H}}{3}\mathcal{H}_{x_2}(\nabla\cdot\bar{\mathbf{u}})_{x_1}.$$

After substituting the last result into formula (A.12) and by multiplying both sides by $\mathcal{H}$ we obtain:

$$\frac{\mathcal{H}}{2}\,\nabla\,(\,|\,\bar{\mathbf{u}}\,|^{\,2}\,) \;=\; \mathcal{H}\,(\bar{\mathbf{u}}\boldsymbol{\cdot}\nabla)\,\bar{\mathbf{u}} - \frac{\mathcal{H}}{2}\,\bigl(\nabla\,\mathcal{H} - 2\,\nabla\,h\bigr)\boldsymbol{\cdot}\bigl(\bar{\mathbf{u}}\boldsymbol{\cdot}\nabla\,\mathscr{D}h + (\nabla\boldsymbol{\cdot}\bar{\mathbf{u}})\,(\bar{\mathbf{u}}\boldsymbol{\cdot}\nabla\,h)\bigr)$$

$$-\Bigl(\tfrac{1}{3}\,\mathcal{H}^{2}\,\nabla\mathcal{H} - \tfrac{1}{2}\,\mathcal{H}^{2}\,\nabla h\Bigr)\,\bar{\mathbf{u}}\boldsymbol{\cdot}\nabla(\nabla\boldsymbol{\cdot}\bar{\mathbf{u}}) + \frac{\mathcal{H}}{2}(\nabla\,\mathscr{D}h + (\nabla\boldsymbol{\cdot}\bar{\mathbf{u}})\,\nabla h)\boldsymbol{\cdot}(\bar{\mathbf{u}}\boldsymbol{\cdot}\nabla\,\mathcal{H} - 2\,\bar{\mathbf{u}}\boldsymbol{\cdot}\nabla h)$$

$$+\;\Bigl(\tfrac{1}{3}\,\mathcal{H}^{2}\,\bar{\mathbf{u}}\cdot\nabla\,\mathcal{H} \;-\; \tfrac{1}{2}\,\mathcal{H}^{2}\,\bar{\mathbf{u}}\cdot\nabla\,h\Bigr)\,\nabla\,(\nabla\cdot\bar{\mathbf{u}})\,.$$

Consequently, for the term (Υ_2) we get:

$$(\Upsilon_2) \;\equiv\; \int_{-h}^{\eta} \tfrac{1}{2}\,\nabla\,(\,|\,\mathbf{u}\,|^{\,2}\,)\,\mathrm{d}y \;\stackrel{(\mathrm{A}.11)}{=}\; \mathcal{H}\,(\bar{\mathbf{u}}\cdot\nabla)\,\bar{\mathbf{u}} \,+$$
$$\frac{\mathcal{H}}{2}\,\bigl(\nabla\,\mathscr{D}h \,+\, (\nabla\cdot\bar{\mathbf{u}})\,\nabla\,h\bigr)\cdot\bigl(\bar{\mathbf{u}}\cdot\nabla\,\mathcal{H} \,-\, 2\,\bar{\mathbf{u}}\cdot\nabla\,h\bigr) \,+$$
$$\Bigl(\tfrac{1}{3}\,\mathcal{H}^{2}\,\bar{\mathbf{u}}\cdot\nabla\,\mathcal{H} \,-\, \tfrac{1}{2}\,\mathcal{H}^{2}\,\bar{\mathbf{u}}\cdot\nabla\,h\Bigr)\,\nabla\,(\nabla\cdot\bar{\mathbf{u}})\,. \qquad (\mathrm{A}.13)$$

Finally, using the asymptotic representation (A.7) for the vertical velocity v, we obtain the required expression of the term (Υ_3):

$$(\Upsilon_3) \;\equiv\; \int_{-h}^{\eta} v\,\nabla v\,\mathrm{d}y \;\stackrel{(\mathrm{A}.7)}{=}\; \frac{\mathcal{H}}{2}\,\bigl(\nabla\,\mathscr{D}h \,+\, (\nabla\cdot\bar{\mathbf{u}})\,\nabla\,h\bigr)\cdot\bigl(\mathcal{H}\,\nabla\cdot\bar{\mathbf{u}} \,+\, 2\,\mathscr{D}h\bigr) \,+$$
$$\Bigl(\tfrac{1}{3}\,\mathcal{H}^{3}\,(\nabla\cdot\bar{\mathbf{u}}) \,+\, \tfrac{1}{2}\,\mathcal{H}^{2}\,\mathscr{D}h\Bigr)\,\nabla\,(\nabla\cdot\bar{\mathbf{u}})\,. \qquad (\mathrm{A}.14)$$

By using one more time the representation (A.6) of the horizontal velocity $\mathbf{u}$, we can express the first term (Υ_1):

$$(\Upsilon_1) \;\equiv\; \int_{-h}^{\eta} \mathbf{u}_t\,\mathrm{d}y \;=\; \mathcal{H}\,\bar{\mathbf{u}}_t \,+\, \frac{\mathcal{H}}{2}\,\bigl(\nabla\,\mathscr{D}h \,+\, (\nabla\cdot\bar{\mathbf{u}})\,\nabla\,h\bigr)\cdot\bigl(\mathcal{H}_t \,-\, 2\,h_t\bigr) \,+$$
$$\Bigl(\tfrac{1}{3}\,\mathcal{H}^{2}\,\mathcal{H}_t \,-\, \tfrac{1}{2}\,\mathcal{H}^{2}\,h_t\Bigr)\,\nabla\,(\nabla\cdot\bar{\mathbf{u}})\,. \qquad (\mathrm{A}.15)$$

Now we can assemble all the elements together:

$$\int_{-h}^{\eta} \bigl(\mathbf{u}_t \,+\, \tfrac{1}{2}\,\nabla\,(\,|\,\mathbf{u}\,|^{\,2}\,) \,+\, v\,\nabla\,v\bigr)\,\mathrm{d}y \;\equiv\; (\Upsilon_1) \,+\, (\Upsilon_2) \,+\, (\Upsilon_3) \;\equiv$$
$$(\mathrm{A}.15) \,+\, (\mathrm{A}.13) \,+\, (\mathrm{A}.14) \;=\; \mathcal{H}\,\bigl(\bar{\mathbf{u}}_t \,+\, (\bar{\mathbf{u}}\cdot\nabla)\,\bar{\mathbf{u}}\bigr)\,,$$

where we used also the mass conservation Equation (1.76). Hence, after dividing both sides of the last equation by $\mathcal{H}$, we recover precisely Equation (1.81), which holds in the PELINOVSKY–ZHELEZNYAK framework. The evolution equation for the variable $\bar{\mathbf{u}}$ is obtained from Equation (1.82) by substituting the asymptotic representation of the pressure p. However, the pressure p in SGN and PELINOVSKY–ZHELEZNYAK models is computed according to the same Eq. (1.80), we conclude that the governing equations of both models coincide identically. Thus, the PELINOVSKY–ZHELEZNYAK model in the case of the moving bottom is given by Eqs. (1.76) and (1.83).

Revisiting the Equation of Motion

We would like to mention here that Eq. (1.83) is not particularly suitable for the theoretical and numerical studies since it involves second derivatives in time of the function $\mathcal{H}$. This is why we propose to recast this equation in an equivalent, more convenient form. By using the equality

$$\nabla \cdot \bar{\mathbf{u}} = -\frac{\mathscr{D}\mathcal{H}}{\mathcal{H}}$$

along with its direct differential consequence:

$$\mathscr{D}^2 \mathcal{H} = \mathcal{H}\big((\nabla \cdot \bar{\mathbf{u}})^2 - \mathscr{D}(\nabla \cdot \bar{\mathbf{u}})\big)$$

we can rewrite Eq. (1.83) in the following equivalent form:

$$\bar{\mathbf{u}}_t + (\bar{\mathbf{u}} \cdot \nabla)\bar{\mathbf{u}} + g\nabla\eta = \frac{1}{\mathcal{H}}\nabla\Big(\tfrac{1}{3}\mathcal{H}^3\mathscr{R}_1 + \tfrac{1}{2}\mathcal{H}^2\mathscr{R}_2\Big) - \nabla h\Big(\tfrac{1}{2}\mathcal{H}\mathscr{R}_1 + \mathscr{R}_2\Big). \qquad \text{(A.16)}$$

The quantities $\mathscr{R}_{1,2}$ have been defined in Eq. (1.87). It is not difficult to see that Eq. (A.16) does not contain any second derivatives of the evolution variables $\mathcal{H}$ and $\bar{\mathbf{u}}$ with respect to the time variable t. At the same time, the System (1.76), (A.16) of the PELINOVSKY–ZHELEZNYAK model is equivalent to Eqs. (1.76) and (1.84) of the Green–Naghdi model.

A.2 Intermediate Conclusions

Hence, in this appendix we have shown that two systems (Green–Naghdi and PELINOVSKY–ZHELEZNYAK) are equivalent in the cases of stationary and moving bottoms. In fact, they are different forms of shallow water equations of the second approximation. We remind that two models are derived under slightly different assumptions and the approximate velocity variable $\bar{\mathbf{u}}$ is not defined in the same way. Despite all these facts the resulting governing equations are identical. Moreover, in Sect. (1.1.5) we demonstrated that the same equations can be obtained without any simplifying irrotationality assumption based solely on the horizontal velocity decomposition (1.30). Thus, these models can be obtained from the base model (1.40), (1.41) derived in this book by taking the simplest possible closure $\tilde{\mathbf{u}} \equiv \mathbf{0}$. However, we have to admit that the potential flow assumption allows to reconstruct much better the velocity $\mathbf{u}$ of the 3D flow from its approximation $\bar{\mathbf{u}}$ given by the SGN model.

Appendix B
Modified Intermediate Weakly Nonlinear Weakly Dispersive Equations

Above in Chap. 1 we presented several families of shallow water models: dispersionless nonlinear shallow water equations (NSWE), fully nonlinear weakly dispersive (FNWS) and weakly nonlinear weakly dispersive (WNWD) equations. WNWD models occupy an intermediate place between NSWE and FNWD. It is considered that they can be applied in practice to model, for example, tsunami propagation far from the coasts, where the flow is not yet fully nonlinear and we may neglect nonlinearities in dispersive terms without losing substantially the accuracy. The advantage of using WNWD models (comparing to NSWE) is that we model eventual dispersive effects in tsunami propagation, which may have quite visible effect on the observed waveforms in the far field. Such operational Boussinesq-type models are well-known [15, 68, 137, 141, 228, 243].

WNWD models can be straightforwardly obtained from the corresponding FNWD models by employing the well-known in the theory Boussinesq approximation consisting to assume that $\varepsilon = \mathcal{O}(\mu^2)$ and neglecting the terms of the order $\mathcal{O}(\mu^4)$ and higher. The simplifications will take place in non-hydrostatic terms $\dfrac{\nabla \mathscr{P}}{\mathcal{H}}$ and $\dfrac{\check{p}}{\mathcal{H}}$ present in Eq. (1.47). As a result of this operation, dispersive terms become linear. This standard method was adopted by several authors. However, this approach has at least one important drawback: by following this purely asymptotic reasoning, some important properties such as the Galilean invariance and some conservation laws of the parent FNWD model may be lost. For example, in the classical Peregrine system, the Galilean invariance is lost due to this asymptotic reasoning. Moreover, this system does not possess any *consistent* energy conservation law even on the flat bottoms. In this appendix we demonstrate another way of thinking which allows to achieve some simplification of the FNWD by preserving important properties of this (parent) model. Here we follow essentially the ideas published earlier in [133].

G. Khakimzyanov et al., *Dispersive Shallow Water Waves*, Lecture Notes in Geosystems Mathematics and Computing, https://doi.org/10.1007/978-3-030-46267-3

B.1 Derivation of the Modified Boussinesq Equations

In order to preserve important *structural* properties of the FNWD model, we propose a modification of the standard approach of obtaining weakly nonlinear models described above. In this approach, some, but not *all*, nonlinear dispersive terms are neglected inside expressions $\frac{\mathscr{P}}{\mathcal{H}}$ and $\frac{\check{p}}{\mathcal{H}}$. We shall demonstrate this approach on the SGN system to obtain finally the *consistent*[3] Peregrine model.

Let us rewrite algebraically the expressions for $\mathscr{P}$ and $\check{p}$ in dimensionless variables by factoring the dispersive terms with $\mathcal{H}$:

$$\mathscr{P} = \frac{\mathcal{H}^2}{2} - \mu^2 \mathcal{H}\left(\frac{\mathcal{H}}{3}\,\mathscr{D}\big(\mathcal{H}\,(\nabla\cdot\bar{\mathbf{u}})\big) + \frac{\mathcal{H}}{2}\,\mathscr{D}^2 h\right),$$

$$\check{p} = \mathcal{H} - \mu^2 \mathcal{H}\left(\frac{1}{2}\,\mathscr{D}\big(\mathcal{H}\,(\nabla\cdot\bar{\mathbf{u}})\big) + \mathscr{D}^2 h\right),$$

where we substituted the following expression for $\mathscr{R}_1 = \frac{\mathscr{D}\big(\mathcal{H}\,(\nabla\cdot\bar{\mathbf{u}})\big)}{\mathcal{H}}$, which was obtained from the continuity Eq. (1.45) seen under a new angle such as

$$\nabla\cdot\bar{\mathbf{u}} = -\frac{\mathscr{D}\mathcal{H}}{\mathcal{H}}. \tag{B.1}$$

Consequently, after the division by a positive wave height function $\mathcal{H}$ we obtain:

$$\frac{\mathscr{P}}{\mathcal{H}} = \frac{\mathcal{H}}{2} - \mu^2\left(\frac{\mathcal{H}}{3}\,\mathscr{D}\big(\mathcal{H}\,(\nabla\cdot\bar{\mathbf{u}})\big) + \frac{\mathcal{H}}{2}\,\mathscr{D}^2 h\right),$$

$$\frac{\check{p}}{\mathcal{H}} = 1 - \mu^2\left(\frac{1}{2}\,\mathscr{D}\big(\mathcal{H}\,(\nabla\cdot\bar{\mathbf{u}})\big) + \mathscr{D}^2 h\right).$$

Now, dispersive terms on the right-hand sides can be effectively simplified using the standard Boussinesq approximation:

$$\frac{\mathscr{P}}{\mathcal{H}} \leftarrow \frac{\mathcal{H}}{2} - \mu^2\underbrace{\left(\frac{h}{3}\,\mathscr{D}\big(h\,(\nabla\cdot\bar{\mathbf{u}})\big) + \frac{h}{2}\,\mathscr{D}^2 h\right)}_{(\mathfrak{M}_1)},$$

[3]We underline the fact that the consistency is understood here in the sense of the energy functional and in the sense of the consistency with fundamental principles of the classical physics, the Galilean invariance being the most important one. Please, do not confuse our *consistency* with the formal notion of the asymptotic consistency.

$$\frac{\check{p}}{\mathcal{H}} \leftarrow 1 - \mu^2 \underbrace{\left(\frac{1}{2}\,\mathscr{D}\big(h\,(\nabla\cdot\bar{\mathbf{u}})\big) + \mathscr{D}^2 h\right)}_{(\mathfrak{M}_2)}.$$

Finally, we come back to simplified expressions of two pressure related quantities $\mathscr{P}$ and $\check{p}$:

$$\mathscr{P} = \frac{\mathcal{H}^2}{2} - \mu^2\left(\frac{h}{3}\,\mathcal{H}\,\mathscr{D}\big(h\,(\nabla\cdot\bar{\mathbf{u}})\big) + \frac{h}{2}\,\mathcal{H}\,\mathscr{D}^2 h\right), \tag{B.2}$$

$$\check{p} = \mathcal{H} - \mu^2\left(\frac{\mathcal{H}}{2}\,\mathscr{D}\big(h\,(\nabla\cdot\bar{\mathbf{u}})\big) + \mathcal{H}\,\mathscr{D}^2 h\right). \tag{B.3}$$

As a result, modified Boussinesq equations have the same form as Eqs. (1.45) and (1.47) of the SGN model. The difference consists only in the expressions of $\mathscr{P}$ and $\check{p}$, whose non-hydrostatic parts $(\mathfrak{M}_{1,\,2})$ differ precisely as prescribed in formulas (B.2), (B.3) above. Namely, the terms $(\mathfrak{M}_{1,\,2})$ are linear functions in the variable $\mathcal{H}$, while the same terms in FNWD models have higher degrees of the nonlinearity in $\mathcal{H}$. Strictly speaking, we cannot call the modified Boussinesq equations a WNWD model since the non-hydrostatic terms $(\mathfrak{M}_{1,\,2})$ are nonlinear in $\bar{\mathbf{u}}$. So, we have to place this model between WNWD and FNWD equations.

One of the most important advantages of the modified Boussinesq system we can mention is the Galilean invariance property. This can be seen directly from expressions (B.2), (B.3) which contain only total derivatives in time $\mathscr{D}$, which is a Galilean invariant operator in contrast to ∂_t. The second advantage consists in the possibility of rewriting the momentum balance equation in the conservative form (1.46). This opens some interesting perspectives regarding the application of robust shock-capturing finite volume schemes to solve these equations numerically [111, 112, 116, 117, 191].

Remark B.1 Exact solitary wave solutions to the modified weakly nonlinear Boussinesq model derived above are described in Sect. 2.1.6.

B.2 Energy Balance in the Modified Boussinesq System

Moreover, the modified Boussinesq system possesses the energy balance equation even in the presence of non-flat and moving bottom. This equation has precisely the form as Eq. (1.50) in the SGN model. The difference consists in the expression of the energy density:

$$\mathscr{E} = \tfrac{1}{2}\,|\bar{\mathbf{u}}|^2 + \mu^2\left[\tfrac{1}{6}\,h^2\,(\nabla\cdot\bar{\mathbf{u}})^2 + \frac{h}{2}\,(\mathscr{D}h)\,(\nabla\cdot\bar{\mathbf{u}}) + \frac{(\mathscr{D}h)^2}{2}\right] + \frac{\mathcal{H} - 2h}{2}.$$

The direct derivation of the energy balance equation for the modified Boussinesq equations is slightly different from the derivation presented in Sect. 1.1.5 for the SGN model. That is why we provide here a brief sketch of this derivation procedure to facilitate the understanding.

First of all, we multiply Eq. (1.46) by $\bar{\mathbf{u}}$ and taking into account the identity $\bar{\mathbf{u}} \cdot (\bar{\mathbf{u}} \cdot \nabla)\,\bar{\mathbf{u}} \equiv \frac{1}{2}\,\bar{\mathbf{u}} \cdot \nabla |\,\bar{\mathbf{u}}\,|^2$, we obtain the following equation:

$$\mathscr{D}\bigl(\tfrac{1}{2}\,|\,\bar{\mathbf{u}}\,|^2\bigr) + \frac{1}{\mathcal{H}}\,\nabla\cdot(\mathscr{P}\,\bar{\mathbf{u}}) - \underbrace{\Bigl(\frac{\mathscr{P}}{\mathcal{H}}\,\nabla\cdot\bar{\mathbf{u}} + \frac{\check{p}}{\mathcal{H}}\,\mathscr{D}h\Bigr)}_{(\mathrm{II})} = -\frac{\check{p}}{\mathcal{H}}\,h_t\,.$$

The sum of two terms (II) can be greatly simplified using identity (B.1) along with formulas (B.2) and (B.3):

$$-(\mathrm{II}) = -\frac{\mathscr{P}}{\mathcal{H}}\,\nabla\cdot\bar{\mathbf{u}} - \frac{\check{p}}{\mathcal{H}}\,\mathscr{D}h = \mathscr{D}\Bigl(\frac{\mathcal{H} - 2h}{2}\Bigr) +$$
$$\mu^2\Bigl[\mathscr{D}\Bigl(\frac{h^2}{6}\,(\nabla\cdot\bar{\mathbf{u}})^2\Bigr) + \mathscr{D}\Bigl(\frac{h}{2}\,(\mathscr{D}h)\,(\nabla\cdot\bar{\mathbf{u}})\Bigr) + \mathscr{D}\Bigl(\frac{(\mathscr{D}h)^2}{2}\Bigr)\Bigr] =$$
$$\mathscr{D}\Bigl(\mathscr{E} - \frac{|\,\bar{\mathbf{u}}\,|^2}{2}\Bigr).$$

Thus, for the modified Boussinesq equations, the energy balance equation in the non-conservative form can be written as

$$\mathscr{E}_t + \bar{\mathbf{u}}\cdot\nabla\mathscr{E} + \frac{1}{\mathcal{H}}\,\nabla\cdot(\mathscr{P}\,\bar{\mathbf{u}}) = -\frac{\check{p}}{\mathcal{H}}\,h_t\,.$$

It is obvious that the last equation can be rewritten in the conservative form (1.50) as well.

This concludes the presentation of the modified Boussinesq equations which occupy a very special place in the hierarchy of long wave models, thanks to their unique properties briefly mentioned hereinabove.

References

1. M.B. Abbott, H.M. Petersen, O. Skovgaard, On the numerical modelling of short waves in shallow water. J. Hydr. Res. **16**(3), 173–204 (1978)
2. R. Abraimi, Modelling the 2010 Chilean Tsunami using the H2Ocean unstructured mesh model. Master thesis, TU Delft, 2014
3. G.B. Alalykin, S.K. Godunov, L.L. Kireyeva, L.A. Pliner, *Solution of One-Dimensional Problems in Gas Dynamics on Moving Grids* (Nauka, Moscow, 1970)
4. Y.Z. Aleshkov, *Currents and Waves in the Ocean* (Saint Petersburg University Press, Saint-Petersburg, 1996)
5. C.J. Ammon, C. Ji, H.-K. Thio, D.I. Robinson, S. Ni, V. Hjorleifsdottir, H. Kanamori, T. Lay, S. Das, D. Helmberger, G. Ichinose, J. Polet, D. Waldm, Rupture process of the 2004 Sumatra-Andaman earthquake. Science **308**, 1133–1139 (2005)
6. D.C. Antonopoulos, V.A. Dougalis, D.E. Mitsotakis, Initial-boundary-value problems for the Bona-Smith family of Boussinesq systems. Adv. Differ. Equ. **14**, 27–53 (2009)
7. D.C. Antonopoulos, V.A. Dougalis, D.E. Mitsotakis, Galerkin approximations of the periodic solutions of Boussinesq systems. Bull. Greek Math. Soc. **57**, 13–30 (2010)
8. D.C. Antonopoulos, V.A. Dougalis, D.E. Mitsotakis, Numerical solution of Boussinesq systems of the Bona-Smith family. Appl. Numer. Math. **30**, 314–336 (2010)
9. J.S. Antunes Do Carmo, F.J. Seabra Santos, E. Barthélemy, Surface waves propagation in shallow water: a finite element model. Int. J. Numer. Meth. Fluids **16**(6), 447–459 (1993)
10. D. Arcas, H. Segur, Seismically generated tsunamis. Phil. Trans. R. Soc. A **370**, 1505–1542 (2012)
11. C. Arvanitis, A.I. Delis, Behavior of finite volume schemes for hyperbolic conservation laws on adaptive redistributed spatial grids. SIAM J. Sci. Comput. **28**(5), 1927–1956 (2006)
12. C. Arvanitis, T. Katsaounis, C. Makridakis, Adaptive finite element relaxation schemes for hyperbolic conservation laws. ESAIM Math. Model. Numer. Anal. **35**(1), 17–33 (2010)
13. S. Assier-Rzadkiewicz, P. Heinrich, P.C. Sabatier, B. Savoye, J.F. Bourillet, Numerical modelling of a landslide-generated Tsunami: the 1979 nice event. Pure Appl. Geophys. **157**(10), 1707–1727 (2000)
14. B.N. Azarenok, S.A. Ivanenko, T. Tang, Adaptive mesh redistribution method based on Godunov's scheme. Commun. Math. Sci. **1**(1), 152–179 (2003)
15. T. Baba, N. Takahashi, Y. Kaneda, K. Ando, D. Matsuoka, T. Kato, Parallel implementation of dispersive tsunami wave modeling with a nesting algorithm for the 2011 Tohoku Tsunami. Pure Appl. Geophys. **172**(12), 3455–3472 (2015)
16. N.S. Bakhvalov, On the optimization of methods of solving boundary value problems with a boundary layer. USSR Comput. Math. Math. Phys. **9**(4), 139–166 (1969)

G. Khakimzyanov et al., *Dispersive Shallow Water Waves*, Lecture Notes in Geosystems Mathematics and Computing, https://doi.org/10.1007/978-3-030-46267-3

17. V.B. Barakhnin, G.S. Khakimzyanov, On the algorithm for one nonlinear dispersive shallow-water model. Russ. J. Numer. Anal. Math. Model. **12**(4), 293–317 (1997)
18. V.B. Barakhnin, G.S. Khakimzyanov, The splitting technique as applied to the solution of the nonlinear dispersive shallow-water equations. Dokl. Math. **59**(1), 70–72 (1999)
19. T.J. Barth, M. Ohlberger, Finite volume methods: foundation and analysis, in *Encyclopedia of Computational Mechanics*, ed. by E. Stein, R. de Borst, and T.J.R. Hughes (Wiley, Chichester, 2004), pp. 439–474
20. E. Barthélémy, Nonlinear shallow water theories for coastal waves. Surv. Geophys. **25**, 315–337 (2004)
21. G.K. Batchelor, *An Introduction to Fluid Dynamics*. (Cambridge University Press, Cambridge, 1967)
22. S.V. Bazdenkov, N.N. Morozov, O.P. Pogutse, Dispersive effects in two-dimensional hydrodynamics. Dokl. Akad. Nauk SSSR **293**(4), 818–822 (1987)
23. J. Beck, S. Guillas, Sequential design with mutual information for computer experiments (MICE): emulation of a Tsunami model. SIAM/ASA J. Uncertainty Quantif. **4**(1), 739–766 (2016)
24. C. Beck, F. Manalt, E. Chapron, P. Van Rensbergen, M. De Batist, Enhanced seismicity in the early post-glacial period: Evidence from the post-würm sediments of lake Annecy, northwestern Alps. J. Geodynamics **22**(1–2), 155–171 (1996)
25. S.A. Beisel, L.B. Chubarov, G.S. Khakimzyanov, Simulation of surface waves generated by an underwater landslide moving over an uneven slope. Russ. J. Numer. Anal. Math. Modelling **26**(1), 17–38 (2011)
26. S.A. Beisel, L.B. Chubarov, D. Dutykh, G.S. Khakimzyanov, N.Y. Shokina, Simulation of surface waves generated by an underwater landslide in a bounded reservoir. Russ. J. Numer. Anal. Math. Modelling **27**(6), 539–558 (2012)
27. S. Beji, K. Nadaoka, A formal derivation and numerical modelling of the improved Boussinesq equations for varying depth. Ocean Eng. **23**(8), 691–704 (1996)
28. S. Beji, K. Nadaoka, Authors' reply to 'Discussion of Schäffer and Madsen on "A formal derivation and numerical modelling of the improved Boussinesq equations for varying depth" (Ocean Engineering, 25 (1998) 497–500)'. Ocean Eng. **25**(7), 615–618 (1998)
29. T.B. Benjamin, J.L. Bona, J.J. Mahony, Model equations for long waves in nonlinear dispersive systems. Philos. Trans. R. Soc. Lond. A **272**, 47–78 (1972)
30. M.J. Berger, D.L. George, R.J. LeVeque, K.T. Mandli, The GeoClaw software for depth-averaged flows with adaptive refinement. Adv. Water Res. **34**(9), 1195–1206 (2011)
31. D. Bestion, The physical closure laws in the CATHARE code. Nucl. Eng. Des. **124**, 229–245 (1990)
32. J.G. Blom, P.A. Zegeling, Algorithm 731; a moving-grid interface for systems of one-dimensional time-dependent partial differential equations. ACM Trans. Math. Softw. **20**(2), 194–214 (1994)
33. N. Bohr, Über die Serienspektra der Element. Z. Phys. **2**(5), 423–469 (1920)
34. J.L. Bona, M. Chen, A Boussinesq system for two-way propagation of nonlinear dispersive waves. Physica D **116**, 191–224 (1998)
35. J.L. Bona, R. Smith, A model for the two-way propagation of water waves in a channel. Math. Proc. Camb. Philos. Soc. **79**, 167–182 (1976)
36. J.L. Bona, V.A. Dougalis, O.A. Karakashian, Fully discrete Galerkin methods for the Korteweg-de Vries equation. Comput. Math. Appl. **12**(7), 859–884 (1986)
37. J.L. Bona, M. Chen, J.-C. Saut, Boussinesq equations and other systems for small-amplitude long waves in nonlinear dispersive media. I: Derivation and linear theory. J. Nonlinear Sci. **12**(4), 283–318 (2002)
38. J.L. Bona, M. Chen, J.-C. Saut, Boussinesq equations and other systems for small-amplitude long waves in nonlinear dispersive media: II. The nonlinear theory. Nonlinearity **17**(3), 925–952 (2004)
39. J.L. Bona, T. Colin, D. Lannes, Long wave approximations for water waves. Arch. Rational Mech. Anal. **178**, 373–410 (2005)

40. P. Bonneton, F. Chazel, D. Lannes, F. Marche, M. Tissier, A splitting approach for the fully nonlinear and weakly dispersive Green-Naghdi model. J. Comput. Phys. **230**, 1479–1498 (2011)
41. J.V. Boussinesq, Théorie générale des mouvements qui sont propagés dans un canal rectangulaire horizontal. C. R. Acad. Sc. Paris **73**, 256–260 (1871)
42. J.V. Boussinesq, Théorie des ondes et des remous qui se propagent le long d'un canal rectangulaire horizontal, en communiquant au liquide contenu dans ce canal des vitesses sensiblement pareilles de la surface au fond. J. Math. Pures Appl. **17**, 55–108 (1872)
43. J.V. Boussinesq, Essai sur la théorie des eaux courantes. Mémoires présentés par divers savants à l'Acad. des Sci. Inst. Nat. France **XXIII**, 1–680 (1877)
44. J.P. Boyd, *Chebyshev and Fourier Spectral Methods*, 2nd edn. (Dover Publications, New York, 2000)
45. M.-O. Bristeau, N. Goutal, J. Sainte-Marie, Numerical simulations of a non-hydrostatic shallow water model. Comput. Fluids **47**(1), 51–64 (2011)
46. M. Brocchini, A reasoned overview on Boussinesq-type models: the interplay between physics, mathematics and numerics. Proc. R. Soc. A **469**(2160), 20130496 (2013)
47. J.G.B. Byatt-Smith, The reflection of a solitary wave by a vertical wall. J. Fluid Mech. **197**, 503–521 (1988)
48. F. Carbone, D. Dutykh, J.M. Dudley, F. Dias, Extreme wave run-up on a vertical cliff. Geophys. Res. Lett. **40**(12), 3138–3143 (2013)
49. M.J. Castro, M. de la Asuncion, J. Macias, C. Parés, E.D. Fernandez-Nieto, J.M. Gonzalez-Vida, T. Morales de Luna, IFCP Riemann solver: Application to tsunami modelling using CPUs, in *Numerical Methods for Hyperbolic Equations: Theory and Applications*, ed. by M.E. Vazquez-Cendon, A. Hidalgo, P. Garcia-Navarro, L. Cea (CRC Press, Boca Raton, 2013), pp. 237–244
50. V. Casulli, A semi-implicit finite difference method for non-hydrostatic, free-surface flows. Int. J. Numer. Meth. Fluids **30**(4), 425–440 (1999)
51. A.-L. Cauchy, Mémoire sur la théorie de la propagation des ondes à la surface d'un fluide pesant d'une profondeur indéfinie. Mém. Présentés Divers Savans Acad. R. Sci. Inst. France **1**, 3–312 (1827)
52. J. Chambarel, C. Kharif, J. Touboul, Head-on collision of two solitary waves and residual falling jet formation. Nonlin. Process. Geophys. **16**, 111–122 (2009)
53. R.K.-C. Chan, R.L. Street, A computer study of finite-amplitude water waves. J. Comput. Phys. **6**(1), 68–94 (1970)
54. K.-A. Chang, T.-J. Hsu, P.L.-F. Liu, Vortex generation and evolution in water waves propagating over a submerged rectangular obstacle. Coastal Eng. **44**(1), 13–36 (2001)
55. E. Chapron, C. Beck, M. Pourchet, J.-F. Deconinck, 1822 earthquake-triggered homogenite in Lake Le Bourget (NW Alps). Terra Nova **11**(2–3), 86–92 (1999)
56. E. Chapron, P. Van Rensbergen, M. De Batist, C. Beck, J.P. Henriet, Fluid-escape features as a precursor of a large sublacustrine sediment slide in Lake Le Bourget, NW Alps, France. Terra Nova **16**(5), 305–311 (2004)
57. J.G. Charney, R. Fjörtoft, J. Neumann, Numerical integration of the barotropic vorticity equation. Tellus **2**(4), 237–254 (1950)
58. F. Chazel, Influence of bottom topography on long water waves. M2AN **41**, 771–799 (2007)
59. F. Chazel, D. Lannes, F. Marche, Numerical simulation of strongly nonlinear and dispersive waves using a Green-Naghdi model. J. Sci. Comput. **48**, 105–116 (2011)
60. Q. Chen, Fully nonlinear Boussinesq-type equations for waves and currents over porous beds. J. Eng. Mech. **132**(2), 220–230 (2006)
61. J.-B. Chen, M.-Z. Qin, and Y.-F. Tang. Symplectic and multi-symplectic methods for the nonlinear Schrödinger equation. Comput. Math. Appl. **43**(8–9), 1095–1106 (2002)
62. A.A. Cherevko, A.P. Chupakhin, Equations of the shallow water model on a rotating attracting sphere. 1. Derivation and general properties. J. Appl. Mech. Tech. Phys. **50**(2), 188–198 (2009)

63. A.A. Cherevko, A.P. Chupakhin, Shallow water equations on a rotating attracting sphere 2. Simple stationary waves and sound characteristics. J. Appl. Mech. Tech. Phys. **50**(3), 428–440 (2009)
64. M. Chhay, D. Dutykh, D. Clamond, On the multi-symplectic structure of the Serre-Green-Naghdi equations. J. Phys. A Math. Gen **49**(3), 03LT01 (2016)
65. J. Choi, J.T. Kirby, S.B. Yoon, Reply to "Discussion to 'Boussinesq modeling of longshore currents in the Sandy Duck experiment under directional random wave conditions' by J. Choi, J. T. Kirby and S.B. Yoon". Coastal Eng. **106**, 4–6 (2015)
66. A. Chorin, Numerical solution of the Navier-Stokes equations. Math. Comput. **22**, 745–762 (1968)
67. C.I. Christov, An energy-consistent dispersive shallow-water model. Wave Motion **34**, 161–174 (2001)
68. L.B. Chubarov, Y.I. Shokin, The numerical modelling of long wave propagation in the framework of non-linear dispersion models. Comput. Fluids **15**(3), 229–249 (1987)
69. L.B. Chubarov, Z.I. Fedotova, Y.I. Shokin, B.G. Einarsson, Comparative analysis of nonlinear dispersive shallow water models. Int. J. Comput. Fluid Dyn. **14**(1), 55–73 (2000)
70. L.B. Chubarov, S.V. Eletsky, Z.I. Fedotova, G.S. Khakimzyanov, Simulation of surface waves by an underwater landslide. Russ. J. Numer. Anal. Math. Model. **20**(5), 425–437 (2005)
71. R. Cienfuegos, E. Barthélemy, P. Bonneton, A fourth-order compact finite volume scheme for fully nonlinear and weakly dispersive Boussinesq-type equations. Part II: boundary conditions and validation. Int. J. Numer. Meth. Fluids **53**(9), 1423–1455 (2007)
72. D. Clamond, D. Dutykh, Practical use of variational principles for modeling water waves. Phys. D **241**(1), 25–36 (2012)
73. D. Clamond, D. Dutykh, Non-dispersive conservative regularisation of nonlinear shallow water (and isentropic Euler equations). Commun. Nonlin. Sci. Numer. Simul. **55**, 237–247 (2018)
74. D. Clamond, D. Dutykh, D. Mitsotakis, Conservative modified Serre–Green–Naghdi equations with improved dispersion characteristics. Commun. Nonlin. Sci. Numer. Simul. **45**, 245–257 (2017)
75. M.J. Cooker, P.D. Weidman, D.S. Bale, Reflection of a high-amplitude solitary wave at a vertical wall. J. Fluid Mech. **342**, 141–158 (1997)
76. R. Courant, K. Friedrichs, H. Lewy, Über die partiellen Differenzengleichungen der mathematischen Physik. Math. Ann. **100**(1), 32–74 (1928)
77. W. Craig, M.D. Groves, Hamiltonian long-wave approximations to the water-wave problem. Wave Motion **19**, 367–389 (1994)
78. A.D.D. Craik, The origins of water wave theory. Ann. Rev. Fluid Mech. **36**, 1–28 (2004)
79. R.A. Dalrymple, S.T. Grilli, J.T. Kirby, Tsunamis and challenges for accurate modeling. Oceanography **19**, 142–151 (2006)
80. M.H. Dao, P. Tkalich, Tsunami propagation modelling – a sensitivity study. Nat. Haz. Earth Syst. Sci. **7**, 741–754 (2007)
81. V.H. Davletshin, Force action of solitary waves on vertical structures, in *Tsunami Meeting* (Institute of Applied Physics, Gorky, 1984), pp. 41–43
82. C. Dawson, V. Aizinger, A discontinuous Galerkin method for three-dimensional shallow water equations. J. Sci. Comput. **22**(1-3), 245–267 (2005)
83. S.J. Day, J.C. Carracedo, H. Guillou, P. Gravestock, Recent structural evolution of the Cumbre Vieja volcano, La Palma, Canary Islands: volcanic rift zone reconfiguration as a precursor to volcano flank instability? J. Volcanol. Geothermal Res. **94**(1–4), 135–167 (1999)
84. A.J.C. de Saint-Venant, Théorie du mouvement non-permanent des eaux, avec application aux crues des rivières et à l'introduction des marées dans leur lit. C. R. Acad. Sci. Paris **73**, 147–154 (1871)
85. P.J. Dellar, Hamiltonian and symmetric hyperbolic structures of shallow water magnetohydrodynamics. Phys. Plasmas **9**(4), 1130–1136 (2002)
86. P.J. Dellar, Dispersive shallow water magnetohydrodynamics. Phys. Plasmas **10**(3), 581–590 (2003)

87. P.J. Dellar, R. Salmon, Shallow water equations with a complete Coriolis force and topography. Phys. Fluids **17**(10), 106601 (2005)
88. B. Delouis, J.-M. Nocquet, M. Vallée, Slip distribution of the February 27, 2010 Mw = 8.8 Maule Earthquake, central Chile, from static and high-rate GPS, InSAR, and broadband teleseismic data. Geophys. Res. Lett. **37**(17), L17305 (2010)
89. F. Dias, D. Dutykh, Dynamics of tsunami waves, in *Extreme Man-Made and Natural Hazards in Dynamics of Structures*, ed. by A. Ibrahimbegovic, I. Kozar (Springer Netherlands, 2007), pp. 35–60
90. F. Dias, D. Dutykh, L. O'Brien, E. Renzi, T. Stefanakis, On the modelling of Tsunami generation and Tsunami inundation. Procedia IUTAM **10**, 338–355 (2014)
91. M.W. Dingemans, *Water Wave Propagation over Uneven Bottom* (World Scientific, Singapore, 1997)
92. V.A. Dougalis, O.A. Karakashian, On some high-order accurate fully discrete Galerkin methods for the Korteweg-de Vries equation. Math. Comput. **45**(172), 329–345 (1985)
93. V.A. Dougalis, D.E. Mitsotakis, Theory and numerical analysis of Boussinesq systems: a review, in *Effective Computational Methods in Wave Propagation*, ed. by N.A. Kampanis, V.A. Dougalis, J.A. Ekaterinaris (CRC Press, Boca Raton, 2008), pp. 63–110
94. V.A. Dougalis, D.E. Mitsotakis, J.-C. Saut, Initial-boundary-value problems for Boussinesq systems of Bona-Smith type on a plane domain: theory and numerical analysis. J. Sci. Comput. **44**, 109–135 (2010)
95. P.G. Drazin, R.S. Johnson, *Solitons: An Introduction* (Cambridge University Press, Cambridge, 1989)
96. A. Durán, D. Dutykh, D. Mitsotakis, On the Galilean invariance of some nonlinear dispersive wave equations. Stud. Appl. Math. **131**(4), 359–388 (2013)
97. A. Durán, D. Dutykh, D. Mitsotakis, Peregrine's system revisited, in *Nonlinear Waves and Pattern Dynamics*, ed. by N. Abcha, E.N. Pelinovsky, I. Mutabazi (Springer International Publishing, Cham, 2018), pp. 3–43
98. D. Dutykh, Mathematical modelling of tsunami waves. Phd thesis, École Normale Supérieure de Cachan, 2007
99. D. Dutykh, D. Clamond, Shallow water equations for large bathymetry variations. J. Phys. A Math. Theor. **44**(33), 332001 (2011)
100. D. Dutykh, D. Clamond, Efficient computation of steady solitary gravity waves. Wave Motion **51**(1), 86–99 (2014)
101. D. Dutykh, D. Clamond, Modified shallow water equations for significantly varying seabeds. Appl. Math. Model. **40**(23–24), 9767–9787 (2016)
102. D. Dutykh, F. Dias, Dissipative Boussinesq equations. C. R. Mécanique **335**(9–10), 559–583 (2007)
103. D. Dutykh, F. Dias, Water waves generated by a moving bottom, in *Tsunami and Nonlinear Waves*, ed. by A. Kundu (Springer, Berlin, 2007), pp. 65–95
104. D. Dutykh, F. Dias, Energy of tsunami waves generated by bottom motion. Proc. R. Soc. Lond. A **465**(2103), 725–744 (2009)
105. D. Dutykh, F. Dias, How does sedimentary layering affect the generation of tsunamis? in *Proceedings of the International Conference on Offshore Mechanics and Arctic Engineering – OMAE*, vol. 6 (2009), pp. 495–503
106. D. Dutykh, F. Dias, Tsunami generation by dynamic displacement of sea bed due to dip-slip faulting. Math. Comput. Simul. **80**(4), 837–848 (2009)
107. D. Dutykh, F. Dias, Influence of sedimentary layering on tsunami generation. Comput. Meth. Appl. Mech. Eng. **199**(21–22), 1268–1275 (2010)
108. D. Dutykh, D. Ionescu-Kruse, Travelling wave solutions for some two-component shallow water models. J. Differ. Equ. **261**(2), 1099–1114 (2016)
109. D. Dutykh, H. Kalisch, Boussinesq modeling of surface waves due to underwater landslides. Nonlin. Process. Geophys. **20**(3), 267–285 (2013)
110. D. Dutykh, F. Dias, Y. Kervella, Linear theory of wave generation by a moving bottom. C. R. Mathématique **343**(7), 499–504 (2006)

111. D. Dutykh, T. Katsaounis, D. Mitsotakis, Dispersive wave runup on non-uniform shores, in *Finite Volumes for Complex Applications VI - Problems and Perspectives, Prague*, ed. by J. Fort (Springer, Berlin, 2011), pp. 389–397
112. D. Dutykh, T. Katsaounis, D. Mitsotakis, Finite volume schemes for dispersive wave propagation and runup. J. Comput. Phys. **230**(8), 3035–3061 (2011)
113. D. Dutykh, R. Poncet, F. Dias, The VOLNA code for the numerical modeling of tsunami waves: generation, propagation and inundation. Eur. J. Mech. B Fluids **30**(6), 598–615 (2011)
114. D. Dutykh, D. Mitsotakis, L.B. Chubarov, Y.I. Shokin, On the contribution of the horizontal sea-bed displacements into the tsunami generation process. Ocean Model. **56**, 43–56 (2012)
115. D. Dutykh, M. Chhay, F. Fedele, Geometric numerical schemes for the KdV equation. Comput. Math. Math. Phys. **53**(2), 221–236 (2013)
116. D. Dutykh, D. Clamond, P. Milewski, D. Mitsotakis, Finite volume and pseudo-spectral schemes for the fully nonlinear 1D Serre equations. Eur. J. Appl. Math. **24**(05), 761–787 (2013)
117. D. Dutykh, T. Katsaounis, D. Mitsotakis, Finite volume methods for unidirectional dispersive wave models. Int. J. Numer. Meth. Fluids **71**, 717–736 (2013)
118. D. Dutykh, D. Mitsotakis, S.A. Beisel, N.Y. Shokina, Dispersive waves generated by an underwater landslide, in *Numerical Methods for Hyperbolic Equations: Theory and Applications*, ed. by E. Vazquez-Cendon, A. Hidalgo, P. Garcia-Navarro, L. Cea (CRC Press, Boca Raton, 2013), pp. 245–250
119. D. Dutykh, D. Mitsotakis, X. Gardeil, F. Dias, On the use of the finite fault solution for tsunami generation problems. Theor. Comput. Fluid Dyn. **27**(1–2), 177–199 (2013)
120. D. Dutykh, D. Clamond, D. Mitsotakis, Adaptive modeling of shallow fully nonlinear gravity waves. RIMS Kôkyûroku **1947**(4), 45–65 (2015)
121. D. Dutykh, Mathematical modeling in the environment: from tsunamis to powder-snow avalanches. Habilitation à Diriger des Recherches, Université de Savoie, (2010). Retrieved from http://tel.archives-ouvertes.fr/tel-00542937/
122. F. Enet, S.T. Grilli, Experimental study of tsunami generation by three-dimensional rigid underwater landslides. J. Waterway Port Coast. Ocean Eng. **133**(6), 442–454 (2007)
123. R.C. Ertekin, W.C. Webster, J.V. Wehausen, Waves caused by a moving disturbance in a shallow channel of finite width. J. Fluid Mech. **169**, 275–292 (1986)
124. Y.D. Evsyukov, Distribution of landslide bodies on the continental slope of the north-eastern part of the Black Sea. Bull. North Caucasus Sci. Center Higher School Nat. Sci. **6**, 100–104 (2009)
125. M.S. Fabien, Spectral methods for partial differential equations that model shallow water wave phenomena. Master, University of Washington, 2014
126. Z.I. Fedotova, On application of the MacCormack difference scheme for problems of long-wave hydrodynamics. Comput. Technol. **11**(5), 53–63 (2006)
127. Z.I. Fedotova, E.D. Karepova, Variational principle for approximate models of wave hydrodynamics. Russ. J. Numer. Anal. Math. Model. **11**(3), 183–204 (1996)
128. Z.I. Fedotova, G.S. Khakimzyanov, Shallow water equations on a movable bottom. Russ. J. Numer. Anal. Math. Model. **24**(1), 31–42 (2009)
129. Z.I. Fedotova, G.S. Khakimzyanov, Nonlinear-dispersive shallow water equations on a rotating sphere. Russ. J. Numer. Anal. Math. Model. **25**(1) (2010)
130. Z.I. Fedotova, G.S. Khakhimzyanov, Full nonlinear dispersion model of shallow water equations on a rotating sphere. J. Appl. Mech. Tech. Phys. **52**(6), 865–876 (2011)
131. Z.I. Fedotova, G.S. Khakimzyanov, Nonlinear dispersive shallow water equations on a rotating sphere and conservation laws. J. Appl. Mech. Tech. Phys. **55**(3), 404–416 (2014)
132. Z.I. Fedotova, V.Y. Pashkova, Methods of construction and the analysis of difference schemes for nonlinear dispersive models of wave hydrodynamics. Russ. J. Numer. Anal. Math. Model. **12**(2) (1997)
133. Z.I. Fedotova, G.S. Khakimzyanov, D. Dutykh, Energy equation for certain approximate models of long-wave hydrodynamics. Russ. J. Numer. Anal. Math. Model. **29**(3), 167–178 (2014)

134. J.D. Fenton, M.M. Rienecker, A Fourier method for solving nonlinear water-wave problems: application to solitary-wave interactions. J. Fluid Mech. **118**, 411–443 (1982)
135. E.D. Fernandez-Nieto, F. Bouchut, D. Bresch, M.J. Castro-Diaz, A. Mangeney, A new Savage-Hutter type models for submarine avalanches and generated tsunami. J. Comput. Phys. **227**(16), 7720–7754 (2008)
136. G.R. Flierl, Particle motions in large-amplitude wave fields. Geophys. Astrophys. Fluid Dyn. **18**(1–2), 39–74 (1981)
137. D.R. Fuhrman, P.A. Madsen, Tsunami generation, propagation, and run-up with a high-order Boussinesq model. Coastal Eng. **56**(7), 747–758 (2009)
138. G. Galilei, *Dialogue Concerning the Two Chief World Systems* (Modern Library, New York, 2001, original date 1632)
139. D.L. George, Finite volume methods and adaptive refinement for tsunami propagation and inundation. PhD thesis, Department of Applied Mathematics, University of Washington, Seattle, 2006
140. P.A. Gilman, Magnetohydrodynamic "shallow water" equations for the solar tachocline. Astrophys. J. **544**(1), L79–L82 (2000)
141. S. Glimsdal, G.K. Pedersen, K. Atakan, C.B. Harbitz, H.P. Langtangen, F. Løvholt, Propagation of the Dec. 26, 2004, Indian Ocean Tsunami: effects of dispersion and source characteristics. Int. J. Fluid Mech. Res. **33**(1), 15–43 (2006)
142. S. Glimsdal, G.K. Pedersen, C.B. Harbitz, F. Løvholt, Dispersion of tsunamis: does it really matter? Nat. Haz. Earth Syst. Sci. **13**(6), 1507–1526 (2013)
143. M.F. Gobbi, J.T. Kirby, G. Wei, A fully nonlinear Boussinesq model for surface waves. Part 2. Extension to $\text{Ø}(kh)^4$. J. Fluid Mech. **405**, 181–210 (2000)
144. E. Godlewski, P.-A. Raviart, *Hyperbolic Systems of Conservation Laws* (Ellipses, Paris, 1990)
145. S.K. Godunov, V.S. Ryabenkii, *Difference Schemes* (North-Holland, Amsterdam, 1987)
146. L. Gosse, *Computing Qualitatively Correct Approximations of Balance Laws: Exponential-Fit, Well-Balanced and Asymptotic-Preserving*. SIMAI Springer Series, 1st edn., vol. 2 (Springer Milan, Milano, 2013)
147. A.E. Green, P.M. Naghdi, A derivation of equations for wave propagation in water of variable depth. J. Fluid Mech. **78**, 237–246 (1976)
148. A.E. Green, N. Laws, P.M. Naghdi, On the theory of water waves. Proc. R. Soc. Lond. A **338**, 43–55 (1974)
149. S.T. Grilli, P. Watts, Tsunami generation by submarine mass failure. I: Modeling, experimental validation, and sensitivity analyses. J. Waterway Port Coast. Ocean Eng. **131**(6), 283 (2005)
150. S. Grilli, S. Vogelmann, P. Watts, Development of a 3D numerical wave tank for modeling tsunami generation by underwater landslides. Eng. Anal. Bound. Elem. **26**, 301–313 (2002)
151. S.T. Grilli, M. Ioualalen, J. Asavanant, F. Shi, J.T. Kirby, P. Watts, Source constraints and model simulation of the December 26, 2004, Indian Ocean Tsunami. J. Waterway Port Coast. Ocean Eng. **133**(6), 414–428 (2007)
152. S.T. Grilli, J.C. Harris, T.S. Tajalli Bakhsh, T.L. Masterlark, C. Kyriakopoulos, J.T. Kirby, F. Shi, Numerical simulation of the 2011 Tohoku Tsunami based on a new transient FEM co-seismic source: comparison to far- and near-field observations. Pure Appl. Geophys. **170**, 1333 (2013)
153. J. Grue, E.N. Pelinovsky, D. Fructus, T. Talipova, C. Kharif, Formation of undular bores and solitary waves in the Strait of Malacca caused by the 26 December 2004 Indian Ocean tsunami. J. Geophys. Res. **113**(C5), C05008 (2008)
154. O.I. Gusev, Algorithm for surface waves calculation above a movable bottom within the frame of plane nonlinear dispersive wave model. Comput. Technol. **19**(6), 19–40 (2014)
155. O.I. Gusev, S.A. Beisel, Tsunami dispersion sensitivity to seismic source parameters. Sci. Tsunami Haz. **35**(2), 84–105 (2016)
156. O.I. Gusev, N.Y. Shokina, V.A. Kutergin, G.S. Khakimzyanov, Numerical modelling of surface waves generated by underwater landslide in a reservoir. Comput. Technol. **18**(5), 74–90 (2013)

157. O.I. Gusev, G.S. Khakimzyanov, L.B. Chubarov, Bulgarian tsunami on 7 May 2007: numerical investigation of the hypothesis of a submarine-landslide origin. Geol. Soc. Lond. Spec. Publ. **477**(1), 303–313 (2019)
158. V.K. Gusyakov, Z.I. Fedotova, G.S. Khakimzyanov, L.B. Chubarov, Y.I. Shokin, Some approaches to local modelling of tsunami wave runup on a coast. Russ. J. Numer. Anal. Math. Model. **23**(6), 551 (2008)
159. E. Hairer, G. Wanner, *Solving Ordinary Differential Equations II. Stiff and Differential-Algebraic Problems*. Springer Series in Computational Mathematics, vol. 14 (Springer, Berlin, 1996)
160. E. Hairer, S.P. Nørsett, G. Wanner, *Solving Ordinary Differential Equations: Nonstiff Problems* (Springer, Berlin, 2009)
161. G.J. Haltiner, R.T. Williams, *Numerical Prediction and Dynamic Meteorology*, 2nd edn. (Wiley, New York, 1980)
162. J. Hammack, D. Henderson, P. Guyenne, M. Yi, Solitary wave collisions, in *Proceedings of the 23rd International Conference on Offshore Mechanics and Arctic Engineering* (2004)
163. F.H. Harlow, J.E. Welch, Numerical calculation of time-dependent viscous incompressible flow of fluid with free surface. Phys. Fluids **8**, 2182 (1965)
164. J.C. Harris, S.T. Grilli, S. Abadie, T.B. Tayebeh, Near-and far-field tsunami hazard from the potential flank collapse of the Cumbre Vieja Volcano, in *Proceedings of the Twenty-second (2012) International Offshore and Polar Engineering Conference, Rhodes* (ISOPE, Mountain View, 2012), pp. 242–249
165. H. Hermes, *Introduction to Mathematical Logic*. Universitext (Springer Berlin, 1973)
166. N.J. Higham, *Accuracy and Stability of Numerical Algorithms*, 2nd edn. (SIAM Philadelphia, Philadelphia, 2002)
167. H. Holden, K.H. Karlsen, K.-A. Lie, N.H. Risebro, *Splitting Methods for Partial Differential Equations with Rough Solutions* (European Mathematical Society, Zürich, 2010)
168. J. Horrillo, Z. Kowalik, Y. Shigihara, Wave dispersion study in the Indian Ocean-Tsunami of December 26, 2004. Marine Geodesy **29**(3), 149–166 (2006)
169. J. Horrillo, W. Knight, Z. Kowalik, Tsunami propagation over the north pacific: dispersive and nondispersive models. Sci. Tsunami Haz. **31**(3), 154–177 (2012)
170. J. Horrillo, S.T. Grilli, D. Nicolsky, V. Roeber, J. Zhang, Performance benchmarking tsunami models for NTHMP's inundation mapping activities. Pure Appl. Geophys. **172**(3–4), 869–884 (2015)
171. S.-C. Hsiao, P.L.-F. Liu, Y. Chen, Nonlinear water waves propagating over a permeable bed. Proc. R. Soc. A **458**(2022), 1291–1322 (2002)
172. W. Huang, Practical aspects of formulation and solution of moving mesh partial differential equations. J. Comput. Phys. **171**(2), 753–775 (2001)
173. W. Huang, R.D. Russell, Adaptive mesh movement – the MMPDE approach and its applications. J. Comput. Appl. Math. **128**(1–2), 383–398 (2001)
174. A.M. Il'in, Differencing scheme for a differential equation with a small parameter affecting the highest derivative. Math. Notes Acad. Sci. USSR **6**(2), 596–602 (1969)
175. F. Imamura, Simulation of wave-packet propagation along sloping beach by TUNAMI-code, in *Long-wave Runup Models*, ed. by H. Yeh, P.L.-F. Liu, C.E. Synolakis (World Scientific, Singapore, 1996), pp. 231–241
176. M. Ioualalen, S. Migeon, O. Sardoux, Landslide tsunami vulnerability in the Ligurian Sea: case study of the 1979 October 16 Nice international airport submarine landslide and of identified geological mass failures. Geophys. J. Int. **181**(2), 724–740 (2010)
177. R.S. Johnson, Camassa-Holm, Korteweg-de Vries and related models for water waves. J. Fluid Mech. **455**, 63–82 (2002)
178. A. Kabbaj, Contribution à l'étude du passage des ondes de gravité et de la génération des ondes internes sur un talus, dans le cadre de la théorie de l'eau peu profonde. Thèse, Université Scientifique et Médicale de Grenoble, 1985

179. H. Kalisch, Z. Khorsand, D. Mitsotakis, Mechanical balance laws for fully nonlinear and weakly dispersive water waves. Phys. D **333**, 243–253 (2016)
180. R.A. Kazantsev, V.V. Kruglyakov, Giant landslide on the Black Sea bottom. Priroda **10**, 86–87 (1998)
181. M. Kazolea, A.I. Delis, A well-balanced shock-capturing hybrid finite volume-finite difference numerical scheme for extended 1D Boussinesq models. Appl. Numer. Math. **67**, 167–186 (2013)
182. A.B. Kennedy, J.T. Kirby, Q. Chen, R.A. Dalrymple, Boussinesq-type equations with improved nonlinear performance. Wave Motion **33**(3), 225–243 (2001)
183. Y. Kervella, D. Dutykh, F. Dias, Comparison between three-dimensional linear and nonlinear tsunami generation models. Theor. Comput. Fluid Dyn. **21**(4), 245–269 (2007)
184. G. Khakimzyanov, D. Dutykh, On supraconvergence phenomenon for second order centered finite differences on non-uniform grids. J. Comp. Appl. Math. **326**, 1–14 (2017)
185. G.S. Khakimzyanov, Y.I. Shokin, V.B. Barakhnin, N.Y. Shokina, *Numerical Simulation of Fluid Flows with Surface Waves.* (Siberian Branch, Russian Academy of Sciences, Novosibirsk, 2001)
186. G.S. Khakimzyanov, O.I. Gusev, S.A. Beizel, L.B. Chubarov, N.Y. Shokina, Simulation of tsunami waves generated by submarine landslides in the Black Sea. Russ. J. Numer. Anal. Math. Model. **30**(4), 227–237 (2015)
187. G.S. Khakimzyanov, N.Y. Shokina, D. Dutykh, D. Mitsotakis, A new run-up algorithm based on local high-order analytic expansions. J. Comp. Appl. Math. **298**, 82–96 (2016)
188. G.S. Khakimzyanov, D. Dutykh, Z.I. Fedotova, Dispersive shallow water wave modelling. Part III: Model derivation on a globally spherical geometry. Commun. Comput. Phys. **23**(2), 315–360 (2018)
189. G.S. Khakimzyanov, D. Dutykh, Z.I. Fedotova, D.E. Mitsotakis, Dispersive shallow water wave modelling. Part I: Model derivation on a globally flat space. Commun. Comput. Phys. **23**(1), 1–29 (2018)
190. G.S. Khakimzyanov, D. Dutykh, O. Gusev, Dispersive shallow water wave modelling. Part IV: Numerical simulation on a globally spherical geometry. Commun. Comput. Phys. **23**(2), 361–407 (2018)
191. G.S. Khakimzyanov, D. Dutykh, O. Gusev, N.Y. Shokina, Dispersive shallow water wave modelling. Part II: Numerical modelling on a globally flat space. Commun. Comput. Phys. **23**(1), 30–92 (2018)
192. G.S. Khakimzyanov, D. Dutykh, D. Mitsotakis, N.Y. Shokina, Numerical simulation of conservation laws with moving grid nodes: application to tsunami wave modelling. Geosciences **9**(5), 197 (2019)
193. G.S. Khakimzyanov, Z.I. Fedotova, O.I. Gusev, N.Y. Shokina, Finite difference methods for 2D shallow water equations with dispersion. Russ. J. Numer. Anal. Math. Model. **34**(2), 105–117 (2019)
194. J.W. Kim, R.C. Ertekin, A numerical study of nonlinear wave interaction in regular and irregular seas: irrotational Green-Naghdi model. Marine Struct. **13**(4–5), 331–347 (2000)
195. K.Y. Kim, R.O. Reid, R.E. Whitaker, On an open radiational boundary condition for weakly dispersive tsunami waves. J. Comput. Phys. **76**(2), 327–348 (1988)
196. J.W. Kim, K.J. Bai, R.C. Ertekin, W.C. Webster, A derivation of the Green-Naghdi equations for irrotational flows. J. Eng. Math. **40**(1), 17–42 (2001)
197. J.T. Kirby, F. Shi, B. Tehranirad, J.C. Harris, S.T. Grilli, Dispersive tsunami waves in the ocean: model equations and sensitivity to dispersion and Coriolis effects. Ocean Model. **62**, 39–55 (2013)
198. N.E. Kochin, I.A. Kibel, N.V. Roze, *Theoretical Hydromechanics* (Interscience Publishers, New York, 1965)
199. R.L. Kolar, W.G. Gray, J.J. Westerink, R.A. Luettich, Shallow water modeling in spherical coordinates: equation formulation, numerical implementation, and application. J. Hydr. Res. **32**(1), 3–24 (1994)

200. D.J. Korteweg, G. de Vries, On the change of form of long waves advancing in a rectangular canal, and on a new type of long stationary waves. Philos. Mag. **39**(5), 422–443 (1895)
201. V.M. Kovenya, N.N. Yanenko, *Splitting method in Gas Dynamics Problems* (Nauka, Novosibirsk, 1981)
202. Z. Kowalik, T.S. Murty, *Numerical Modeling of Ocean Dynamics* (World Scientific, Singapore, 1993)
203. H.O. Kreiss, G. Scherer, Method of lines for hyperbolic equations. SIAM J. Numer. Anal. **29**, 640–646 (1992)
204. E.A. Kulikov, P.P. Medvedev, S.S. Lappo, Satellite recording of the Indian Ocean tsunami on December 26, 2004. Dokl. Earth Sci. A **401**, 444–448 (2005)
205. A.A. Kurkin, S.V. Semin, Y.A. Stepanyants, Transformation of surface waves over a bottom step. Izvestiya Atmos. Ocean. Phys. **51**(2), 214–223 (2015)
206. O.A. Ladyzhenskaya, N.N. Uraltseva, *Linear and Quasilinear Elliptic Equations* (Nauka, Moscow, 1973)
207. Z. Lai, C. Chen, G.W. Cowles, R.C. Beardsley, A nonhydrostatic version of FVCOM: 1. Validation experiments. J. Geophys. Res. **115**(C11), C11010 (2010)
208. L.D. Landau, E.M. Lifshitz, *Mechanics* (Elsevier Butterworth-Heinemann, Amsterdam, 1976)
209. D. Lannes, *The Water Waves Problem: Mathematical Analysis and Asymptotics*. (American Mathematical Society, AMS, Philadelphia, 2013)
210. D. Lanser, J.G. Blom, J.G. Verwer, Spatial discretization of the shallow water equations in spherical geometry using Osher's scheme. Technical report, Centrum voor Wiskunde en Informatica, Amsterdam, Netherlands, 1999
211. T. Lay, H. Kanamori, C.J. Ammon, M. Nettles, S.N. Ward, R.C. Aster, S.L. Beck, S.L. Bilek, M.R. Brudzinski, R. Butler, H.R. DeShon, G. Ekstrom, K. Satake, S. Sipkin, The great Sumatra-Andaman earthquake of 26 December 2004. Science **308**, 1127–1133 (2005)
212. O. Le Métayer, S. Gavrilyuk, S. Hank, A numerical scheme for the Green-Naghdi model. J. Comput. Phys. **229**(6), 2034–2045 (2010)
213. D.Y. Le Roux, Spurious inertial oscillations in shallow-water models. J. Comput. Phys. **231**(24), 7959–7987 (2012)
214. D.Y. Le Roux, V. Rostand, B. Pouliot, Analysis of numerically induced oscillations in 2D finite-element shallow-water models part I: inertia-gravity waves. SIAM J. Sci. Comput. **29**(1), 331–360 (2007)
215. R.J. LeVeque, *Numerical Methods for Conservation Laws*, 2nd edn. (Birkhäuser Basel, Basel, 1992)
216. V. Liapidevskii, D. Dutykh, On the velocity of turbidity currents over moderate slopes. Fluid Dyn. Res. **51**, 035501 (2019)
217. V.Y. Liapidevskii, D. Dutykh, M. Gisclon, On the modelling of shallow turbidity flows. Adv. Water Res. **113**, 310–327 (2018)
218. E.K. Lindstrøm, G.K. Pedersen, A. Jensen, S. Glimsdal, Experiments on slide generated waves in a 1:500 scale fjord model. Coastal Eng. **92**, 12–23 (2014)
219. R. Liska, B. Wendroff, Shallow water conservation laws on a sphere, in *Hyperbolic Problems: Theory, Numerics, Applications* (Birkhäuser Basel, Basel, 2001), pp. 673–682
220. P.L.-F. Liu, T.-R. Wu, F. Raichlen, C.E. Synolakis, J.C. Borrero, Runup and rundown generated by three-dimensional sliding masses. J. Fluid Mech. **536**(1), 107–144 (2005)
221. J.W.S. Lord Rayleigh, On waves. Philos. Mag. **1**, 257–279 (1876)
222. F. Løvholt, G. Pedersen, Instabilities of Boussinesq models in non-uniform depth. Int. J. Numer. Meth. Fluids **61**(6), 606–637 (2009)
223. F. Løvholt, G. Pedersen, G. Gisler, Oceanic propagation of a potential tsunami from the La Palma Island. J. Geophys. Res. **113**(C9), C09026 (2008)
224. F. Løvholt, G. Pedersen, S. Glimsdal, Coupling of dispersive tsunami propagation and shallow water coastal response. Open Oceanogr. J. **4**(1), 71–82 (2010)
225. P. Lynch, *The Emergence of Numerical Weather Prediction: Richardson's Dream* (Cambridge University Press, Cambridge, 2014)

226. P. Lynett, P.L.F. Liu, A numerical study of submarine-landslide-generated waves and run-up. Proc. R. Soc. A **458**(2028), 2885–2910 (2002)
227. P.J. Lynett, T.R. Wu, P.L.-F. Liu, Modeling wave runup with depth-integrated equations. Coastal Eng. **46**(2), 89–107 (2002)
228. P.A. Madsen, D.R. Fuhrman, Run-up of tsunamis and long waves in terms of surf-similarity. Coastal Eng. **55**(3), 209–223 (2008)
229. O.S. Madsen, C.C. Mei, The transformation of a solitary wave over an uneven bottom. J. Fluid Mech. **39**(04), 781–791 (1969)
230. P.A. Madsen, O.R. Sørensen, A new form of the Boussinesq equations with improved linear dispersion characteristics. Part 2. A slowly-varying bathymetry. Coastal Eng. **18**, 183–204 (1992)
231. P.A. Madsen, R. Murray, O.R. Sørensen, A new form of the Boussinesq equations with improved linear dispersion characteristics. Coastal Eng. **15**, 371–388 (1991)
232. P.A. Madsen, H.B. Bingham, H. Liu, A new Boussinesq method for fully nonlinear waves from shallow to deep water. J. Fluid Mech. **462**, 1–30 (2002)
233. S.V. Manoylin, Some experimental and theoretical methods of estimation of tsunami wave action on hydro-technical structures and seaports. Technical report, Siberian Branch of Computing Center, Krasnoyarsk, 1989
234. G.I. Marchuk, *Numerical Solution of Problems of Atmosphere and Ocean Dynamics* (Gidrometeoizdat, Leningrad, 1974)
235. Y. Matsuno, Hamiltonian formulation of the extended Green-Naghdi equations. Phys. D **301–302**, 1–7 (2015)
236. T. Maxworthy, Experiments on collisions between solitary waves. J. Fluid Mech. **76**, 177–185 (1976)
237. J. McCloskey, A. Antonioli, A. Piatanesi, K. Sieh, S. Steacy, S. Nalbant, M. Cocco, C. Giunchi, J.D. Huang, P. Dunlop, Tsunami threat in the Indian Ocean from a future megathrust earthquake west of Sumatra. Earth Planet. Sci. Lett. **265**, 61–81 (2008)
238. C.C. Mei, B. Le Méhauté, Note on the equations of Long waves over an uneven bottom. J. Geophys. Res. **71**(2), 393–400 (1966)
239. J.W. Miles, R. Salmon, Weakly dispersive nonlinear gravity waves. J. Fluid Mech. **157**, 519–531 (1985)
240. J.W. Milnor, *Topology from the Differentiable Viewpoint*, rev. edition (Princeton University Press, Princeton, 1997)
241. N.R. Mirchina, E.N. Pelinovsky, Nonlinear and dispersive effects for tsunami waves in the open ocean. Int. J. Tsunami Soc. **2**(4), 1073–1081 (1982)
242. S.M. Mirie, C.H. Su, Collision between two solitary waves. Part 2. A numerical study. J. Fluid Mech. **115**, 475–492 (1982)
243. D.E. Mitsotakis, Boussinesq systems in two space dimensions over a variable bottom for the generation and propagation of tsunami waves. Math. Comput. Simul. **80**, 860–873 (2009)
244. D. Mitsotakis, B. Ilan, D. Dutykh, On the Galerkin/finite-element method for the Serre equations. J. Sci. Comput. **61**(1), 166–195 (2014)
245. T. Miyoshi, T. Saito, D. Inazu, S. Tanaka, Tsunami modeling from the seismic CMT solution considering the dispersive effect: a case of the 2013 Santa Cruz Islands tsunami. Earth Planets Space **67**(1), 4 (2015)
246. D. Moldabayev, H. Kalisch, D. Dutykh, The Whitham equation as a model for surface water waves. Phys. D **309**, 99–107 (2015)
247. M. Moreno, M. Rosenau, O. Oncken, 2010 Maule earthquake slip correlates with pre-seismic locking of Andean subduction zone. Nature **467**(7312), 198–202 (2010)
248. N. Mori, T. Takahashi, T. Yasuda, H. Yanagisawa, Survey of 2011 Tohoku earthquake tsunami inundation and run-up. Geophys. Res. Lett. **38**(7), L00G14 (2011)
249. T.S. Murty, Storm surges – meteorological ocean tides. Technical report, National Research Council of Canada, Ottawa, 1984
250. T.S. Murty, A.D. Rao, N. Nirupama, I. Nistor, Numerical modelling concepts for tsunami warning systems. Curr. Sci. **90**(8), 1073–1081 (2006)

251. A. Nayfeh, *Perturbation Methods*, 1st edn. (Wiley-VCH, New York, 2000)
252. H. Nersisyan, D. Dutykh, E. Zuazua, Generation of 2D water waves by moving bottom disturbances. IMA J. Appl. Math. **80**(4), 1235–1253 (2015)
253. M.A. Nosov, G.N. Nurislamova, A.V. Moshenceva, S.V. Kolesov, Residual hydrodynamic fields after tsunami generation by an earthquake. Izv. Atmos. Ocean. Phys. **50**(5), 520–531 (2014)
254. O. Nwogu, Alternative form of Boussinesq equations for nearshore wave propagation. J. Waterway Port Coast. Ocean Eng. **119**, 618–638 (1993)
255. Y. Okada, Surface deformation due to shear and tensile faults in a half-space. Bull. Seism. Soc. Am. **75**, 1135–1154 (1985)
256. Y. Okada, Internal deformation due to shear and tensile faults in a half-space. Bull. Seism. Soc. Am. **82**, 1018–1040 (1992)
257. P.J. Olver, Unidirectionalization of Hamiltonian waves. Phys. Lett. A **126**, 501–506 (1988)
258. G. Paranas-Carayannis, The earthquake and tsunami of 27 February 2010 in Chile – evaluation of source mechanism and of near and far-field tsunami effects. Sci. Tsunami Haz. **29**(2), 96–126 (2010)
259. G.K. Pedersen, F. Løvholt, Documentation of a global Boussinesq solver. Technical report, University of Oslo, Oslo, 2008
260. J. Pedlosky, *Geophysical Fluid Dynamics* (Springer, New York, 1992)
261. E.N. Pelinovsky, *Tsunami Wave Hydrodynamics* (Institute of Applied Physics Press, Nizhny Novgorod, 1996)
262. E. Pelinovsky, B.H. Choi, T. Talipova, S.B. Wood, D.C. Kim, Solitary wave transformation on the underwater step: asymptotic theory and numerical experiments. Appl. Math. Comput. **217**(4), 1704–1718 (2010)
263. D.H. Peregrine, Calculations of the development of an undular bore. J. Fluid Mech. **25**(02), 321–330 (1966)
264. D.H. Peregrine, Long waves on a beach. J. Fluid Mech. **27**, 815–827 (1967)
265. D.H. Peregrine, Water-wave impact on walls. Annu. Rev. Fluid Mech. **35**, 23–43 (2003)
266. S.D. Poisson, Mémoire sur la théorie des ondes. Mém. Acad. R. Sci. Inst. France **1816**(1), 70–186 (1818)
267. B. Ranguelov, S. Tinti, G. Pagnoni, R. Tonini, F. Zaniboni, A. Armigliato, The nonseismic tsunami observed in the Bulgarian Black Sea on 7 May 2007: was it due to a submarine landslide? Geophys. Res. Lett. **35**(18), L18613 (2008)
268. S.C. Reddy, L.N. Trefethen, Stability of the method of lines. Numer. Math. **62**(1), 235–267 (1992)
269. J.S. Russell, Report on waves. Technical report, Report of the fourteenth meeting of the British Association for the Advancement of Science, York, September 1844, London, 1845
270. O.B. Rygg, Nonlinear refraction-diffraction of surface waves in intermediate and shallow water. Coastal Eng. **12**(3), 191–211 (1988)
271. G. Sadaka, Solution of 2D Boussinesq systems with FreeFem++: the flat bottom case. J. Numer. Math. **20**(3–4), 303–324 (2012)
272. A.A. Samarskii, *The Theory of Difference Schemes* (CRC Press, New York, 2001)
273. P. Sammarco, E. Renzi, Landslide tsunamis propagating along a plane beach. J. Fluid Mech. **598**, 107–119 (2008)
274. H.A. Schäffer, P.A. Madsen, Further enhancements of Boussinesq-type equations. Coastal Eng. **26**(1-2), 1–14 (1995)
275. H.A. Schäffer, P.A. Madsen, Discussion of "A formal derivation and numerical modelling of the improved Boussinesq equations for varying depth". Ocean Eng.neering **25**(6), 497–500 (1998)
276. H.A. Schäffer, P.A. Madsen, Further discussion related to 'A formal derivation and numerical modeling of the improved Boussinesq equations for varying depth' by Beji and Nadaoka. Ocean Eng. **26**(11), 1057–1062 (1999)
277. W.E. Schiesser, Method of lines solution of the Korteweg-de Vries equation. Comput. Math. Appl. 28(10–12), 147–154 (1994)

278. F.J. Seabra-Santos, D.P. Renouard, A.M. Temperville, Numerical and experimental study of the transformation of a solitary wave over a shelf or isolated obstacle. J. Fluid Mech. **176**, 117–134 (1987)
279. F.J. Seabra-Santos, A.M. Temperville, D.P. Renouard, On the weak interaction of two solitary waves. Eur. J. Mech. B Fluids **8**(2), 103–115 (1989)
280. L.I. Sedov, *Mechanics of Continuous Media*, vol. 1 (World Scientific, Singapore, 1997)
281. H. Segur, Waves in shallow water, with emphasis on the tsunami of 2004, in *Tsunamis and Nonlinear Waves*, ed. by A. Kundu (Springer, New York, 2007), pp. 3–29
282. F. Serre, Contribution à l'étude des écoulements permanents et variables dans les canaux. La Houille blanche **8**, 830–872 (1953)
283. F. Serre, Contribution à l'étude des ondes longues irrotationnelles. La Houille Blanche, (3), 375–390 (1956). https://doi.org/10.1051/lhb/1956033
284. L.F. Shampine, ODE solvers and the method of lines. Numer. Meth. Part. Differ. Equ. **10**(6), 739–755 (1994)
285. F. Shi, J.T. Kirby, J.C. Harris, J.D. Geiman, S.T. Grilli, A high-order adaptive time-stepping TVD solver for Boussinesq modeling of breaking waves and coastal inundation. Ocean Model. **43–44**, 36–51 (2012)
286. F. Shi, J.T. Kirby, B. Tehranirad, Tsunami benchmark results for spherical coordinate version of FUNWAVE-TVD (Version 2.0). Technical report, University of Delaware, Newark, Delaware, USA, 2012
287. Y.I. Shokin, Y.V. Sergeeva, G.S. Khakimzyanov, Predictor-corrector scheme for the solution of shallow water equations. Russ. J. Numer. Anal. Math. Model. **21**(5), 459–479 (2006)
288. Y.I. Shokin, Z.I. Fedotova, G.S. Khakimzyanov, L.B. Chubarov, S.A. Beisel, Modelling surface waves generated by a moving landslide with allowance for vertical flow structure. Russ. J. Numer. Anal. Math. Model. **22**(1), 63–85 (2007)
289. Y.I. Shokin, V.V. Babailov, S.A. Beisel, L.B. Chubarov, S.V. Eletsky, Z.I. Fedotova, V.K. Gusiakov, Mathematical modeling in application to regional tsunami warning systems operations, in *Computational Science and High Performance Computing III* (2008), pp. 52–68
290. Y.I. Shokin, Z.I. Fedotova, G.S. Khakimzyanov, Hierarchy of nonlinear models of the hydrodynamics of long surface waves. Dokl. Phys. **60**(5), 224–228 (2015)
291. N.Y. Shokina, To the problem of construction of difference schemes on movable grids. Russ. J. Numer. Anal. Math. Model. **27**(6), 603 (2012)
292. G. Simarro, Energy balance, wave shoaling and group celerity in Boussinesq-type wave propagation models. Ocean Model. **72**, 74–79 (2013)
293. G. Simarro, A. Orfila, A. Galan, Linear shoaling in Boussinesq-type wave propagation models. Coastal Eng. **80**, 100–106 (2013)
294. G. Söderlind, L. Wang, Adaptive time-stepping and computational stability. J. Comput. Appl. Math. **185**(2), 225–243 (2006)
295. O.R. Sørensen, H.A. Schäffer, L.S. Sørensen, Boussinesq-type modelling using an unstructured finite element technique. Coastal Eng. **50**(4), 181–198 (2004)
296. J.J. Stoker, *Water Waves: The Mathematical Theory with Applications* (Wiley, Hoboken, 1992)
297. C.H. Su, C.S. Gardner, KdV equation and generalizations. Part III. Derivation of the Korteweg-de Vries equation and Burgers equation. J. Math. Phys. **10**, 536–539 (1969)
298. C.H. Su, R.M. Mirie, On head-on collisions between two solitary waves. J. Fluid Mech. **98**, 509–525 (1980)
299. C.E. Synolakis, E.N. Bernard, Tsunami science before and beyond Boxing Day 2004. Phil. Trans. R. Soc. A **364**, 2231–2265 (2006)
300. C.E. Synolakis, E.N. Bernard, V.V. Titov, U. Kânoglu, F.I. González, Validation and verification of tsunami numerical models. Pure Appl. Geophys. **165**, 2197–2228 (2008)
301. L. Tang, V.V. Titov, E.N. Bernard, Y. Wei, C.D. Chamberlin, J.C. Newman, H.O. Mofjeld, D. Arcas, M.C. Eble, C. Moore, B. Uslu, C. Pells, M. Spillane, L. Wright, E. Gica, Direct energy estimation of the 2011 Japan tsunami using deep-ocean pressure measurements. J. Geophys. Res. Oceans **117**(C8), 8008 (2012)

302. Y. Tanioka, Analysis of the far-field tsunamis generated by the 1998 Papua New Guinea Earthquake. Geophys. Res. Lett. **26**(22), 3393–3396 (1999)
303. D.R. Tappin, P. Watts, S.T. Grilli, The Papua New Guinea tsunami of 17 July 1998: anatomy of a catastrophic event. Nat. Hazards Earth Syst. Sci. **8**, 243–266 (2008)
304. D.R. Tappin, S.T. Grilli, J.C. Harris, R.J. Geller, T. Masterlark, J.T. Kirby, F. Shi, G. Ma, K. Thingbaijam, P.M. Mai, Did a submarine landslide contribute to the 2011 Tohoku tsunami? Marine Geol. **357**, 344–361 (2014)
305. P.D. Thomas, C.K. Lombart, Geometric conservation law and its application to flow computations on moving grid. AIAA J. **17**(10), 1030–1037 (1979)
306. A.N. Tikhonov, A.A. Samarskii, Homogeneous difference schemes. Zh. vych. mat. **1**(1), 5–63 (1961)
307. A.N. Tikhonov, A.A. Samarskii, Homogeneous difference schemes on non-uniform nets. Zh. vych. mat. **2**(5), 812–832 (1962)
308. S. Tinti, E. Bortolucci, C. Vannini, A block-based theoretical model suited to gravitational sliding. Nat. Hazards **16**(1), 1–28 (1997)
309. V.V. Titov, F.I. González, Implementation and testing of the method of splitting tsunami (MOST) model. Technical Report ERL PMEL-112, Pacific Marine Environmental Laboratory, NOAA, 1997
310. V.V. Titov, C.E. Synolakis, Numerical modeling of 3-D long wave runup using VTCS-3, in *Long Wave Runup Models*, ed. by H. Yeh, P. L.-F. Liu, C.E. Synolakis (World Scientific, Singapore, 1996), pp. 242–248
311. M. Tort, T. Dubos, F. Bouchut, V. Zeitlin, Consistent shallow-water equations on the rotating sphere with complete Coriolis force and topography. J. Fluid Mech. **748**, 789–821 (2014)
312. J. Touboul, E. Pelinovsky, Bottom pressure distribution under a solitonic wave reflecting on a vertical wall. Eur. J. Mech. B Fluids **48**, 13–18 (2014)
313. L. Umlauf, H. Burchard, Second-order turbulence closure models for geophysical boundary layers. A review of recent work. Cont. Shelf Res. **25**, 795–827 (2005)
314. F. Ursell, The long-wave paradox in the theory of gravity waves. Proc. Camb. Philos. Soc. **49**, 685–694 (1953)
315. A. van Dam, P.A. Zegeling, A robust moving mesh finite volume method applied to 1D hyperbolic conservation laws from magnetohydrodynamics. J. Comput. Phys. **216**(2), 526–546 (2006)
316. I.N. Vekua, *Fundamentals of Tensor Analysis and Theory of Covariants* (Nauka, Moscow, 1978)
317. C. Vigny, A. Socquet, S. Peyrat, J.-C. Ruegg, M. Metois, R. Madariaga, S. Morvan, M. Lancieri, R. Lacassin, J. Campos, D. Carrizo, M. Bejar-Pizarro, S. Barrientos, R. Armijo, C. Aranda, M.-C. Valderas-Bermejo, I. Ortega, F. Bondoux, S. Baize, H. Lyon-Caen, A. Pavez, J.P. Vilotte, M. Bevis, B. Brooks, R. Smalley, H. Parra, J.-C. Baez, M. Blanco, S. Cimbaro, E. Kendrick, The 2010 Mw 8.8 Maule Megathrust Earthquake of Central Chile, monitored by GPS. Science **332**(6036), 1417–1421 (2011)
318. I. Vilibic, J. Sepic, B. Ranguelov, N.S. Mahovic, S. Tinti, Possible atmospheric origin of the 7 May 2007 western Black Sea shelf tsunami event. J. Geophys. Res. **115**(C7), C07006 (2010)
319. N.E. Voltsinger, R.V. Pyaskovsky, *Theory of Shallow Water. Oceanological Problems and Numerical Methods* (Gidrometeoizdat, Leningrad, 1977)
320. M. Walkley, M. Berzins, A finite element method for the two-dimensional extended Boussinesq equations. Int. J. Numer. Meth. Fluids **39**(10), 865–885 (2002)
321. S.N. Ward, Landslide tsunami. J. Geophysical Res. **106**, 11201–11215 (2001)
322. S.N. Ward, S.J. Day, Cumbre Vieja Volcano – potential collapse and tsunami at La Palma, Canary Islands. Geophys. Res. Lett. **28**(17), 3397 (2001)
323. P. Watts, Tsunami features of solid block underwater landslides. J. Waterway Port Coast. Ocean Eng. **126**(3), 144–152 (2000)
324. P. Watts, F. Imamura, S.T. Grilli, Comparing model simulations of three benchmark tsunami generation cases. Sci. Tsunami Hazards **18**(2), 107–123 (2000)

325. P. Watts, S.T. Grilli, J.T. Kirby, G.J. Fryer, D.R. Tappin, Landslide tsunami case studies using a Boussinesq model and a fully nonlinear tsunami generation model. Nat. Hazards Earth Syst. Sci. **3**(5), 391–402 (2003)
326. C.F. Waythomas, P. Watts, J.S. Walder, Numerical simulation of tsunami generation by cold volcanic mass flows at Augustine Volcano, Alaska. Nat. Hazards Earth Syst. Sci. **6**(5), 671–685 (2006)
327. G. Wei, J.T. Kirby, Time-dependent numerical code for extended Boussinesq equations. J. Waterway Port Coast. Ocean Eng. **121**(5), 251–261 (1995)
328. G. Wei, J.T. Kirby, S.T. Grilli, R. Subramanya, A fully nonlinear Boussinesq model for surface waves. Part 1. Highly nonlinear unsteady waves. J. Fluid Mech. **294**, 71–92 (1995)
329. Y. Wei, E.N. Bernard, L. Tang, R. Weiss, V.V. Titov, C. Moore, M. Spillane, M. Hopkins, U. Kanoglu, Real-time experimental forecast of the Peruvian tsunami of August 2007 for U.S. coastlines. Geophys. Res. Lett. **35**(4), L04609 (2008)
330. R. Wen, Y. Ren, X. Li, R. Pan, Comparison of two great Chile tsunamis in 1960 and 2010 using numerical simulation. Earthquake Sci. **24**(5), 475–483 (2011)
331. G.B. Whitham, Variational methods and applications to water waves. Proc. R. Soc. Lond. A **299**(1456), 6–25 (1967)
332. D.L. Williamson, J.B. Drake, J.J. Hack, R. Jakob, P.N. Swarztrauber, A standard test set for numerical approximations to the shallow water equations in spherical geometry. J. Comput. Phys. **102**(1), 211–224 (1992)
333. T.Y. Wu, Long waves in ocean and coastal waters. J. Eng. Mech. **107**, 501–522 (1981)
334. D.M. Young, Iterative methods for solving partial difference equations of elliptic type. Phd, Harvard University, 1950
335. N.N. Zagryadskaya, S.V. Ivanova, I.S. Nudner, A.I. Shoshin, Action of long waves on a vertical obstacle. Bull. VNIIG **138**, 94–101 (1980)
336. A.I. Zaitsev, A.A. Kurkin, B.V. Levin, E.N. Pelinovsky, A.C. Yalciner, Y.I. Troitskaya, S.A. Ermakov, Numerical simulation of catastrophic tsunami propagation in the Indian Ocean. Dokl. Earth Sci. **402**(4), 614–618 (2005)
337. V. E. Zakharov (Ed.), *What is Integrability?* Springer Series in Nonlinear Dynamics (Springer, Berlin, 1991)
338. P.A. Zegeling, J.G. Blom, An evaluation of the gradient-weighted moving-finite-element method in one space dimension. J. Comput. Phys. **103**(2), 422–441 (1992)
339. P.A. Zegeling, J.G. Verwer, J.C.H. Van Eijkeren, Application of a moving grid method to a class of 1D brine transport problems in porous media. Int. J. Numer. Meth. Fluids **15**(2), 175–191 (1992)
340. V. Zeitlin (ed.), *Nonlinear Dynamics of Rotating Shallow Water: Methods and Advances* (Elsevier B.V., Amsterdam, 2007)
341. Y. Zhang, A.B. Kennedy, N. Panda, C. Dawson, J.J. Westerink, Boussinesq-Green-Naghdi rotational water wave theory. Coastal Eng. **73**, 13–27 (2013)
342. B.B. Zhao, R.C. Ertekin, W.Y. Duan, A comparative study of diffraction of shallow-water waves by high-level IGN and GN equations. J. Comput. Phys. **283**, 129–147 (2015)
343. M.I. Zheleznyak, Influence of long waves on vertical obstacles, in *Tsunami Climbing a Beach*, ed. by E.N. Pelinovsky (Applied Physics Institute Press, Gorky, 1985), pp. 122–139
344. M.I. Zheleznyak, E.N. Pelinovsky, Physical and mathematical models of the tsunami climbing a beach, in *Tsunami Climbing a Beach*, ed. by E.N. Pelinovsky (Applied Physics Institute Press, Gorky, 1985), pp. 8–34
345. V.A. Zorich, *Mathematical Analysis I*, 2nd edn. (Springer, Berlin, 2008)

Index

G. Khakimzyanov et al., *Dispersive Shallow Water Waves*, Lecture Notes in Geosystems Mathematics and Computing, https://doi.org/10.1007/978-3-030-46267-3

G

T